Palaeomagnetism and Diagenesis in Sediments

GEOLOGICAL SOCIETY SPECIAL PUBLICATION NO. 151

Palaeomagnetism and Diagnesis in Sediments

EDITED BY

DONALD H. TARLING
University of Plymouth

AND

PETER TURNER
University of Birmingham

1999

Published by

The Geological Society

London

THE GEOLOGICAL SOCIETY

The Society was founded in 1807 as The Geological Society of London and is the oldest geological society in the world. It received its Royal Charter in 1825 for the purpose of 'investigating the mineral structure of the Earth'. The Society is Britain's national society for geology with a membership of around 8500. It has countrywide coverage and approximately 1500 members reside overseas. The Society is responsible for all aspects of the geological sciences including professional matters. The Society has its own publishing house, which produces the Society's international journals, books and maps, and which acts as the European distributor for publications of the American Association of Petroleum Geologists, SEPM and the Geological Society of America.

Fellowship is open to those holding a recognized honours degree in geology or cognate subject and who have at least two years' relevant postgraduate experience, or who have not less than six years' relevant experience in geology or a cognate subject. A Fellow who has not less than five years' relevant postgraduate experience in the practice of geology may apply for validation and, subject to approval, may be able to use the designatory letters C Geol (Chartered Geologist).

Further information about the Society is available from the Membership Manager, The Geological Society, Burlington House, Piccadilly, London W1V 0JU, UK. The Society is a Registered Charity, No. 210161.

Published by The Geological Society from:
The Geological Society Publishing House
Unit 7, Brassmill Enterprise Centre
Brassmill Lane
Bath BA1 3JN
UK
(*Orders*: Tel. 01225 445046
  Fax 01225 442836)

First published 1999

The publishers make no representation, express or implied, with regard to the accuracy of the information contained in this book and cannot accept any legal responsibility for any errors or omissions that may be made.

© The Geological Society of London 1999. All rights reserved. No reproduction, copy or transmission of this publication may be made without written permission. No paragraph of this publication may be reproduced, copied or transmitted save with the provisions of the Copyright Licensing Agency, 90 Tottenham Court Road, London W1P 9HE. Users registered with the Copyright Clearance Center, 27 Congress Street, Salem, MA 01970, USA: the item-fee code for this publication is 0305-8719/99/$15.00.

British Library Cataloguing in Publication Data
A catalogue record for this book is available from the British Library.

ISBN 1-86239-028-2

Distributors

USA
AAPG Bookstore
PO Box 979
Tulsa
OK 74101-0979
USA
(*Orders*: Tel. (918) 584-2555
  Fax (918) 560-2652)

Australia
Australian Mineral Foundation
63 Conyngham Street
Glenside
South Australia 5065
Australia
(*Orders*: Tel. (08) 379-0444
  Fax (08) 379-4634)

India
Affiliated East-West Press PVT Ltd
G-1/16 Ansari Road
New Delhi 110 002
India
(*Orders*: Tel. (11) 327-9113
  Fax (11) 326-0538)

Japan
Kanda Book Trading Co.
Cityhouse Tama 204
Tsurumaki 1-3-10
Tama-Shi
Tokyo 0206-0034
Japan
(*Orders*: Tel. (0423) 57-7650
  Fax (0423) 57-7651)

Typeset by Aarontype Ltd, Unit 47, Easton Business Centre, Felix Road, Bristol BS5 0HE, UK

Printed by WBC, Bridgend, UK.

Contents

TARLING, D. H. Introduction: sediments and diagenesis — 1

BATT, C. M. Preliminary investigations into the acquisition of remanence in archaeological sediments — 9

PISAREVSKY, S. A. Studies of post-depositional remanent magnetization and their relevance to the palaeomagnetic record — 21

BORRODALE, G. J. Viscous remanent magnetization of high thermal stability in limestone — 27

MAHER, B. A. & HOUNSLOW, M. W. The significance of magnetotactic bacteria for the palaeo- and rock magnetic record of Quaternary sediments and soils — 43

MARCO, S., RON, H., MCWILLIAMS, M. O. & STEIN, M. The locking-in of remanence in late Pleistocene sediments of Lake Lisan (palaeo Dead Sea) — 47

DINARÉS-TURELL, J. & DEKKERS, M. J. Diagenesis and remanence acquisition in the early Pliocene Trubi Marks at Punta di Maiata (southern Sicily): palaeomagnetic and rock magnetic observations — 53

VIGLIOTTI, L., CAPOTONDI, L. & TORII, M. Magnetic properties of sediments deposited in suboxic–anoxic environments: relationships with biological and geochemical proxies — 71

URBAT, M. DEKKERS, M. J. & VRIEND, S. P. The isolation of diagenetic groups in marine sediments using fuzzy c-means cluster analyses — 85

WILSON, G. S. & ROBERTS, A. P. Diagenesis of magnetic mineral assemblages in multiply redeposited siliclastic marine sediments, Wanganui basin, New Zealand — 95

TURNER, P., CHANDLER, P., ELLIS, D., LEVEILLE, G. P. & HEYWOOD, M. L. Remanence acquisition and magnetostratigraphy of the Leman Sandstone Formation: Jupiter Fields, southern North Sea — 109

HAILWOOD, E. A., BOWEN, D., DING, F., CORBETT, P. W. M. & WHATTLER, P. Characterizing pore fabrics in sediments by anisotropy of magnetic susceptibility analyses — 125

HROUDA, F. & JEZEK, J. Magnetic anisotropy indications of deformations associated with diagenesis — 127

BORRADAILE, G. J., FRALICK, P. W. & LAGROIX, F. Acquisition of anhysteretic remanence and tensor subtraction from AMS isolates true palaeocurrent grain alignments — 139

D'ARGENIO, B., FERRERI, V., IORIO, M., RASPINI, A. & TARLING, D. H. Diagenesis and remanence acquisition in the cretaceous carbonates of Monte Raggeto, southern Italy — 147

MÁRTON, E. Diagenesis in platform carbonates: a palaeomagnetic study of a late Triassic–early Jurassic section, Tata (Hungary) — 157

SHOGENOVA, A. The influence of dolomitization on the magnetic properties of early Palaeozoic carbonate rocks in Estonia — 167

HAUBOLD, H. Alteration of magnetic properties of Palaeozoic platform carbonates during burial diagenesis (Lower Ordovician, Texas, USA) — 181

Glossary — 205

Index — 209

Introduction: sediments and diagenesis

D. H. TARLING

Department of Geological Sciences, University of Plymouth, Plymouth PL4 8AA, UK

This introduction outlines the history of the palaeomagnetic study of the diagenesis of sediments, followed by a very general description of the main concepts of diagenetic process involved as loose sediments become increasing lithified into true rocks. This is followed by a discussion of how this collection of articles originated and the basis for the way they are organized. A glossary of most relevant palaeomagnetic and diagenetic terms, with some minerals, is given at the end of the book (**Tarling**, this volume).

As sediments are almost always more weakly magnetized than igneous rocks, the earliest palaeomagnetic studies were mostly made on igneous rocks (Delesse 1849; Folgerhaiter 1894) and fired clays (Melloni 1853; Gheradi 1862), although the discovery of thermal remanence was made by Boyle (1691) using sedimentary iron ores. Studies of the magnetization of Quaternary varved sediments and marine samples were reported by McNish & Johnson (1938), followed shortly by Ising (1943) working on varved sediments in Sweden. Granar (1958) expanded on the Swedish work and developed the early principles of how magnetic grains, as they are deposited, become aligned by the ambient geomagnetic field and that this alignment can be preserved after deposition. Studies of unconsolidated sediments in the United States was expanded by Johnson *et al.* (1948) and further studies of the Swedish varves by Griffiths (1953) and Griffiths *et al.* (1960). Pioneering magnetic studies of laboratory deposited materials were made by King (1955), Griffiths *et al.* (1957), Rees (1961), Griffiths *et al.* (1962) and Hamilton (1963) in which the basic understanding of the origin and nature of magnetic fabrics in sediments were also established. These showed that elongated grains, when deposited, became slightly flattened into the bedding plane. As such grains tend to be magnetized along their long axes, this depositional tilt necessarily rotated the net magnetization towards the bedding plane, thereby shallowing the inclination and creating an 'inclination error' of up to 4–5°. Improvements in coring systems, following Mackereth (1958, 1969), enabled samples to be collected with little or no disturbance, leading to a vast expansion in palaeomagnetic studies and of the processes occurring within lake and marine sediments, as well as how such information can be used to determine previous environments (Thompson & Oldfield 1986).

Early studies of sedimentary rocks, as opposed to sediments, were made by Graham (1949) in America, and Khramov (1958) in the USSR. However, the expansion in the magnetic study of consolidated sedimentary rocks mostly followed Blackett's suggestion that his new, highly sensitive astatic magnetometer (Blackett 1952) could be used for the measurement of the weak magnetization of sediments. This was followed shortly by results from mainly red sandstones in Britain and Europe (Creer *et al.* 1954; Clegg *et al.* 1954; Nairn 1956; Creer 1957; Irving 1957a; Collinson *et al.* 1957). These were then extended rapidly throughout the geological time scale and through most continents, e.g. Creer (1958) in South America, Irving (1957b) in Australia; Doell (1955), Runcorn (1955) and Balsley & Buddington (1958) in the United States; Clegg *et al.* (1956) in India; Graham & Hales (1957), Nairn (1960) in southern Africa, etc. In general, such studies assumed that these sediments carried a depositional remanence acquired at the time of deposition, which could be isolated, using partial thermal and alternating magnetic field demagnetization, from later magnetic overprints, such as viscous remanences, and magnetic changes associated with burial, uplift and exposure. It was generally assumed that diagenetic changes were minor or occurred so soon after deposition that there was no meaningful differences in age between deposition and diagenesis.

Most early concerns were whether the depositional remanence in a sediment had been affected by the depositional 'inclination error' found in laboratory deposits. Comparison of the magnetic properties of sedimentary rocks with coeval igneous rocks largely discounted this idea, but it is still suspected to occur in unconsolidated sediments. The absence of this effect in lithified sediments implies that any 'inclination error' is eliminated during post-depositional processes, i.e. essentially during diagenesis. Concurrently

TARLING, D. H. 1999. Introduction: sediments and diagenesis. *In*: TARLING, D. H. & TURNER, P. (eds) *Palaeomagnetism and Diagenesis in Sediments*, Geological Society, London, Special Publications, **151**, 1–8.

with such comparisons, sedimentological studies demonstrated that many lithified 'red beds' had necessarily undergone complex chemical changes between the time of their original deposition and their present lithified state (Turner 1980). Consequently, hematitic reddening is generally a post-depositional phenomenon, but could have occurred almost immediately after deposition (Steiner & Helsley 1974a, b) or have been delayed, possibly for hundreds of millions of years (Tarling *et al.* 1976). With increasing time, the importance of such chemical changes, diagenesis, has become increasingly recognized in all fields of geology. One of the earliest meetings to specifically consider diagenesis and palaeomagnetism was held at the University of Newcastle (Tarling 1976). Since then, the knowledge in all fields – in studies of diagenesis, rock magnetism and mineral magnetism – has improved enormously. While such studies have often served to demonstrate the complexities of diagenetic processes, rather than explain them, there is now considerably greater understanding of the processes. There has also been an increasing realization by sedimentologists of how magnetic analyses can assist in the understanding of sedimentological processes. It thus seemed timely to consider the current 'state of the art' in terms of the relationship between palaeomagnetism and diagenesis in sediments and sedimentary rocks.

There are numerous definitions of diagenesis. Most of these are only partially satisfactory because of the complexity of the interacting biochemico-physical processes involved. In theory, diagenetic processes comprize all physical and chemical events that affect sediments from the time they are initially deposited to when they show identifiable signs of the very earliest stages of tectonic metamorphism (ancho-metamorphism). However, the boundaries are obscure. While grains are still being deposited, they are re-acting with their environment, both physically and chemically, so some diagenetic processes have actually commenced prior to deposition. Similarly, the distinction between deep-burial diagenetic processes and physical-chemical changes associated with the onset of tectonic deformation are more semantic than real. Both processes usually operate simultaneously. Nonetheless, most sedimentologists recognize two phases of diagenesis. 'Early' diagenesis (or syndiagenesis) includes all those processes associated with deposition and shallow burial, while 'late' diagenesis (or anadiagenesis) are those associated with deeper burial. Unfortunately, 'shallow' and 'deep' cannot be quantified because it is impossible to make any meaningful generalization; the difference between shallow and deep process depends on the characteristics of the sediments and their precise environment. Carbonates, for example, can form well-cemented rocks at depths which can be only centimetres deep, or even at the sediment/water interface, while sands may exceptionally escape cementation altogether or lithification may be delayed until cementing fluids pass through them at depths which could exceed a kilometre.

In clastic rocks, such as shales, siltstones and sandstones, three main phases of diagenesis are generally recognized. The *redoxomorphic* stage is associated with compaction and dewatering in zones that may be either oxidising or reducing. Above the water-sediment interface, the conditions are largely oxic, so that organic and sulphur compounds tend to react with iron, destroying much, but not necessarily all, of the detrital ferric minerals to form soluble iron sulphates which can react and be re-deposited as the iron compounds elsewhere. Under such circumstances, many detrital magnetic grains are likely to be lost during prolonged diagenesis, and the original remanence may only be carried by magnetic grain inclusions within relatively inert grains, such as quartz and ilmenite (Hounslow *et al.* 1995). Below the interface, the low oxygen concentrations tend to preserve the organic compounds and the iron tends to link with sulphur to form pyrite. Usually underlying this phase, but not necessarily so, the *locomorphic phase* occurs where the main diagenetic cementation takes place. This cementation is usually of silica or carbonate, depending on the composition of migrating fluids, and can be preceded by authigenic growth of pre-existing minerals, such as quartz and calcite overgrowths. Both of these predominantly chemically defined phases (redoxomorphic and locomorphic) are accompanied by changes in the physical properties, in particular changes in porosity. Clays, for example, are usually deposited with 50–80% porosity. Pore fluids are then mostly lost during the earliest changes from clays to claystones. Cementation generally further reduces the permeability, although some mineralogical changes, particularly any inversion of calcite to dolomite, may increase the permeability. Within this complex of physical effects and inorganic reactions, the presence of organisms and organic materials may have crucial effects. Macro-organisms do not merely disturb the sediment by their motions (bioturbation), but also actively comminute the material which they digest. Living organisms can also directly promote the formation and changes in the magnetic mineralogy. For example, the microbe, *Desulfovibrio*, promotes the formation of iron sulphides in anoxic conditions and has a profound influence on the magnetic mineralogy of lake and marine sediments, particularly if organic material is present. The knowledge of much of the magnetic mineralogy is relatively new; the existence of the iron sulphide mineral, greigite, was largely unknown until the studies of Skinner et al. (1964), partially because of its instability on exposure to air, yet this mineral provides a major link in the pathway of iron minerals during the evolution of many lake and marine sediments. The biochemical significance of wide variety of such organisms similarly remains poorly understood, even when they are a direct agent in affecting the magnetic mineralogy. Magnetotactic bacteria are now known to produce chains of single domain magnetite grains, apparently in order to determine 'way up' in the interface between fluid and sediment (Blakemore 1975; Kirschvink & Lowenstam 1979; Frankel & Blakemore 1990). On death, such chains or the individual grains may be preserved to provide a high stability magnetic mineral whose remanence may be locked in as the grains become cemented. However, bacteria can fix iron minerals within their structure in many different ways (Vali & Kirschvink 1990) and are currently being cultivated for their potential ability to fix heavy metal pollutants. The current knowledge of the magnitude of the effects of these and other bacteria is still poor, but they are likely to be major factors in numerous diagenetic processes in anoxic and sub-oxic environments, even at depths down to 6–7 km. However, in more oxidizing environments, such biogenetic grains are likely to be oxidized.

In carbonates, diagenetic processes are far more complex and rapid because the initial deposits have a high porosity, 40–70%, and much of the initial carbonate mineralogy may be in the hexagonal form, aragonite, which is metastable under non-marine conditions and inverts readily to the more stable orthorhombic form, calcite. This crystallization results in a decrease in the primary porosity. In high salinity conditions, such as in restricted marine basins, the higher concentration of magnesium in bottom waters can react with the aragonitic muds, forming early diagenetic ('syngenetic') dolomite. These complex carbonate interactions between aragonite, calcite and dolomite (polymorphism) are also accompanied by re-crystallization, resolution and growth of pre-existing carbonates, all forming the neomorphic phase of diagenesis. Such reactions can take place in a variety of locations, depending on the source for the Mg-enriched fluids. This can occur at very shallow, sub-tidal depths, under conditions of high marine salinity in silled-basins, or at sea-level where such higher density brines interface with fresh-water lenses, such as in sabkha conditions. Such Mg-enriched brines can also enter into the evolving system at somewhat later stages, resulting in new or further dolomitization; this later, but still 'early' diagenetic dolomite may be recognized because it is usually related to more permeable horizons, such as unconformities, or to tectonic features that cut across depositional surfaces. Conversely, if relatively fresh water, i.e. with low Mg/Ca, penetrates the system, de-dolomitization can result, accompanied by a decrease in porosity as the dolomite crystals invert to calcite which is up to 13% more voluminous. This also occurs if sea level fell and exposed the rocks to subaerial or partial subaerial conditions. It is also possible that silica-rich fluids can enter the system resulting in the formation of a silica cement and formation of cherts. Within this complex, organic material can drastically change the local geochemical environment and can accumulate to form major anoxic organic deposits such as those that form future hydrocarbon sources. Within such deposits, complex biochemical reactions will occur in a variety of reactions – as outlined above for clastic deposits. In carbonates, however, the reactions are even more complex because the deposits themselves are generally more chemically active than those of the clastic deposits. Such reactions with organic matter usually convert iron oxides and hydroxides into sulphides or highly soluble sulphates. Again, as for the clastic sediments, the role of microorganisms is fundamental, yet poorly understood; even dolomitization may well be enhanced in the presence of such organisms. It seems probable that magnetite-bearing bacteria may also be of far greater importance than in clastic rocks. This is partially because of the very low concentrations of any other magnetic minerals in carbonates, which are often diamagnetic, but also because the chemical reactivities in the carbonates are far greater and such grains are far more likely to become cemented in the sediment very shortly after death. This not only locks in any magnetic orientation, but also preserves the grains against subsequent oxidation. However, even now, little is known of any of the complex inorganic and organic re-actions in such circumstances. Finally, drastic chemical changes can also occur as a result of the migration of, for example, hydro-carbon-rich fluids through the system at any stage (Machel 1995).

During even greater burial of all lithologies, clastics and carbonates, the increasing pressure and temperature induce further diagenetic changes, termed the phyllomorphic phase, which involves the recrystallization of clays, in particular, and the recrystallization of less stable phyllosilicates, such as muscovite and biotite to form chlorites. While the net orientation of these flaky minerals remains close to the horizontal, such changes are still recognized as diagenetic, but if the crystal growths are under the influence of oblique-to-horizontal, tectonic forces, they are recognized as an early metamorphic process, ancho-metamorphism. In reality, the boundaries between diagenesis and metamorphism are poorly defined and partially lithologically dependent. At any stage, accumulating sediments may well be raised, by tectonic forces, from one regime to another. This can cause a reversal of some of the diagenetic processes as minerals that are stable at greater pressures equilibrate with the new conditions. Such reactions are termed *epidiagenesis* as they are superimposed on previous diagenetic features. As, or sometimes more, important is the entry of new fluids into the sedimentary basins due to tectonic activity on their margins. The actual pathways and composition of these basinal fluids are largely controlled by the permeability characteristics of the sediments. These fluids may completely destroy all pre-existing iron-bearing minerals, although new authigenic ferromagnetic minerals may then date the time that such fluids have passed. Many of the Jurassic limestones in Britain, for example, appear to have been altered by prolonged water circulation from the overlying Cretaceous seas. Only very rarely can minute specks of blue limestone be seen in deep quarries. Such limestones still show, superimposed on this deep 'weathering', a few centimetres of present-day weathering. Such chemical activity will be enhanced if the fluids are at higher temperatures, commonly up to 200–300°C according to fluid inclusion studies in many sedimentary basins. If such temperatures are maintained for a few million years, then they can physically cause the complete remagnetisation of the ferromagnetic minerals in the basin, irrespective of any chemical effects although these are usually more locally deleterious.

In this book, the evidence from sediments deposited over archaeological time scales are considered first by **Batt** (this volume) who shows that many undisturbed archaeological sediments have acquired remanence directions related to the Earth's magnetic field direction only shortly after deposition, suggesting that such remanences are probably not depositional, but related to post-deposition rotations that have taken place from a few months to a few years after deposition. Poor results are usually associated with bioturbated sites and those with higher organic content. **Pisarevky** (this volume) considers experiments on a variety of Quaternary continental sediments to simulate the natural wetting and drying that will have occurred to them as a result of changes in the local height of water table. Not surprisingly, these confirm the models suggested by Noël (1980) and confirm that the magnetic properties of permeable sediments of any age may have been drastically altered by this process. More importantly, any sediment samples that have dried out, or been soaked, since collection may have had their properties changed. In some sense, such processes are similar to weathering, but without chemical changes. Therefore the effects of weathering and viscous remanences are considered next as these can affect the results of all of the techniques used to isolate the original, primary magnetization (depositional or diagenetic). Such methods involve fundamental assumptions concerning the properties of viscous remanence acquired by rocks while simply lying in the Earth's magnetic field prior and subsequent to collection. Viscous remanence in limestones is considered by **Borradaile** (this volume), although such behaviour may well occur in other lithologies. He demonstrates, using stones from buildings, that viscous remanences are not necessarily so readily removed during standard demagnetization tests as is generally supposed, and poses the problem of how such stable remanences can arise and be recognized. (It is interesting to note that the technique used by Borradaile was originally used by Melloni in 1853 who used stones from the Roman amphitheatre at Pompeii to test the magnetic stability of lavas from Vesuvius.) The importance of biologically produced magnetite is considered by **Maher & Hounslow** (this volume), predominantly for Quaternary sediments and soils. Such magnetite is potentially of great palaeomagnetic importance as organisms have optimized the grain-size to be that of single-domain sizes with very high magnetic stability. When present, they may well carry a remanence relating to the time that the grain became locked into the rock, usually during cementation. As is pointed out, the separation and identification of such materials is not simple. Where ferromagnetic grains of detrital origin are present, these may well swamp the signal of the biogenic material, but in areas of very low detrital input, as on some carbonate platforms, these may well be a significant carrier of remanence. However, as with all other iron-bearing minerals, they are subject to potential oxidation or sulphidization during later diagenesis, particularly if organic matter is present.

Unconsolidated lake sediments have usually accumulated over time scales that cover conventional archaeological studies and extend back to even earlier times. However, most lacustrine environments are complicated by the effects of high organic activity in them and would require an entire publication on such aspects alone. Consequently, the next contribution considers a lacustrine situation where biological activity can be largely excluded, in this case because of extremely high salinities. **Marco et al.** (this volume) present studies of the 'chemical' deposits of the Lisan Formation in Israel. They show that the remanences preserve short-term features (secular variations) of the geomagnetic field from shortly after deposition, possibly within a few weeks or months. However, being in a highly active seismic region close the the Dead Sea Fault, violent seismic shocks can briefly fluidize such sediments, causing a remagnetisation down to depths of about 1 m below which the sediments were presumably already sufficiently consolidated that their remanences were preserved. Mörner (1996) has attributed similar seismically induced fluidization effects in the Swedish Quaternary varves where locally some 2 to 4 m may be affected in an region generally regarded as seismically quiet! The reason for this difference is, presumably, because the Lisan Formation is likely to become more cohesive because of its composition than the more inert varved silts and clays in Sweden. Such studies also highlight the hazards of handling poorly consolidated sediments which could clearly result in remagnetization. On the bright side, however, it is interesting to note that the magnetic effects of palaeo-seismic disturbances provides a new method for studying palaeo-seismicity.

Turning to somewhat more lithified sediments, **Dinares-Turell & Dekkers** (this volume) consider palaeomagnetic results from the lower Pliocene Trubi Marls in Sicily. They show that it can be extremely difficult to assess the extent to which exposed sediments are truly pristine and also demonstrate the drastic differences that can occur as a result of different depositional environments. In this particular case, identification of the correct components is also vital to determining the structural evolution of the area. Further evidence of the particular complexity in the diagenetic evolution of organic matter is given by **Vigliotti et al.** (this volume) in both lacustrine and marine sediments. They show that careful rock magnetic analyses may provide a vital key to assessing environmental conditions, particular if these can be combined with palaeontological and geochemical data. The two previous papers have clearly demonstrated the complexities of diagenesis, particularly when organic materials are involved, and suggest any statistical parameters attempting to identify different diagenetic zones must consider a host of physico-chemico-biological properties. However, standard variate analyses of various partially related palaeomagnetic, rock-magnetic, geochemical and palaeontological parameters are obviously tedious and most such statistical techniques require class boundaries to be defined before assessing the means and variances. Such sharp classifications seems inappropriate for properties that merge and overlap over a range of scales. **Urbat & Dekkers** (this volume) propose that fuzzy-c statistics appear to be particularly appropriate to such a situation. The examples which they use relate mostly to early diagenetic situations, but the techniques are equally applicable to later stages of diagenesis. Fuzzy-c statistical analysis are shown to be particularly appropriate when considering overlapping diagenetic groupings which, in nature, are very fuzzy! There is also the advantage of speed of comparative analysis and the avoidance of subjectively selecting classes, as required by more routine multivariate analyses – although outliers still have to be manually excluded! (While this may mean starting on yet another initial learning curve, the potential does seem high!)

Wilson & Roberts (this volume) consider an unusual situation where sediments have been repeatedly re-cycled through erosion, diagenesis and re-deposition within sulphate-reducing conditions. Such processing essentially removed most of the 'standard' magnetite and hematite that normally carry the remanence in sedimentary rocks, demonstrating the extreme of repeated early diagenetic processes. They demonstrate that the remanence appears to be carried almost entirely by hemo-ilmenites, with a very minor contribution from chromite. This seems strange as such ilmenite would normally be considered to have paramagnetic properties at room temperature, but it suggests that such remanence-carrying hemo-ilmenites may be a significant component in other sediments where diagenetic changes have largely removed most magnetite and hematite. **Turner et al.** (this volume) demonstrate how palaeomagnetic techniques, combined with petrological studies, can be used to evaluate the reliability in borecores from sediments that cannot be studied in outcrop. This restricts many of the standard tests used in palaeomagnetic studies, but such combined studies enable a magnetostratigraphy to be established, indicating different rates of deposition in different parts of the Jupiter oil field. **Hailwood** *et al.* (this volume) then shows how the fluid permeability anisotropy of such bore-samples can be determined more quickly and efficiently after impregnation by high magnetic susceptibility fluids. Such a technique is not, of course, confined to bore-cores but could be invaluable when, as for **Turner** *et al.* (this volume), there are no outcrop exposures to enable three-dimensional assessment of the routes through which fluids have passed. Such fluids can obviously drastically affect the magnetic properties of the rocks through which they pass, whether they are oxic or reducing, but, conversely, the remaining minerals provide a clue to the nature of such fluids, while their magnetization enables the date of their passage to be assessed.

The problem of defining the change from deep diagenesis to tectonic metamorphism is considered by **Hrouda & Jezek** (this volume). A variety of sediments bordering the Alps have fabrics that range from almost entirely sedimentary to strongly tectonized. As the distinction between deep diagenetic and tectonic processes is based fundamentally on whether the phyllosilicate minerals have bedding-parallel or cross-bedding orientation, it is clearly fabric techniques that define which process is dominant. Hrouda and Jezek demonstrate how magnetic anisotropy studies pick out the earlier phases of grain alignments by these two different orientation process faster, cheaper and more effectively than most standard petrological techniques. **Borradaile** (this volume) finds that it is still possible to isolate fabrics with sedimentary characteristics, by separating the fabrics associated with diamagnetic quartz from those of ferromagnetic minerals. This clearly requires considerable care in undertaking the analyses in the correct sequence, but also indicates how strictly diagenetic fabrics can occur together with tectonic fabrics - thereby demonstrating the arbitrary nature of the distinction between very late diagenesis and ancho-metamorphism.

The carbonates are considered last as having different reactivities from clastic rocks. **D'Argenio** *et al.* (this volume) deal with Cretaceous shallow-water carbonates which, on sedimentological grounds, have retained their early diagenetic features. These carbonates show cyclical depositional as well as early diagenetic changes related to periodic variations in the Earth's orbital parameters. Early diagenetic features developed almost immediately after deposition in a shallow, sub-tidal marine realm which was only briefly interrupted by rare exposures to a meteoric regime when relative sea-level dropped at the top of the more major planetary orbital cycles. The palaeomagnetic properties show same cyclicities as the sedimentary parameters, implying that the Earth's magnetic field is also affected by such perturbations. They also show the patterns and rates of change that suggest that geomagnetic secular variations have been preserved. Such observations mean that the remanence must have been locked-in during very early diagenesis of these sediments. The combined sedimentological and palaeomagnetic evidence can therefore also be used to date diagenetic features with a remarkably high precision, averaging some 360 years. **Márton** (this volume) describes an unusual example at the Triassic/Jurassic boundary where Triassic carbonates were exposed prior to being infilled and succeeded by Jurassic pelagic carbonates. Careful examination of these exposures and comparison with the data from a bore-core and neighbouring localities show how the late Triassic remanences had been re-set, without evident chemical change, during the Jurassic – a re-setting that would have been almost impossible to distinguish under normal circumstances. In Ordovician and Silurian carbonates in Estonia, **Shogenova** (this volume) finds that dolomites have different ages and genesis in different part of the region. Geochemical observations combined with palaeomagnetic analyses show that many are of very early diagenetic origin, but others are different, both geochemically and magnetically. These atypical dolomites are spatially related to deep basement faults that provided conduits for the fluids that caused the dolomitization at later times. In Texas, **Haubold** (this volume) found analogous results in similarly aged carbonates, again combining geochemical and magnetic observations. In this case, he was able to palaeomagnetically date the effects of the passage of the fluids responsible for the changes. These were clearly orogenic fluids generated in the nearby Ouachita Orogen which then passed through the more permeable beds and fissures. Palaeomagnetic dating enabled both primary (early diagenesis) and secondary re-magnetizations to be dated, showing that impermeable beds, and other lithologies more distant from the orogenic front, still retained their original properties.

The following people, in alphabetical order, were asked to referee one or more of the submitted papers: C. M. Batt, G. Borradaile, J. Dinares-Turrell, N. Hamilton, H. Haubold, R. Ixer, S. Marco, E. Marton, M. Noël, S. Openshaw, W. Owens, S. Pisarevsky, G. Potts,

A. P. Roberts, J. Sahota, J. Shaw, A. A. Shogenova, Stephenson, A. Turner, L. Vigliotti, H. Vizan and G.S. Wilson.

Their help is very gratefully acknowledged. I also wish to acknowledge my indebtedness to all of my colleagues in the department for releasing me from virtually all teaching and administrative duties while undertaking the task of editing – and to all those contributors who not only followed the recommendations of the referees, but also tolerated my extensive re-writes. Bruno D'Argenio and Marina Iorio also helped to check the introduction and glossary. Not least, I thank my wife, Nanette, for her tolerance and support.

References

BALSLEY, J. R. & BUDDINGTON, A. F. 1958. Iron-titanium oxide minerals, rocks and aero-magnetic anomalies of the Adirondack area, New York. *Econic Geology*, **53**, 777–805.

BLACKETT, P. M. S. 1952. A negative experiment relating to magnetism and the Earth's rotation. *Philosophical Transaction Royal Society, London*, **A245**, 309–370.

BLAKEMORE, R. 1975. Magnetotactic bacteria *Science*, **190**, 377–379.

BOYLE, R. 1691. Chymico magnetical experiments and observations. *In: Experimenta & Observationes Physicae*, London, Chapter 1.

CLEGG, J. A., ALMOND, A. & STUBBS, P. H. S. 1954a. Remanent magnetism of some sedimentary rocks in Great Britain. *Philosophical Magazine*, **45**, 583–598.

——, DEUTSCH, E. R. & GRIFFITHS, D. H. 1956. Rock magnetism in India. *Philosophical Magazine, Series 8*, **1**, 419–531.

COLLINSON, D. W., CREER, K. M., IRVING, E. & RUNCORN, S. K. 1957. Palaeomagnetic measurements in Great Britain – 1. Measurement of the permanent magnetization of rocks. *Philosophical Transactions of the Royal Society, London*, **A250**, 73–82.

CREER, K. M. 1957. Palaeomagnetic investigations in Great Britain – IV. The natural remanence magnetization for certain stable rocks from Great Britain. *Philosophical Transactions of the Royal Society, London*, **A250**, 111–129.

——1958. Preliminary palaeomagnetic measurements from South America. *Annales Geophysica*, **15**, 373–390.

——, IRVING, E. & RUNCORN, S. K. 1954. The direction of the geomagnetic field in remote epochs in Great Britain. *Journal Geomagnetism and Geoelectricity*, **6**, 163–168.

DELESSE, A. 1849. Sur le magnétisme polaire dans les mineraux et dans les roches. *Annales Chimique et Physique*, **25**, 194–209.

DOELL, R. R. 1955. Paleomagnetic study of rocks from the Grand Canyon of the Colorado River. *Nature*, **176**, 1167.

FOLGERHAITER, G. 1894. Origine del magnetismo nelle roccie vulcaniche del Lazio *et al. Attai. della Reala Accadema Lincei*, **3**, 53, 117 & 165.

FRANKEL, R. B. & BLAKEMORE, R. P. (eds) 1990. *Iron Biominerals*. Plenum Press, New York

GHERADI, S. 1862. **Title to be supplied in proof** *Nuovo Cimento*, **16**, 384.

GRAHAM, J. W. 1949. The stability and significance of magnetism in sedimentary rocks. *Journal of Geophysical Research*, **54**, 131–167.

GRAHAM, K. W. & HALES, A. L. 1961. Preliminary palaeomagnetic measurements on Silurian sediments from South Africa. *Geophysical Journal Royal Astronomical Society*, **5**, 318–325.

GRANAR, L. 1958. Magnetic measurements on Swedish varved sediments. *Arkive fur Geofysik*, **3**, 1–40.

GRIFFITHS, D. H. 1953. Remanent magnetism of varved clays from Sweden. *Nature*, **172**, 539.

——, KING, R. F. & REES, A. I. 1962. The relevance of magnetic measurements on some fine grained silts to the study of their depositional process. *Sedimentology*, **1**, 134–144.

——, —— & WRIGHT, A. E. 1960. The remanent magnetism of some recent varved sediments. *Proceedings of the Royal Society, London*, **A256**, 359–383.

——, —— & WRIGHT, A. E. 1957. Some field and laboratory studies of the depositional remanence of Recent sediments. *Advances in Physics*, **6**, 306–316.

HAMILTON, N. 1963. Susceptibility anisotropy measurements on some Silurian siltstones. *Nature*, **197**, 170–171.

HOUNSLOW, M. W., MAHER, B. & THISTLEWOOD, L. 1995. Magnetic mineralogy of sandstones from the Lunde Formation late Triassic, northern North Sea, UK: origin of the palaeomagnetic signal. *In*: TURNER, P. & TURNER, A. (eds) *Palaeomagnetic Applications in Hydrocarbon Exploration*. Geological Society, London, Special Publications, **98**, 119–147.

IRVING, E. 1957a. The origin of the palaeomagnetism of the Torridonian Sandstones of northwest Scotland. *Philosophical Transactions Royal Society, London*, **A250**, 1–10.

——1957b. Directions of magnetization in the Carboniferous glacial varves of Australia. *Nature*, **180**, 280–281.

ISING, G. 1943. On the magnetic properties of varved clay. *Arkive fur Matematik, Astronomie Fysik*, **29A**, 1–37.

JOHNSON, E. A., MURPHY, T. & TORRESON, O. W. 1948. The prehistory of the Earth's magnetic field. *Terrestrial Magnetism Atmospheric Electricity*, **53**, 349–372.

KHRAMOV, A. N. 1958. *Palaeomagnetism and Stratigraphic Correlation*, Gostoptechizdat, Leningrad English version Ed. E. Irving, Trans. A. J. Lojkine, A.N.U. Australia, p. 204.

KING, R. F. 1955. The remanent magnetism of artificially deposited sediments. *Monthly Notices Royal Astronomical Society*, **7**, 115–134.

KIRSCHVINK, J. L. & LOWESTAM, H. A. 1979. The role of biogenic minerals in the magnetization of sediments. *Eos*, **60**, 247.

MACHEL, H. G. 1995. Magnetic mineral assemblages and magnetic contrasts in diagenetic environments – with implications for studies of palaeomagnetism, hydrocarbon migration and exploration. *In*: TURNER, P. & TURNER, A. (eds) *Palaeomagnetic Applications in Hydrocarbon Exploration*. Geological Society, London, Special Publications, **98**, 9–29.

MACKERETH, F. J. H. 1958. A portable core sampler for lake deposits. *Limnology Oceanography*, **3**, 181–191.

—— 1969. A short core sampler for subaqueous deposits. *Limnology Oceanography*, **14**, 145–151.

MCNISH, A. E. & JOHNSON, E. A. 1938. Magnetization of unmetamorphosed varves and marine sediments. *Journal of Terrestrial Magnetism*, **43**, 401–407.

MELLONI, M. 1853. Richerche intorno al Magnetismo delle Rocce: sulla Polarità magnetica delle laves e rocce affini *Memorie della Reale Academi Scienze, Napoli*, **1**, 117–164.

MÖRNER, N.-A. 1996. Liquefaction and varve deformation as evidence of paleoseismic events and tsunamis: The autumn 10,430 BP case in Sweden. *Quaternary Science Review*, **15**, 939–948.

NAIRN, A. 1956. Relevance of palaeomagnetic studies of Jurassic rocks to continental drift. *Nature*, **178**, 935–936.

—— 1960. A palaeomagnetic survey of the Karroo System. *Overseas Geology Mineral Resources*, **7**, 398–410.

NOËL, M. 1980. Surface tension phenomena in the magnetization of sediments. *Geophysical Journal Royal Astronomical Society*, **62**, 15–25.

REES, A. I. 1961. The effect of water currents on the magnetic remanence and aniso tropy of susceptibility of some sediments. *Geophysical Journal Royal Astronomical Society*, **5**, 235–251.

RUNCORN, S. K. 1955. Palaeomagnetism of sediments from the Colorado Plateau. *Nature*, **176**, 505.

SKINNER, B. J., ERD, R. C. & GRIMALDI, F. S. 1964. Greigite, the thio spinel of iron; a new mineral. *American Mineralogist*, **49**, 543–555.

STEINER, M. B. & HELSLEY, C. E. 1974a. Magnetic polarity sequence of the Upper Triassic Kayenta Formation, *Geology*, **2**, 191–194.

—— & —— 1974b. Reproducible Anomalous Upper Triassic Magnetization. *Geology*, **2**, 195–198.

TARLING, D. H. 1976. Magnetic dating *Nature*, **262**, 8.

——, DONOVAN, R. N., ABOU-DEEB, J. M. & EL-BATROUK, S. I. 1976. Palaeomagnetic dating of hematite genesis in Orcadian Basin sediments. *Scottish Journal of Geology*, **12**, 125–134.

THOMPSON, R. & OLDFIELD, F. 1986. *Environmental Magnetism*. Allen & Unwin, London, p. 227.

TURNER, P. 1980. *Continental Red Beds* Elsevier, Amsterdam.

VALI, H. & KIRSCHVINK, J. L. 1990. Observations of magnetosome organization, surface structure, and iron biomineralization of undescribed magnetic bacteria: evolutionary speculations. *In*: FRANKEL, R. B. & BLAKEMORE, R. P. (eds) *Iron Biominerals*. Plenum Press, New York, 97–117.

ZIJDERVELD, J. D. A. 1967. A.C. demagnetization of rocks: analysis of results. *In*: RUNCORN, CREER & COLLINSON, D. W. (eds) *Methods in Palaeomagnetism*. Elsevier, Amsterdam, 254–286.

Preliminary investigations into the acquisition of remanence in archaeological sediments

C. M. BATT

Department of Archaeological Sciences, University of Bradford, Bradford BD7 1DP, UK
(e-mail: c.m.batt@bradford.ac.uk)

Abstract: Investigations of magnetic remanence, anisotropy of magnetic susceptibility and magnetic mineralogy of sediments from over 30 depositional environments within British archaeological excavations are compared with archaeological evidence provided by excavation and analysis of associated artefacts and ecofacts. The sites include canals, wells, river floods, ditches, moats and estuarine deposits. A comparison of the magnetic and archaeomagnetic determinations allows the characterization of those deposits that appear to retain a record of the past magnetic field and are therefore datable by comparison with the UK archaeomagnetic calibration curve. Most water-lain sediments have a measurable magnetization, but finer grain sizes tend to have higher intensities, although there is no relationship between depositional environment and intensity of magnetization. The sediments with a higher organic content have a lower consistency index probably as they were more likely to have been bioturbated. Although the magnetic fabrics did not always correspond to the expected 'primary depositional style', almost all sites seemed to have directions of remanence that corresponded well to the dates indicated by the archaeological evidence.

Archaeomagnetic dating is a familiar and well-established technique for the dating of *in situ* fired materials, based on a comparison of the direction of the remanent magnetization of the material with a calibration curve showing changes in the direction of the geomagnetic field over time (Aitken 1970; Clark *et al.* 1988). The possibility that water-lain sediments in archaeological contexts might also retain a magnetization reflecting the geomagnetic field at deposition, and therefore be datable, has generated considerable interest (e.g. Tarling 1983, p.154; Hammo-Yassi 1984; Clark *et al.* 1988; Batt & Noël 1991; Eighmy & Howard 1991). For the archaeologist, dating sediments rather than fired materials has many attractions, as water-lain sediments occur frequently in archaeological sites and rarely need to be retained. It may be possible to date depositional sequences rather than a single firing event, and such investigations may provide large numbers of magnetic directions to improve the fidelity of the calibration curve (Batt 1997). In addition, magnetic studies of sediments may give information about sediment source and mode of deposition. However, for the date obtained to be of value, it must relate to a specific archaeological event, usually that of sediment deposition. Such studies of archaeological sediments can also contribute to an understanding of remanence acquisition and diagenesis processes because they allow the magnetic data to be compared with archaeological evidence to investigate the fidelity of the magnetic direction recorded, the event that it relates to, and the effect of depositional conditions on remanence. The magnetic remanence, magnetic fabric and magnetic mineralogy of over 30 sediments from known depositional environments of known date are reported here as part of a study to establish the reliability of magnetic remanence as a record of date of deposition.

Investigative methods

The archaeological contexts were chosen according to specific criteria. They had to be water-lain, as identified from their lithology, sedimentary structures, biota and location on the site; and to appear minimally bioturbated. The deposits, or material within them, were required to be independently dated by methods such as documentary sources, stratigraphic relationships, radiocarbon or ceramic typologies, and an indication of the depositional environment had to be identifiable from their archaeological context. Deposits were selected to be representative of the wide range of archaeological contexts in which water-lain sediments occur, including fluviatile, lacustrine, static water and marine environments. Finding suitable contexts proved surprisingly difficult, partly because many had suffered heavy bioturbation and reworking, and partly because of the difficulty in identifying water-lain deposits during excavation.

BATT, C. M. 1999. Preliminary investigations into the acquisition of remanence in archaeological sediments. *In*: TARLING, D. H. & TURNER, P. (eds) *Palaeomagnetism and Diagenesis in Sediments*, Geological Society, London, Special Publications, **151**, 9–19.

Sediments were sampled by pushing 2.5 cm diameter plastic tubes into a cleaned vertical or horizontal sediment face (Clark et al. 1988). This small sample size allowed obvious inclusions within the sediment to be avoided and trials in which sample holders were pushed in different directions into the same sediment suggested that such insertion caused negligible deformation of the sediment. Overlapping samples were taken from sediment sequences and at least 15 repeat samples were taken from deposits expected to represent a single date. Orientation was by magnetic compass or, in the presence of steel shoring, a gyrotheodolite. Care was taken to avoid sample drying during storage, which could have resulting grain reorientations (particularly in coarser-grained deposits), by sealing sample tubes and keeping them in a cool, damp environment.

The first stage of investigation was to establish whether the sediment contained magnetized particles which retained a stable magnetization. Natural remanent magnetization (NRM) was measured using a spinner fluxgate magnetometer (Molyneux 1971), and alternating field demagnetization (As & Zijderveld 1958) was used to ascertain the stability of the magnetization and the presence of any viscous remanence and other overprint magnetic directions arising from factors such as chemical change in the sediment or lightning strike. The consistency of the remanence was quantified using the consistency index of Tarling & Symons (1967) and, where appropriate, a small viscous component was removed.

Crucial to the use of archaeomagnetic dates obtained from sediments is whether the magnetization is acquired on deposition, by alignment of magnetized grains with the geomagnetic field as they settle through water (Nagata 1962), or post-depositionally, by the alignment or realignment of magnetized grains within water-filled pore spaces in the sediment matrix (Irving 1957), and subsequently 'locked-in' when the sediment is consolidated. It was hoped to investigate these alternatives by comparing the archaeological date for the deposit with the archaeomagnetic date. The latter was obtained in the conventional manner by visually comparing the mean stable magnetic remanence direction, with error bars at α_{95}, with the existing British archaeomagnetic calibration curve (Clark et al. 1988), after correction to a central location to reduce the effects of spatial variation of the geomagnetic field (Noël & Batt 1990). In this study, α_{95} is the radius of the cone of 95% confidence within which the true mean direction lies (Fisher 1953).

Whereas most archaeomagnetic studies of sediments are concerned with their dating, measurement of the anisotropy of magnetic susceptibility (AMS) can give a useful insight into the orientation of the magnetizable grains within a material, and therefore aid understanding of the magnetic, gravitational and hydrodynamic forces acting on the grains during deposition (Hamilton & Rees 1970). For example, laboratory experiments have suggested that values of q, the ratio of lineation to foliation, in the range 0.06–0.67 are characteristic of a 'primary depositional style' magnetic fabric (Hamilton & Rees 1970). Measurements of AMS were carried out using a Molspin balanced air cored transformer and a Molspin anisotropy delineator (Collinson 1983). To understand the origin of the magnetization of a sediment, and therefore which sediments are suitable for archaeomagnetic study, it is helpful to know the mineralogy and grain size of the magnetic paticles present. Although there are a number of techniques available to determine these, including chemical, X-ray, optical and thermal studies, the use of laboratory induced isothermal remanent magnetization (IRM) has proved particularly informative (Oldfield et al. 1985). In these studies an artificial IRM was produced by exposing a number of pilot samples to a steady magnetic field at room temperature using a pulse magnetizer (Collinson 1983). The remanence acquired depends upon the strength of the applied field and the grain size, composition and concentration of the magnetic minerals in the sample, and was measured using the spinner magnetometer. Particular note was taken of the percentage of the saturating IRM, acquired in fields up to 1 T, to give an indication of the concentration of remanence-carrying magnetite in the sample, and of the IRM acquired when a previously saturated sample was placed in reverse fields of increasing strength. The coercivity of remanence, $B_{0\,cr}$, the reverse field required to reduce the initial saturation IRM to zero, was used as an indicator of grain size for a material of single mineralogy.

The sediments studied occurred either as a direct result of human activity, for example sediments accumulated in a well; or as a result of natural processes associated with archaeological remains, such as a flood deposit covering a site. In all, 30 distinct contexts were sampled (Batt 1992), representing a variety of depositional environments and archaeological dates, and including most of the types of water-lain sediments commonly encountered in British archaeological excavations, for example, ditches, river floods, wells, moats, silted river channels and estuarine deposits. To discuss the information available from magnetic studies of sediments and their limitations, this paper will initially consider two specific examples; sediment samples from overbank flooding in Roman deposits at the Stakis Hotel excavations in York, and recent overbank floods from the same region.

Two case studies: overbank flood deposits in York

Roman overbank deposits

Excavations were carried out by York Archaeological Trust on the southwest bank of the River Ouse, between May 1988 and August 1989 (Ottaway 1988). The site was less than 20 m from the present course of the River Ouse and within the Roman colonia area of York. A number of trenches were opened to provide information about the Roman and Medieval town. One trench (Trench 4) located the Roman road that passed through the civilian town and into the legionary fortress, and excavations revealed a number of horizons showing the initial construction and repeated resurfacings of the road (Ottaway 1988). Context 4159 within this sequence consisted of a grey brown (Munsell No. 10YR 5/2) silt, 0.5 m thick, which was interpreted as being a flood deposit from the nearby river (P. Ottaway, pers. comm.). The deposit was situated over what appeared to be an early road surface consisting of brushwood and cobbles, built up over a sandy material to provide a camber, and overlain by a substantial (0.6 m thick) horizon of cobbles and a series of remetallings. Thirty-three samples (SK 1–33) were taken from this context. The archaeological interpretation of the deposits was that they represented various stages in the construction of the Roman road. This commenced with a raft of brushwood and cobbles to create a stable foundation on marshy ground, which had then been covered by a silt layer from flooding of the river. The response to this was to construct a mound of cobbles on which subsequent road surfaces were built. Hence the origins and date of the flood deposit were of considerable archaeological interest.

The remanence of the samples had a consistency index of 5.3 (very stable, Tarling & Symons 1967). After removal of a soft viscous component of magnetization by demagnetizing all samples in a field of 7.5 mT (Batt 1992), the mean of the samples' remanence had an α_{95} of very low value, 1.3°, comparable with the best results from the dating of fired material (Clark et al. 1988) and implying that all samples recorded a similar magnetic field and that there had been no appreciable change in the geomagnetic field during the period of deposition. The mean magnetic direction appeared to give a date for the sediment in the range AD 20 to AD 80 (Fig. 1). Ebor ware pottery, a red earthenware associated with the earliest Roman pottery assemblages in York (P. Ottaway, pers. comm.), was found under the horizon sampled and late first to early second century AD pottery was found in the cobbled layer overlying the silt, indicating that the archaeomagnetic date corresponded well to the archaeological evidence. The episode of flooding would appear, therefore, to have been around the time of the founding of the Roman fortress, an event dated by documentary sources to AD 71, and implying that the magnetization was acquired on deposition. However, caution must be urged in the use of this procedure for dating the recorded magnetic direction, as discussed later in this paper.

Measurements of the AMS of the samples taken from the Stakis Hotel excavations (Fig. 2a) revealed that the maximum axes of

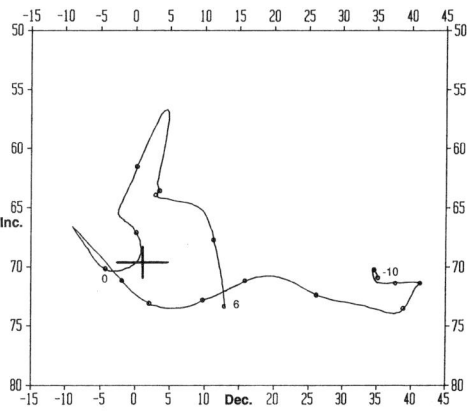

Fig. 1. Mean magnetic direction after partial demagnetization of samples SK 1–33. The direction is corrected to Meriden and fitted to the British calibration curve covering 1000 BC–AD 600 digitized from Clark et al. (1988). It should be noted that this curve was shallowed by 1.7° to allow for magnetic refraction in fired archaeological materials. Declination (Dec.) and inclination (Inc.) are in degrees, with associated 95% confidence errors. Circles indicate centuries and the figures along the curve are in centuries.

susceptibility were clustered, suggesting that elongated grains have been aligned with each other, producing a magnetic lineation. One possible explanation of this is that the sediments have been deposited grain by grain with water currents acting on them (Hamilton & Rees 1970). The close grouping of the minimum axes at shallow inclinations is not typical of water-lain sediments. It could be attributable to post-depositional compaction of the sediment or to sediment deformation during sampling, but further studies of similar sediments are necessary before firm conclusions can be drawn. On the assumption that the alignment of the axes of maximum susceptibility was due to fluid stress and that the magnetic remanence measured had been acquired on deposition, the direction of current flow during deposition can be inferred (Fig. 2b). Using the algorithm developed by Noël & Rudnicki (1988), the remanence directions were connected to the directions of maximum susceptibility by a small circle path which represents the rotation of magnetic grains by the current, away from the direction of the geomagnetic field at the time of deposition. As the current varied locally the small circle paths for all the samples were not identical, but the zone in which they intersected was taken to represent the best estimate of the ancient field direction and the mean direction of the small

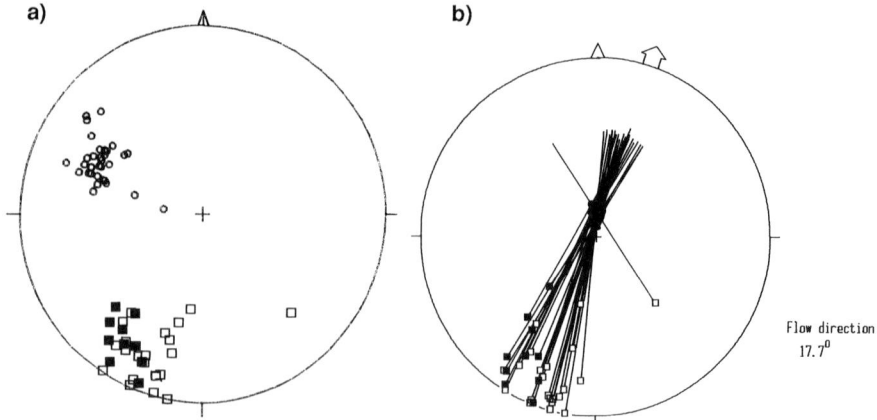

Fig. 2. (a) The directions of the principal axes of susceptibility of samples SK 1–33. Minimum axes are represented by circles and maximum axes by squares; closed symbols represent positive inclinations and open symbols represent negative inclinations. (b) Flow direction from the axes of maximum susceptibility and the directions of magnetization (SK 1–33). Arrow indicates mean flow direction.

circles to indicate the direction of current flow (small circle paths will appear as straight lines on this projection). In the case of the Stakis samples, this analysis suggested that the mean direction of the current flow at the time of deposition had been $c.18°$ NE. This direction indicated current flow roughly at right angles to, and towards, the River Ouse, consistent with the hypothesis that the sediment had been deposited as flood waters subsided back into the river. If the magnetic fabric arose from post-depositional compaction or deformation on sampling, such analysis of current flow would not be valid.

IRM studies were carried out on a number of pilot examples from the Stakis samples; 98% of the saturation value was attained in a field of 1 T, indicating that the majority of the magnetic minerals present were easily saturated. The $B_{0\,cr}$, was 30–35 mT, suggesting that the magnetic carrier was in the form of small, single-domain (SD) grains (Thompson & Oldfield 1986, p. 32). Thus the IRM experiments suggested that the predominant magnetic mineral present in the Stakis samples was small-grained, (SD) magnetite. Magnetite is common in detrital sediments and these results gave no indication of significant post-depositional chemical change to the magnetic minerals.

Modern overbank deposits near York

Flooding is a common occurrence in spring in York and so it was possible to compare the results obtained from the investigation of the supposed archaeological flood deposit with a modern overbank flood deposit. In March 1991 the combination of heavy rain and melting snow caused the River Ouse to rise 5 m above its normal level, burst its banks and flood adjacent properties. A large quantity of sediment was deposited on the flood plain as the waters receded and 21 samples were taken from two apparently undisturbed deposits in a sediment trap for comparison with the archaeological equivalent. The deposits were underlain by autumn leaves and were clearly the result of that season's flooding. Two samples disintegrated on measurement and are therefore not included in the results. The deposits were very dark brown (Munsell No. 10YR 2/2) silty clay and dark yellowish brown (Munsell No. 10YR 4/4) fine sand with occasional silty clay mottling. The NRM intensities fell in a much wider range than for the Stakis samples and the directions of magnetization were widely dispersed, with α_{95} values of 59.5° (Fig. 3a) and 48.3° (Fig. 3b). Demagnetization of pilot samples showed a variety of multicomponent behaviour and no single stable direction could be isolated. The magnetic fabric q values fell in the range 0.17–0.73, 78% of them in the range characteristic of an undisturbed depositional fabric (Fig. 4a and b). The minimum axes were slightly scattered, but most appeared to be consistent with a magnetic foliation, whereas the maximum axes were widely scattered, with no clear lineations. The lack of lineation suggested that the magnetic grains had not been aligned at the time of deposition by water currents, the geomagnetic field or grain interactions, or that the magnetic fabric had been subsequently disturbed. Almost

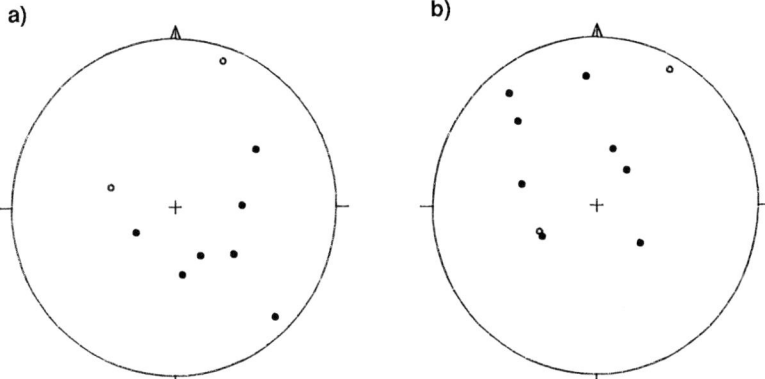

Fig. 3. Stereographic plot of the directions of NRM of two modern river flood deposits. (**a**) The Roman flood deposits; (**b**) their modern analogues. Closed symbols represent positive inclinations and open symbols represent negative inclinations.

all, 96%, of the saturation isothermal remanent magnetization (SIRM) value was reached in an applied field of 1 T and $B_{0\,cr}$ was 22 mT, suggesting that although the magnetic mineral carrying the remanence was predominantly magnetite, there was a larger contribution from a harder magnetic mineral, probably hematite or goethite, than was the case with the Stakis samples.

Comparison

Results from the modern analogue contexts suggested that the deposits contained magnetic grains capable of carrying a remanence within the sediment matrix, but that either the magnetic grains had not been systematically orientated during or after deposition, or that such an alignment had been destroyed. This raises the question of why this deposit does not record the geomagnetic field at the time of deposition, whereas the sediment from the Stakis excavations, which appears to have a similar magnetic grain size, concentration, mineralogy and depositional conditions, appears to do so. One possible explanation is that, as the modern analogue deposits are much shallower than the archaeological deposit, bioturbation is likely to be more significant, although there was no visible evidence of disturbance. An alternative explanation is that insufficient time had elapsed for magnetic grains to orient themselves in the geomagnetic field and become fixed in the matrix, thus implying that the remanence acquired by such sediments is post-depositional in origin. Such a possibility has serious implications for the dating of the Stakis sediments and needs to be re-examined in the light of results from a larger group of sediments.

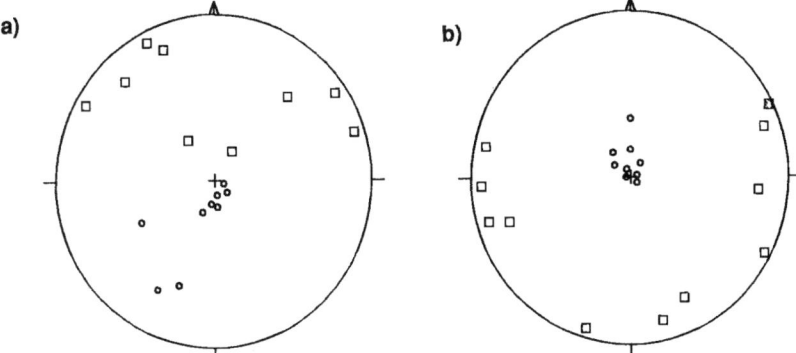

Fig. 4. The directions of the principal axes of susceptibility of modern river flood deposits. Minimum axes are represented by circles and maximum axes by squares.

Results from the entire sediment group

Results from similar studies on the remaining contexts have generated a substantial body of data which can be used to answer a number of questions. The primary objective of this study was to indicate which environments might be most suitable for routine archaeomagnetic dating. This involves the following aspects.

Which contexts retain a record of the geomagnetic field at the time of deposition?

The two key factors considered to be most likely to affect remanence acquisition and retention are grain size and depositional environment. Quantitative measurements of grain size were not attempted; broad groupings were obtained by the qualitative tests commonly used in field archaeology (Keeley & Macphail 1981), with sands having grain sizes of 2.0–0.06 mm, silts with grain sizes of 0.06–0.002 mm and clays with grain sizes less than 0.002 mm. Grouping contexts according to their depositional environment must be done with caution because the categories 'river flood' and 'silted stream–ditch' include 16 and 7 contexts, respectively, whereas the categories 'estuarine', 'static', 'dump' and 'modern analogue' include only two contexts each. Such broad categorizations give little information about the specific depositional conditions affecting a particular deposit and can only be of limited value with such an unevenly distributed number of contexts, but do serve to allow general conclusions to be drawn.

Do the samples have a measurable remanent magnetization?

In the majority of cases the NRM of the sediment samples was easily measurable in a spinner fluxgate magnetometer, with some intensities as high as $4 \times 10^{-4}\,Am^2\,kg^{-1}$, comparable with the intensities of some fired materials. On this evidence, use of a more sensitive cryogenic magnetometer would be unnecessary for most sediments, even given the small sample size, although it would have the advantage of increasing the speed with which samples could be measured. The intensities of NRM were generally in a narrow range; samples with intensities an order of magnitude different from the main group usually also had anomalous remanence directions or showed atypical demagnetization behaviour, often because of a pebble or pottery inclusion in the sample revealed by dissection. Such anomalous remanences were rare, possibly because the sediments were chosen for their homogeneity and visible inclusions were easily avoided with the small sample holders. The problems of sample inhomogeneity encountered in earlier archaeomagnetic studies (Hammo-Yassi 1984) were largely avoided by carefully selecting undisturbed, homogeneous sediments with no visible evidence of iron redeposition or panning.

Is intensity of magnetization related to depositional environment or grain size?

The range of intensities of magnetization and sediment type, in order of increasing grain size, including the sediments with high organic content for comparison, are shown in Fig. 5. Excluded from this graph are samples shown, by dissection, to be anomalous (see above). Intensity tended to be higher in the finer-grained deposits of clay, silty clay or silt than in the sandier material, such as the sand horizons. This might be expected, as in the finer-grained deposits the magnetized grains are likely to have had more time in which to align in the geomagnetic field and to be less disturbed by

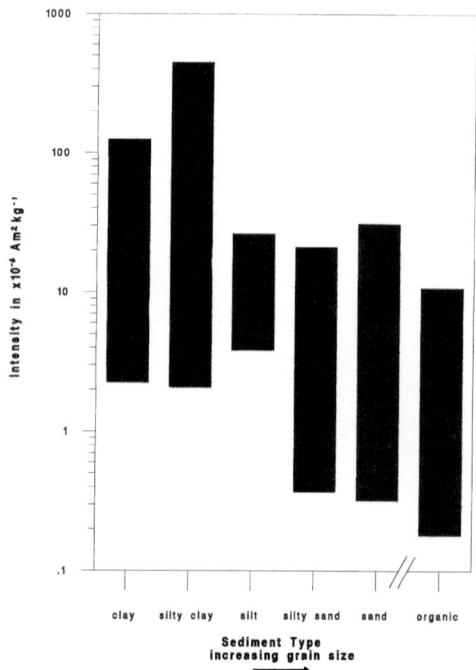

Fig. 5. The intensity of magnetization as a function of sediment grain size.

strong water currents. However, it must be remembered that the magnetic mineralogy and the concentration of these minerals are also significant factors. The plot also showed that sediments with high organic content had noticeably lower intensities of magnetization, which might arise from lower concentrations of magnetic minerals or disturbance of the remanence by bioturbation. The broad conclusions that can be drawn from these results are that most water-lain sediments have a measurable magnetization, but that finer grain sizes tend to have higher values. There appeared to be no relationship between depositional environment and intensity of magnetization.

Which contexts preserve a stable remanent magnetization?

The consistency index was used in the identification of the primary component of magnetization and assessing its consistency. Sixty-two per cent of samples had stable or very stable magnetizations (CI > 4.0) and there was typically a small, soft overprinted magnetization removed by demagnetization in fields of 7.5–15 mT, consistent with a viscous overprint of a primary magnetization. Lower consistency indices were generally associated with coarser-grained, sandy deposits or deposits with a high organic content. It could be suggested that the magnetization of sandy deposits was less stable as there would have been less time for remanence acquisition during deposition or the sample may have been more prone to grain reorientation caused by moisture loss after sampling. The organic sediments have a lower consistency index probably as they are more likely to have been bioturbated.

Which contexts preserve a consistently stable magnetic direction?

The distribution of the direction of the primary components of magnetization (where it was possible to isolate them), quantified by the α_{95} value, was of particular importance, as it could be taken as an indication of whether a group of samples retained a record of the geomagnetic field, although this did not demonstrate the time that the record was retained. A large α_{95} value indicated that the material was undatable, whereas a small value suggested that all samples from that context recorded the same field and meant that dating might be feasible, although the possibility of later overprinting had also to be considered. The broad indications were that small α_{95} values were associated with the finer-grained deposits (clay, silty clay and silt). In some cases, the α_{95} values were extremely small (for example, the Stakis Hotel samples described above), less than those commonly expected for fired materials. Sands and deposits with high organic content had consistently higher α_{95} values. In fact, all contexts with α_{95} values under 4° were deposits of clay, silty clay or silt, whereas 85% of contexts with α_{95} values of 4° or above were silty sand, sand or organic deposits. Bearing in mind the limitations of grouping contexts according to their depositional environment, as discussed above, it appeared (Fig. 6) that a wide range of α_{95} values occurred in river floods; estuarine, silted river–stream and static deposits gave consistently low values, whereas dump and modern analogue deposits were generally significantly higher. These observations suggested that contexts where there had been lowest energy of deposition, smallest grain size and least disturbance by bioturbation or water currents were most likely to yield a small α_{95} value, and therefore are more likely to be datable. It was also clear that low consistency indices (<3) were associated with high α_{95} values (>9°) whereas

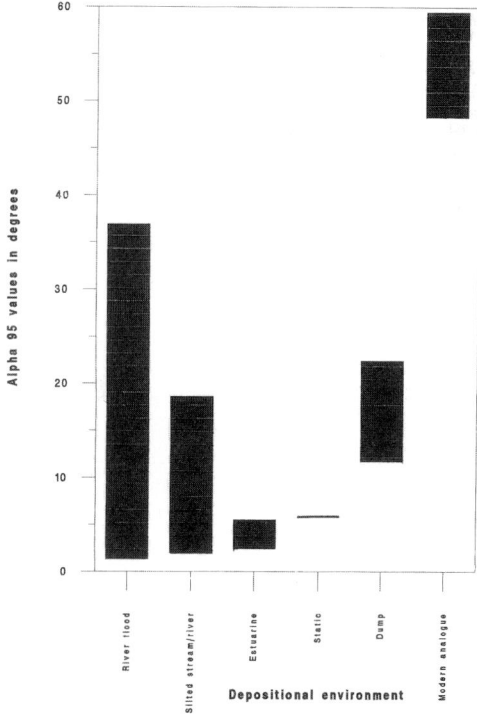

Fig. 6. α_{95} of stable remanence directions as a function of the depositional environment.

Table 1. *Summary of archaeological and archaeomagnetic information for contexts considered datable*

Site details	Context	Dec.	Inc.	α95	Archaeological date	Dating evidence	Archaeomagnetic date
ABC Cinema excavations, York							
Context 2307	river flood	359.9	66.6	2.0	late 1st–2nd century AD	pottery stratigraphy	AD 50–150
Context 2311	river flood	359.8	65.5	3.3	after context 2311 and 2313 AD 160–200 stratigraphy	pottery	AD 40–180
Context 2313	river flood	2.2	68.5	1.5	before context 2307 and 2311	after context 2313, before 2307 stratigraphy	AD 20–AD 100
Stakis Hotel excavations, York							
Context 4159	river flood	1.0	69.6	1.3	late 1st century AD	pottery	AD 20–80
Caldicot							
Context 174	silted-up stream	long section			before 837–993 BC after context 173	radiocarbon stratigraphy	350–1950 BC
Context 173	silted-up stream	long section			before 800–1260 BC 1650–1850 BC	radiocarbon archaeomagnetism	1950–2500 BC
Farnley	overbank river flood	43.7	78.1	3.1	late–mid-Bronze Age after 4600–4940 BC	standing stones above radiocarbon	550–1850 BC
Hartlepool							
Upper	estuarine	348.6	71.2	2.4	before 4330–4470 BC after 4750–5010 BC	radiocarbon radiocarbon	6000–6500 BC
West Heslerton	primary silt in well	2.8	63.7	5.8	AD 450–850	pottery from nearby structures	AD 50–500
Wood Hall Moated Manor							
Context 805	primary silt in ditch	353.4	61.2	1.9	AD 1200–1325 before AD 1327	nation-wide survey of moats documentary sources	AD 1300–1350

All magnetic directions are corrected to Meriden (Batt 1992).

consistency indices over three were usually, but not exclusively, associated with α_{95} values below 6°. It would seem reasonable to expect samples with low consistency to have poorly grouped magnetic directions.

Is the magnetization datable?

In all cases where the magnetization was shown to be stable and the magnetic directions well grouped ($\alpha_{95} < 6°$), an attempt was made to date the sediment by a visual comparison of the mean magnetic direction, and its error, with the British archaeomagnetic calibration curve (Clark et al. 1988). For material expected to date to before 1000 BC, the limit of the calibration curve, a comparison with the lake sediment secular variation (SV) record (Turner & Thompson 1982) was attempted. The comparison of a sequence of magnetic directions recording SV with the calibration curve, possibly using statistical pattern matching methods, should be more satisfactory than comparison of a single magnetic direction. Unfortunately, suitable sequences were rarely found in this study as the sedimentary deposits were frequently of short duration, often representing a single depositional event, or disturbed, with magnetic directions too dispersed for SV to be discerned. Even with a smooth, precise SV record, comparison with the calibration curve was extremely difficult as the deposition rate was unknown and hence the record of direction could be expanded or contracted over time, not necessarily uniformly. The visual comparison of mean magnetic directions with the calibration curve and the construction of the British calibration curve itself have a number of shortcomings and limitations, which have been discussed elsewhere (Batt 1997). These include lack of error representation on the calibration curve, uneven distribution of calibration data through time, subjective fitting of the curve itself by hand and the lack of statistical comparison of values. Bearing in mind these difficulties, dates were proposed for nine of the contexts studied, and these closely matched those suggested by the archaeological evidence (Table 1). The dated contexts included river floods, estuarine deposits, silted-up streams and primary silts in ditches and wells. Although the dates obtained in this way often corresponded well to the dates indicated by the archaeological evidence, suggesting a depositional model for the acquisition of remanence, the limitations of the calibration curve and the uncertainties in both the archaeomagnetic and the archaeological date, make conclusive statements impossible.

Is there any relationship between magnetic fabric and depositional environment?

As expected, specific susceptibility had no clear relationship to grain size or depositional environment. In the majority of cases, the magnetic fabric appeared to be similar to those found previously in natural environments and laboratory experiments (MacDonald & Elwood 1987). The distribution of minimum and maximum axes of susceptibility apparently provided useful indication of environment of deposition and sometimes indicated flow directions consistent with the archaeological evidence. However, in a number of cases the results were difficult to interpret. For example, laboratory experiments have suggested that values of q, the ratio of lineation to foliation, can be used as a characteristic of a 'primary depositional style' magnetic fabric (Hamilton & Rees 1970), but there is no indication from this study that small α_{95} values, which would suggest an undisturbed deposit, were associated with a large number of samples in the range indicated and vice versa (Fig. 7). Hence, it would appear that the definition of a sediment as having a 'primary depositional style' fabric on the basis of the q values is inappropriate for the sediments investigated here. The percentage of q values falling

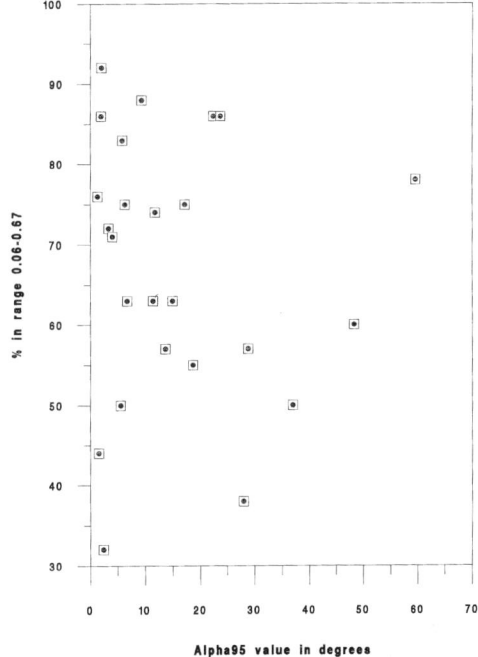

Fig. 7. The scatter of sample directions (α_{95}) as a function of their magnetic fabrics (q).

in the given range varied widely for river flood and silted stream–river deposits, probably simply a reflection of the number of different contexts included in these broad categories. A lower proportion of estuarine samples fell in the range thought to be associated with undisturbed depositional conditions, whereas a much higher proportion of samples from static environments fell into that range, as might be expected. However, caution must be exercised when drawing such conclusions, as a high proportion of samples from dump and modern analogue deposits appeared to exhibit characteristics of primary depositional conditions! It must be borne in mind that laboratory experiments and theoretical models are very simplified. Archaeological sediments reflect particularly complex environments, with both human and natural influences acting on the sediment. The sediments are composed of a number of magnetic and non-magnetic minerals of a wide variety of grain sizes and the magnetic fabric of such materials is not well characterized.

What is the magnetic mineralogy of the archaeological sediments?

Magnetic mineral studies proved useful in the identification of remanence carriers and in characterizing their behaviour. A comparison of SIRM and initial intensity was used to indicate whether low intensity of remanence was a result of absence of remanence-carrying grains or non-alignment of existing magnetized grains. SIRM values are affected by too many factors, including magnetic grain size, concentration and mineralogy, so that comparison between different contexts is difficult. A more informative measurement is S, which gave a normalized measurement of the concentration of magnetic minerals that saturated in low fields, in comparison with those saturating in fields over 1 T. In detrital material, this is likely to be a reflection of the ratio of concentration of magnetite to goethite–hematite. In this study it was noticeable that almost all the samples had high S values (>7.5) suggesting that magnetite was the dominant material. The coercivity of remanence is determined solely by magnetic mineralogy and magnetic grain size and is therefore independent of concentration of magnetic minerals. In this study the values ranged between 13 and 37 mT, generally in the range indicative of SD grains. As expected, there is no correlation between magnetic mineralogy and depositional environment, as so many other factors also influence depositional environment.

Conclusions and cautions

This study has indicated that some water-lain sediments from archaeological contexts appear to retain a stable record of the geomagnetic field at the time they were deposited. The most suitable appear to be fine-grained, undisturbed deposits, with low organic content, laid down in slow-moving or static water conditions. For such sediments, the results obtained are at least as good as those from fired materials, and it appears that bioturbation and strongly magnetic inclusions do not cause significant problems. If this record is of the geomagnetic field at the time of deposition, the sediment can therefore potentially be dated by archaeomagnetism. However, the calibration of remanence direction to obtain a date is a problem common to the study of both fired materials and sediments. Although useful information can potentially be obtained from other magnetic properties, such as the anisotropy of magnetic susceptibility and IRM values, this is sometimes difficult to interpret in such complex depositional environments. It is not possible to use archaeomagnetic dates to investigate whether the magnetization recorded is depositional or post-depositional in origin, as the precision of the archaeomagnetic dates is not sufficient, although the results from the modern analogue deposit would suggest that the magnetization is post-depositional in origin. However, until this issue is resolved archaeomagnetic methods cannot be used to date sediments, as the event to be dated is unclear (Verosub 1977).

In summary, archaeomagnetic studies have the potential to date water-lain archaeological sediments and to determine their environment of deposition and magnetic mineralogy. However, a number of problems remain with the underlying methods of the calibration of magnetic directions for the archaeomagnetic dating of both fired materials and sediments, which must be addressed if the technique is to reach its full potential. An investigation of modern analogue sediments might contribute, as the field direction and date of deposition are known and laboratory experiments cannot satisfactorily replicate the depositional conditions. In particular, sequences of recent secular variation would prove very valuable, if it were possible to locate them.

This work was carried out in the Department of Archaeology, University of Durham, in collaboration with M. Noël, and was made possible by a grant from SERC. I would like to thank York Archaeological Trust in general and P. Addyman and P. Ottaway in particular for access to the Stakis site and many helpful comments throughout the research.

References

AITKEN, M. J. 1970. Dating by archaeomagnetic and thermoluminescent methods. *Philosophical Transactions of the Royal Society of London, Series A*, **A269**, 77–88.

AS, J. A. & ZIJDERVELD, J. D. A. 1958. Magnetic cleaning of rocks in palaeomagnetic research. *Geophysical Journal of the Royal Astronomical Society*, **1**, 308–319.

BATT, C. M. 1992. *Archaeomagnetic dating: investigating new materials and techniques.* PhD thesis, University of Durham.

—— 1997. The British archaeomagnetic calibration curve: an objective treatment. *Archaeometry*, **39**, 153–168.

—— & NOËL, M. 1991. Developing new themes in archaeomagnetism. *In*: PERNICKA, E. & WAGNER, G. A. (eds) *Archaeometry '90: International Symposium on Archaeometry*. Birkhäuser, Basel, 541–550.

CLARK, A. J., TARLING, D. H. & NOËL, M. 1988. Developments in archaeomagnetic dating in Britain. *Journal of Archaeological Science*, **15**, 645–667.

COLLINSON, D. W. 1983, *Methods in Palaeomagnetism and Rock Magnetism*. Chapman and Hall, London.

EIGHMY, J. L. & HOWARD, J. B. 1991. Direct dating of prehistoric canal sediments using archaeomagnetism. *American Antiquity*, **56**, 88–102.

FISHER, R. A. 1953. Dispersion on a sphere. *Proceedings of the Royal Society of London, Series A*, **A217**, 295–305.

HAMILTON, N. & REES, A. I. 1970. The uses of magnetic fabric in palaeocurrent estimation. *In*: RUNCORN, S. K. (ed.) *Palaeogeophysics*. Academic Press, London, 445–464.

HAMMO-YASSI, N. 1984. *Archaeomagnetic work in Britain and Iraq.* PhD thesis, University of Newcastle upon Tyne.

IRVING, E. 1957. Origin of the palaeomagnetism of the Torridonian sandstones of northwest Scotland. *Philosophical Transactions of the Royal Society of London, Series A*, **A250**, 100–110.

KEELEY, H. C. M. & MACPHAIL, R. I. 1981. A soil handbook for archaeologists. *Institute of Field Archaeologists Bulletin*, **18**, 225–241.

MACDONALD, W. D. & ELWOOD, B. B. 1987 Anisotropy of magnetic susceptibility. *Reviews of Geophysics*, **25**, 905–909.

MOLYNEUX, L. 1971. A complete result magnetometer for measuring the remanent magnetization of rocks. *Geophysical Journal of the Royal Astronomical Society*, **24**, 429–433.

NAGATA, T. 1962. Notes on detrital remanent magnetization of sediments. *Journal of Geomagnetism and Geoelectricity*, **14**, 99–106.

NOËL, M. & BATT, C. M. 1990. A method for correcting geographically separated remanence directions for the purpose of archaeomagnetic dating. *Geophysical Journal International*, **102**, 753–756.

—— & RUDNICKI, M. D. 1988. A computer program for determining current directions from rock magnetic data. *Computers and Geosciences*, **14**, 321–338.

OLDFIELD, F., KRAWIECKI, A., MAHER, B., TAYLOR, J. J. & TWIGGER, S. 1985. The role of mineral magnetic measurements in archaeology. *In*: FIELLER, N. R. J., GILBERTSON, D. D. & RALPH, N. G. A. (eds) *Palaeoenvironmental Investigations: Research Design, Methods and Data Analysis.*, British Archaeology Reports. International Series, **258**, 29–43.

OTTAWAY, P. 1988. According to plan – the Stakis Hotel site. *Interim: Archaeology in York*, **13.4**, 13–22.

TARLING, D. H. 1983. *Palaeomagnetism*. Chapman and Hall, London.

—— & SYMONS, D. T. A 1967. A stability index of remanence in palaeomagnetism. *Geophysical Journal of the Royal Astronomical Society*, **12**, 443–448.

THOMPSON, R. & OLDFIELD, F. 1986. *Environmental Magnetism*. Allen and Unwin, London.

TURNER, G. M. & THOMPSON, R. 1982. Detransformation of the British geomagnetic secular variation record for Holocene times. *Geophysical Journal of the Royal Astronomical Society*, **70**, 789–792.

VEROSUB, K. L. 1977. Depositional and post-depositional processes in the magnetization of sediments. *Reviews of Geophysics and Space Physics*, **15**, 129–143.

Studies of post-depositional remanent magnetization and their relevance to the palaeomagnetic record

S. A. PISAREVSKY

All-Russian Petroleum Scientific Research and Geological Exploration Institute (VNIGRI), Liteiny 39, 191104 St Petersburg, Russia (e-mail: psa@oloy.spb.su)

Abstract: The results of experiments on the watering of different types of continental Quaternary sediments in the anomalous magnetic field show that it can cause strong changes in the direction of remanence. The direction of acquired post-depositional remanent magnetization is stable to thermal and alternating field (AF) demagnetization. These changes are not related to the chemical remagnetization and are probably caused by mechanical rotations of magnetic particles in the pore spaces. Post-depositional remagnetization, both in nature and after collection of sediment samples, can distort the palaeomagnetic record of geomagnetic excursions, events and polarity transitions, as well as altering the fine-structure features, such as the palaeomagnetic record of secular variaitons.

Post-depositional remanent magnetization (PDRM) is one of the most important sources for the natural remanent magnetization (NRM) of unconsolidated sediments (Irving & Major 1964). Several studies of PDRM has been carried out since those of Noël (1980), including those of Arason & Levi (1990), Lovlie (1994) and Meynadier & Valet (1996). These have mostly dealt with marine and artificial (laboratory deposited) sediments, although continental deposits are also palaeomagnetically important. For example, numerous records of geomagnetic events and excursions have been obtained from Quaternary continental sediments, including loess and drift deposits (e.g. Zubakov et al. 1982; Tretyak et al. 1989; Coe & Liddicoat 1994). However, such records are not very internally consistent, i.e. records from parallel sections may show the same palaeomagnetic anomaly in particular layers, but this signature may be different in different sections. It can also differ from that in marine sediments and lava sections, even during the Brunhes epoch. The cause of such differences may reflect some additional distortion of the palaeomagnetic record in continental sediments, possibly relating to the irregularities in the post-depositional remagnetization of continental sediments because of the penetration of atmospheric and ground water, whereas most marine sediments have been deposited in more consistent subaqueous conditions. Noël (1980) proposed a model of the

Table 1. *Angular changes in the NRM of continental samples*

N	Sample ref.	γ	α	$\beta_{1/2}$	β_1
Loess, Tashkent area					
1	ks-7	136	13	0	26
2	ks-90	174	1	4	5
Loess, Taman peninsula					
3	605-10	155	15	–	42
4	605-31	128	59	14	31
Varved clay, St Petersburg					
5	lg-35	117	1	2	7
Lake sediments, Voronezh					
6	21–76	71	8	23	64
7	21–91	148	18	94	–
8	21–74a	151	2	–	31
9	21–74b	147	2	–	58
Drift sediments, Voronezh					
10	c-7	162	5	77	112
11	c-12	166	5	21	90
12	c-31	158	13	13	32
13	c-33	152	18	13	18
14	c-43	144	25	–	59
15	c-10a	159	9	–	33
16	c-10b	161	9	–	25

γ is the angle between the direction of local geomagnetic field and that of NRM; α is the angle between the directional changes of NRM caused by the viscosity; $\beta_{1/2}$ is the directional change of NRM caused by watering to half of the 'critical' value and β_1 to the actual 'critical' wetting value.

behaviour of the remanence during the drying-out of sediments, such as occurs in changes in the water table. His experiments with artificial sediments showed that strong changes in the intensity and direction of remanence occurred during both wetting and drying processes. These were not attributable to magnetic viscosity, but could be explained by surface-tension forces changing the orientation of interstitial grains. Such a process is likely to operate, not only under natural changes in the level of the water table, but also in samples extracted from sediments if their interstitial fluid or gases change. Although Noël demonstrated that some features in studies of Scandinavian Quaternary varved sediments could be explained by such processes, it is very difficult to simulate natural compositions and processes in the laboratory correctly, and one of the main aims of this study is to examine the post-depositional remagnetization in a range of continental sediments (drift deposits, varved clays and lake sediments) during their watering and drying, and to estimate the magnitude of such changes. The main questions were: (1) Does wetting and drying in the presence of magnetic field of different direction lead to similar changes of the NRM direction in natural continental sediments, as was shown for the artificial sediment by Noël (1980)? (2) Are magnitudes of these changes large enough to have a sufficient influence on the palaeomagnetic records? (3) Do these changes also exceed the effects of magnetic viscosity? (4) Are these changes in natural sediments caused predominantly by mechanical rotation of magnetic particles, or by chemical changes? Preliminary answers to some of

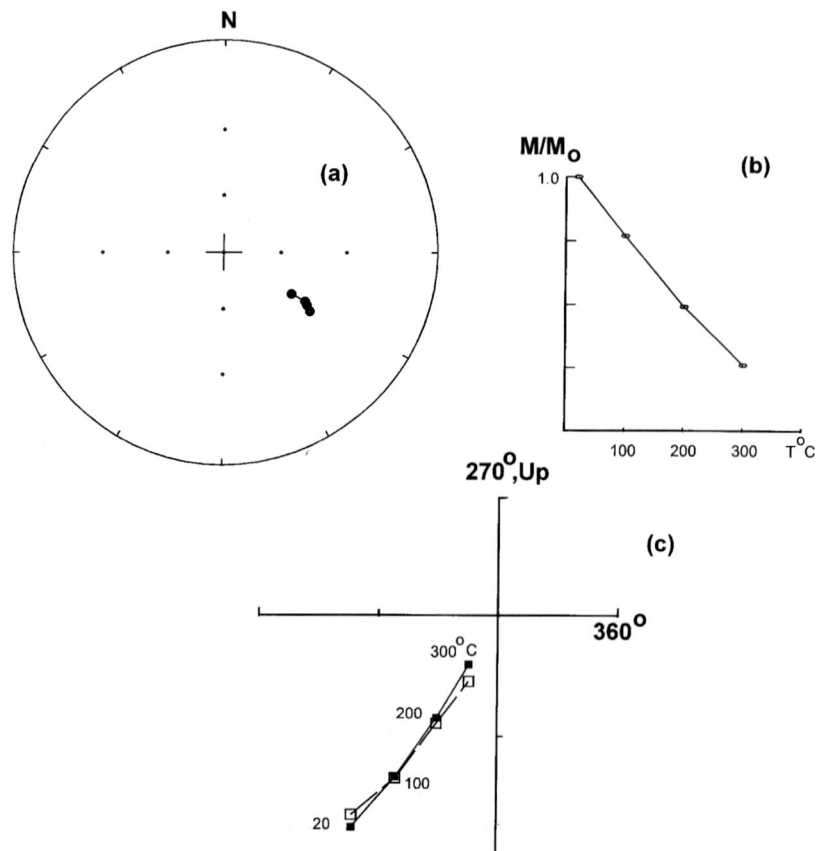

Fig. 1. An example of thermal demagnetization of the PDRM. (**a**) A stereographic projection of characteristic remanent magnetization (ChRM) directions; filled and open symbols represent downward and upward directions, respectively. (**b**) The demagnetization curve. (**c**) Zijderveld (1967) diagram; filled and open symbols denote horizontal and vertical projections, respectively.

these questions have been published (Pisarevsky 1983), and confirmed by Minyuk (1986); new results and conclusions are presented here.

Samples

The following sediments have been studied: (1) Upper Quaternary varved clays from the St Petersburg area, Russia; (2) Upper Quaternary drift deposits and lake sediments from the Voronezh area, Russia; (3) Middle Quaternary loess and drift deposits sediments from the Taman peninsula, Black Sea, Russia; (4) Upper Quaternary loess from the Tashkent area, Uzbekistan. All these samples have been taken from large collections made for palaeomagnetic studies (Pisarevsky & Toychiev 1981; Praslov et al. 1981; Zubakov et al. 1982; Pisarevsky 1986). Thermomagnetic analyses had shown they contained a variety of magnetic minerals from iron hydroxides to hematite. Four cubic (2.4 mm) specimens were prepared from each sample. One of these was used to determine the maximum (critical) value of water that could be added without destroying it. After that the

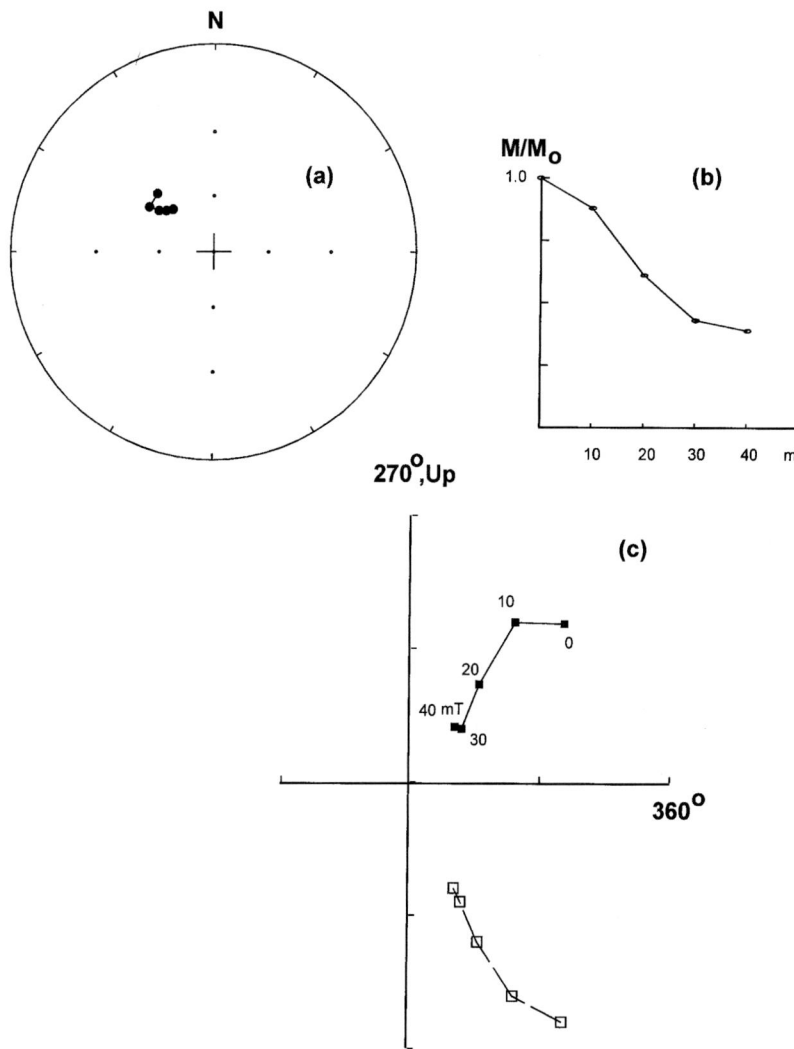

Fig. 2. An example of AF demagnetization of the PDRM. Symbols as for Fig. 1.

remaining specimens were placed in non-magnetic holders. The NRM and low-field susceptibility of 48 remaining specimens were then measured using an astatic magnetometer in the VNIGRI palaeomagnetic laboratory (St Petersburg). The porosity of some of the sediments used was also measured in the VNIGRI laboratory of physical properties of rocks and found to vary between 33 and 51%. (Future work will also measure the permeability.)

Procedure and results

Three specimens from 16 samples were placed within a steady field of the order of the local geomagnetic field (about 0.05 mT) provided by a square coil system used normally for the compensation of the Earth's magnetic field during palaeomagnetic studies, with their NRM directions at random, but large, angles to the applied field (see Table 1). Immediately after placing them in the coils, one of specimen from each sample was watered with the critical amount of water. Another was watered with one-half of this value, and the third specimen was not watered at all. These were then left for few weeks for drying out before remeasuring their NRM and susceptibility. The directional changes in the NRM were then evaluated for three angular properties: α, the angle between vector of remanence before and after experiment for the untreated specimen (i.e. the specimen that was not subjected to watering), to allow the magnetic viscosity to be calculated; $\beta = \beta' - \alpha$ the corresponding angle for the watered sample, β', less the angle, α, determined for magnetic viscosity; and γ, the angle between the direction of local geomagnetic field and that of NRM in the position of the sample during watering and drying.

All specimens showed an ability to acquire a strong remanent magnetization as a result of secondary watering, but its value varied strongly for different sediments (Table 1), being smallest in the varved clays and some loess from the Tashkent area. Only in two specimens were the viscous components large (Table 1) but even in these specimens, the PDRM was clearly recognizable. The change in susceptibility before and after the experiment was also small, up to a maximum of 5%, suggesting the absence of significant chemical changes in magnetic carriers of the studied sediments. Consequently, the observed rotations in the remanence vector because of a mechanical rotation of magnetic particles in pores seem to be the explanation for this remagnetization. Thermal (up to 300°C) and alternating field (AF) (up to 40 mT) stepwise demagnetizations of six specimens (Figs 1 and 2) showed that the PDRM was stable to both forms of demagnetization.

As continental sediments are subjected to natural changes in moisture content on repeated, irregular time scales, repeating wetting and drying was attempted using the drift deposits from the Voronezh area. These showed (Fig. 3) that although complete remagnetization could be achieved after three repeat cycles, the main part of the remagnetization occurred during the first step. The next experiment simulated a major change in the ambient field, such as could occur during a geomagnetic reversal or event. In this case, four specimens of Voronezh drift sediments and Taman loess that had already been watered, dried and measured were placed at large angles (80°–180°) to the direction of the present-day field and then rewatered to the critical value. In some cases, the record of 'excursion' was been erased completely (Fig. 4a); in other cases, partially (Fig. 4b). An additional experiment was undertaken using 19 specimens from outcrops in one part of an Upper Quaternary section (Kostyonki 14) in the Voronezh area. The same procedure was used, but this time the specimens were turned so that the angles between the NRM vector and the local geomagnetic field were between 145° and 175° (Fig. 5a).

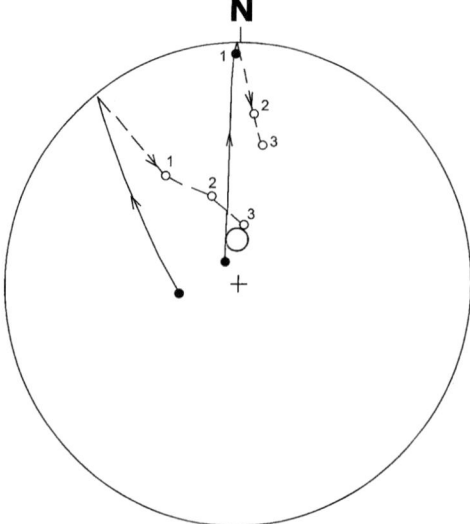

Fig. 3. Examples of repeated wetting of continental sediment samples. Filled and open symbols represent downward and upward directions, respectively. Larger open circle shows the direction of the local field during experiment (the spatial position of the stereogram is arbitrary).

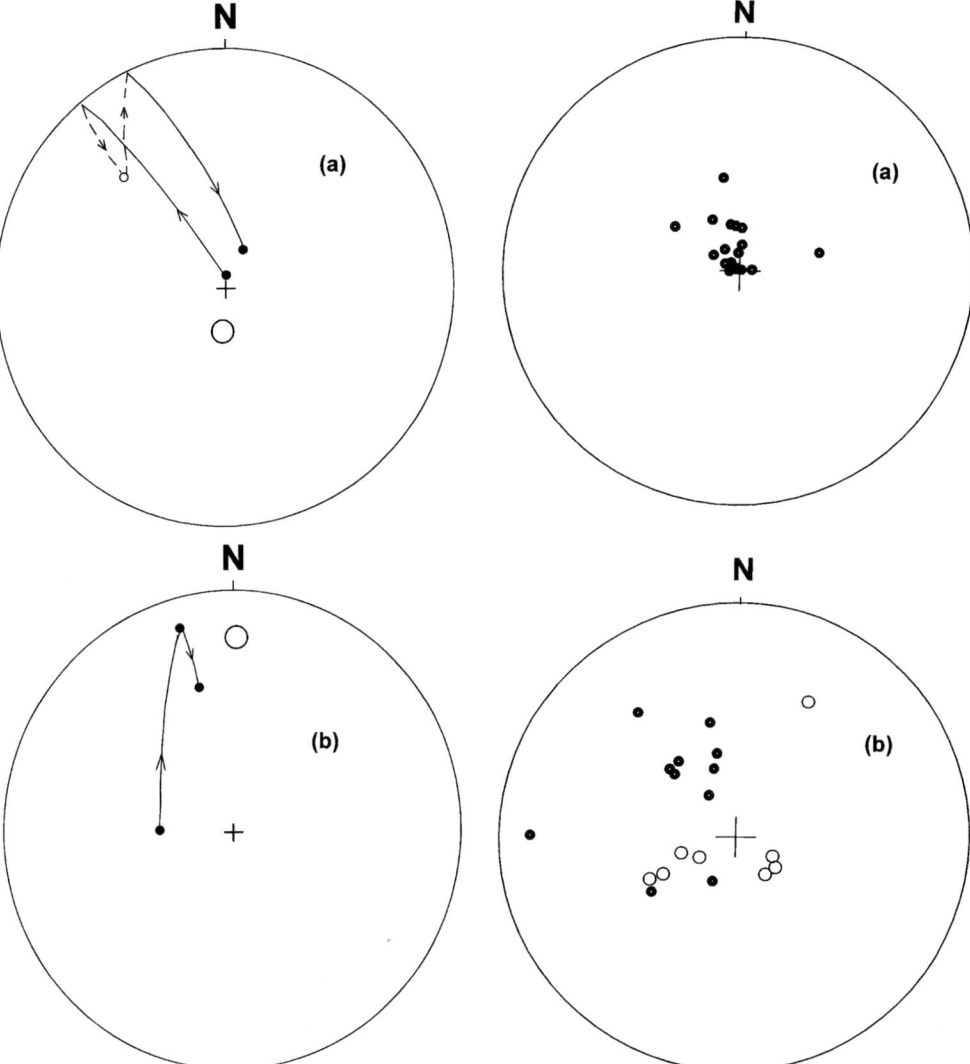

Fig. 4. Examples of the emulation of the excursion of geomagnetic field. Symbols as for Fig. 3. (**a**) The complete destruction of the 'record' of the geomagnetic event, and (**b**) its partial destruction.

Fig. 5. Stereoplots of remanence directions of drift sediments from the Voronezh area (**a**) initial; (**b**) after watering and drying in the presence of a reversed magnetic field.

These samples were then watered, dried and remeasured (Fig. 5b). The behaviour of the remanence varied strongly. In some, it did not change at all, whereas in others if had been reversed completely, with other samples showing 'intermediate' behaviour. The viscous component in these specimens did not exceed a few per cent of NRM, and no correlation was found between the post-depositional changes of remanence and porosity.

Discussion and conclusions

These experiments, using different types of continental sediments, confirm the experimental evidence of strong directional changes during wetting and drying obtained by Noël (1980). It is therefore clear that such post-depositional processes can be a fundamental influence on the observed properties of many types of continental sediments, and that these are not readily removed during routine thermal or AF demagnetization procedures. The experiments were not designed to examine the mechanism of this type of PDRM, but demonstrated that it was of a mechanical rather than chemical origin, and there are no reasons for not accepted the surface-tension effects proposed by Noël (1980). The oldest sediments studied here have an age of about 400 ka, and these were still able to be remagnetized by changes in the fluid and gaseous environment within their pore spaces. This also means that similar such effects may occur in all sediments that are allowed to dry out after collection and before measurement.

An obvious requirement for watering to have any effect on the direction of remanence is that the direction of the ambient field must have changed. If the directional change in the field is small, then the remagnetization effect is also small. Consequently, such changes are particularly significant if wetting takes place after there has been a major change in the orientation of the geomagnetic field, such as during a large-scale geomagnetic event, excursion or polarity transition. It is possible that its record of such behaviour will be 'smeared' over time intervals of a few hundred years if there are changes in the water table or inadequate care is taken to avoid drying out before measurement.

The author is grateful to A. N. Khramov and V. V. Kochegura for detailed discussions and advice, and to M. A. Kovalevsky for measurements of porosity.

References

ARASON, P. & LEVI, S. 1990. Models of inclination shallowing during sediment compaction. *Journal of Geophysical Research*, **95**, 4481–4499.

COE, R. S. & LIDDICOAT, J. C. 1994. Overprinting of natural magnetic remanence in lake sediments by a subsequent high-intensity field. *Nature*, **367**, 57–59.

IRVING, E. & MAJOR, A. 1964. Post-depositional detrital remanent magnetization in a synthetic sediment. *Sedimentology*, **3**, 135–143.

LOVLIE, R. 1994. Field-dependent postdepositional grain realignment in the PDRM process: experimental evidence and implications. *Physics of the Earth and Planetary Interiors*, **85**, 101–111.

MEYNADIER, L. & VALET, J.-P. 1996. Post-depositional realignment of magnetic grains and asymmetrical saw-tooth patterns of magnetization intensity. *Earth and Planetary Science Letters*, **140**, 123–132.

MINYUK, P. S. 1986. The influence of freezing processes on the magnetization of deposits. *In: Geomagnetic investigations in the East of the USSR*. DVNC of Academy of Sciences of the USSR, Magadan, 1–31 (in Russian).

NOËL, M. 1980. Surface tension phenomena in the magnetization of sediments. *Geophysical Journal of the Royal Astronomical Society*, **62**, 15–25.

PISAREVSKY, S. A. 1983. Postpositional remanent magnetization of the continental deposits. *Fizika Zemli*, **N12**, 76–81 (in Russian).

—— 1986. The confirmation of reality of Gothenburg excursion by results of research of varved clays of Leningrad area. *III All-Union Congress in Geomagnetism, Kiev*, 125–126 (in Russian).

—— & TOYCHIEV, H. A. 1981. About an excursion of geomagnetic field found during study of Holocene deposits in Pre-Tashkent area. *XI Congress in Geomagnetism, Tbilisi*, 143 (in Russian).

PRASLOV, M. A., IVANOVA, L. A., GUGALINSKAYA, M. M. & PISAREVSKY, S. A. 1981. Kostyonki 21 (Gmelinskaya stoyanka). *In: Archeology and Paleogeography of the Late Paleolite in Russian Plain*, Leningrad, 6–18 (in Russian).

TRETYAK, A. N., VIGILYANSKAYA, L. I., MAKARENKO, V. N. & DUDKIN, V. P. 1989. *Fine Structure of the Geomagnetic Field in the Late Cenozoic*. Naukova Dumka, Kiev (in Russian).

ZIJDERVELD, J. D. A. 1967. A.C. demagnetization of rocks: analysis of results. *In:* COLLINSON, D. W., CREER, K. M. & RUNCORN, S. K. (eds) *Methods in Palaeomagnetism*. Elsevier, Amsterdam, 254–286.

ZUBAKOV, V. A., PISAREVSKY, S. A. & BOGATINA, N. B. 1982. Detail stratification, stratigraphic thickness and the age of the Karangat horizon in Black Sea area. *Doklady Akademii Nauk SSSR*, **267**, 26–429 (in Russian)

Viscous remanent magnetization of high thermal stability in limestone

GRAHAM J. BORRADAILE

Geophysics Department, Lakehead University, Thunder Bay, Ont., Canada, P7B 5E1
(e-mail: borradaile@lakeheadu.ca)

Abstract: Undisturbed Mesozoic limestones of eastern England show univectorial magnetizations of high thermal stability that appear, on first inspection, to be primary geological remanences. However, viscous remanent magnetization (VRM) of anomalously high thermal stability is an alternative possibility. VRM acquired elsewhere in laboratory experiments, or in a few hundred years as in the examples studied here, may require unblocking temperatures >300°C to demagnetize. Normally, such demagnetization temperatures are required to remove only very ancient geological remanences. The classical 'conglomerate test' of palaeomagnetism fails to discriminate between a VRM and a primary remanence. Any lithology sensitive to VRM will yield a negative result, suggesting that the remanence is secondary. However, tests of physical phenomena that caused multiple reorientations of clasts, at rotation rates comparable with the viscous remagnetization rate, provide a more suitable field test for VRM. In this study, these latter kinds of test reveal that some *in situ* Mesozoic limestones of eastern England have VRM, despite stability to temperatures >300°C. The high thermal stability VRM of the bedrock must have been acquired in the last 0.7 Ma, since the last geomagnetic reversal. Two new field tests detected the viscous nature of the high stability remanence. The first was a 'debris flow test', in which limestone blocks spin slowly with respect to the geomagnetic field. The blocks record variably tilted or helicoidal, multicomponent vectors. The second was a 'masonry test', in which samples of limestone from buildings of known age were exposed to the geomagnetic field for known periods, up to 1600 years in this study. This shows that unusually stable VRM may be common in carbonates. Finally, the masonry test permits an empirical estimate of viscous remagnetization rates over hundreds of years. This established chronometric curves that date ground disturbances of geotechnical or geomorphological interest, and archaeological features of unknown age.

Theory of viscous remanent magnetization

Viscous remanent magnetization (VRM) is a time-dependent magnetization with acquisition (S_a) and decay coefficients (S_d) that predict, from short-term laboratory experiments, a log(time) dependence of remanence intensity. Thus, where M is magnetization and t is time,

$$S_a = \frac{dM}{d(\log t)} \quad \text{and} \quad \frac{dM}{d(\log t)} \tag{1}$$

S_d may be less than S_a, depending on the nature of the grain-size distribution, and the coefficients increase with temperature proportional to T/M_s or $\sqrt{(T/M_s)}$ (Walton 1980; Dunlop 1989; Dunlop & Özdemir 1997). For magnetite, the coefficients have their largest values for grain sizes near 30–40 nm. Dunlop showed that for short laboratory time scales, $M \propto \log t$ yielded the above formulae, except where grain interactions are particularly strong (Walton & Dunlop 1985). Dating methods in archaeology or geomorphology can use VRM because it may have a predictable acquisition rate after some rock sample has been disturbed by slumping or construction (Heller & Markert 1973; Borradaile 1996; Borradaile & Brann 1997). As the acquisition and decay coefficients are not constant (contrary to the assumption of Heller and Markert (1973)) the acquisition rate or remagnetization rate must be calibrated using another dating method, e.g. radiocarbon or historical documentation (Borradaile & Brann 1997; Borradaile et al. 1998). Empirically determined acquisition rates reveal a power-law relationship for VRM acquisition over the last 2000 years in some English limestones bearing pseudo-single-domain (PSD) magnetite and will be discussed later.

The time-dependent acquisition of thermoviscous remanent magnetization (TVRM) was given by Néel (1955):

$$\tau = \frac{1}{C} \exp \left[\frac{VH_c M_s}{2kT} \right] \tag{2}$$

The relaxation time (τ) of a viscous-remanence component depends on the following rock magnetic properties: coercivity (H_c), saturation magnetization (M_s) and effective grain volume (V). Absolute temperature (T) is the most significant variable. One also requires Boltzmann's constant (k) and a frequency factor (C) that may vary from $10^9 \, \text{s}^{-1}$ (Butler 1992) to $10^{11} \, \text{s}^{-1}$ (Pullaiah et al. 1975). The choice of these parameters can affect the shape of the nomograms significantly, and the validity of published curves to one's own study should be justified before use.

Néel's equation shows the relationship of VRM and thermal remanent magnetization (TRM) as end-members of a continuous spectrum of behaviour. All TRMs require some finite time to develop, the acquisition constant being higher at higher temperatures. This principle led Pullaiah et al. (1975) and Kent (1985) to estimate acquisition ages from the temperatures required to demagnetize samples on a laboratory time scale. The principle uses theoretically derived temperature–time nomograms that relate the long relaxation time of a VRM at surface temperatures (e.g. 5000 years at 30°C) to the much shorter laboratory relaxation time when the sample is demagnetized thermally in the laboratory at, for example, 250°C during an experiment lasting <1 h. However, for remanences acquired over geological time scales, several workers found unblocking temperatures higher than predicted by Néel, single-domain (SD) theory (e.g. Kent 1985; Jackson & Van der Voo 1986). Moreover, for historically documented buildings I found that neither the nomograms of

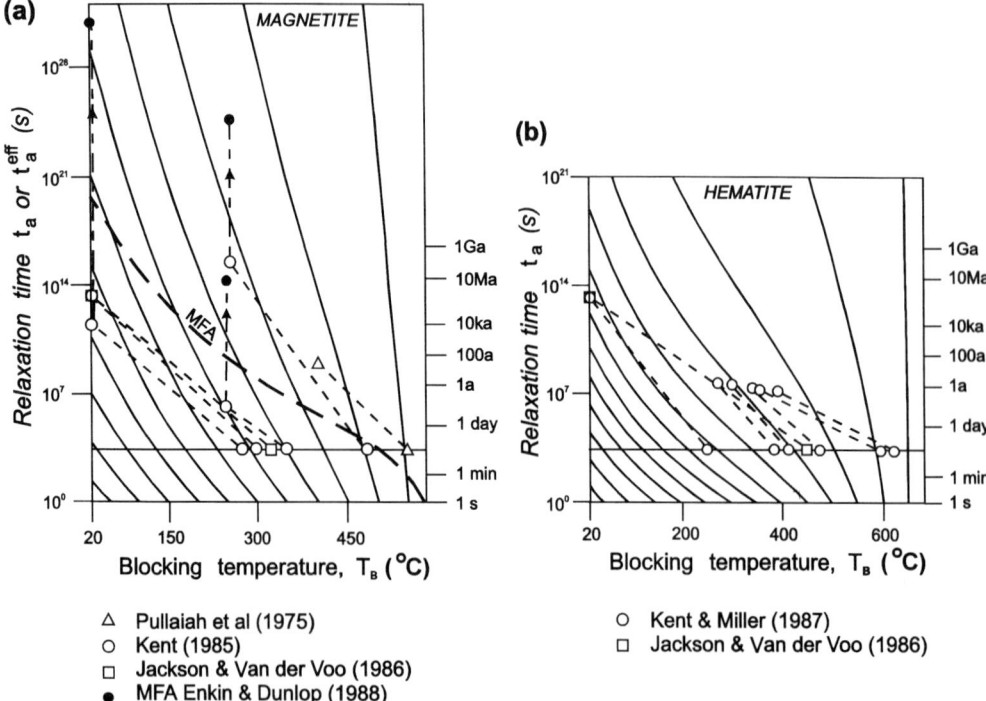

Fig. 1. Nomograms relating the demagnetization or unblocking temperature (T_{UB}) necessary for a laboratory test. Laboratory test times, typically lasting less than 1 h, are those that will remove geological remanence acquired at a lower temperature over geological time. Curves for (**a**) magnetite and (**b**) hematite are shown from Enkin & Dunlop (1988) derived from Néel's SD theoretical relationship. For a given experiment, tie-lines connecting (t_{LAB}, T_{LAB}) with (T_{GEOL}, T_{GEOL}) should follow the nomogram curves. In some cases a much higher laboratory demagnetization temperature (T_{LAB}) is required, and examples from the literature are illustrated. Such samples are considered to have anomalously high thermal stability, the subject of this paper. Strictly speaking, the graphs are valid only for SD, non-interacting grains at low fields. Better estimates of the laboratory unblocking temperatures are obtained when one assumes that acquisition and decay rates are not equal, for which a medium field approximation (MFA; see (**a**)) provides a correction factor that adjusts some data points to more satisfactory positions on the curves (Enkin & Dunlop 1988).

Pullaiah et al. (1975) nor those of Middleton & Schmidt (1982) could account for the high unblocking temperatures. Of course, my studies involved archaeological time scales and PSD magnetite.

Alternating field (AF) demagnetization of VRM may also be difficult in certain, especially hematite-bearing sedimentary rocks (red beds) and some rocks with fine-grained titanomagnetite (Biquand & Prévot 1971, Dunlop & Özdemir 1990, 1997). Dunlop & Stirling (1977) showed that high coercivity is the fundamental cause of this phenomenon but that fine grain size is a prerequisite for the acquisition of the VRM. Because of this early recognition of the failure of AF demagnetization to distinguish VRM from characteristic remanences, thermal demagnetization is the preferred routine treatment by the palaeomagnetic community (Dunlop & Özdemir 1997). However, this paper is preoccupied with VRM that is resistant to thermal cleaning.

The time–temperature nomograms first presented by Pullaiah et al. (1975) are here shown in the form reviewed by Enkin & Dunlop (1988) as Fig. 1a and b.

From Néel's equation, Dunlop & West (1969) derived time–temperature relationships that Pullaiah et al. (1975) used to generate nomograms from the following equation:

$$\frac{T_d \log(Ct_d)}{J_s(T_d)H_c(T_d)} = \frac{T_a \log(Ct_a)}{J_s(T_a)H_c(T_a)} \quad (3)$$

Here, a magnetization acquired in time t_a at temperature T_a should demagnetize in a relaxation period t_d at temperature T_d in the laboratory. J_s and H_c depend on temperature and the relationship may not be well understood. The nomograms provide a simple graphical means of estimating geological relaxation or acquisition times (assuming for the moment that they are equal) from the much shorter laboratory times t_d at temperature T_d. The theory is well justified for strictly SD grains (Williams & Walton 1988). However, field studies on limestones bearing PSD magnetite and on red beds with VRM acquired over either geological (Fig. 1) or archaeological time scales (Fig. 2) show that the nomograms may be inaccurate (e.g. Kent 1985; Kent & Miller 1987). Enkin & Dunlop (1988) drew attention to the fact that VRM acquisition in the Earth's field (equal to the medium field approximation (MFA) ~50 µT of Enkin & Dunlop (1988)) is faster than decay in near-zero field in the laboratory experiment. Thus, they determined and MFA correction factor (Fig. 1a) that adjusted some previously reported anom-

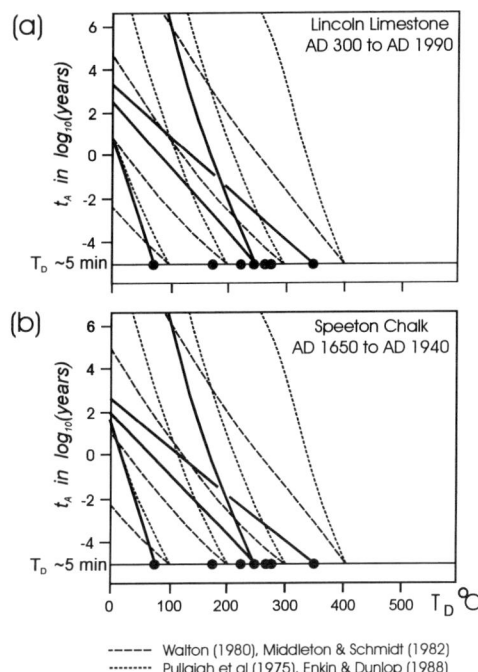

Fig. 2. Tests of the nomograms over archaeological time scales. These are derived from precisely dated historical buildings and show that anomalously high laboratory T_{UB} (>300°C) may be needed to erase young remanences only a few hundred years old.

alously high unblocking temperatures to yield more reasonable geological ages that would result from acquisition in a geomagnetic field (Fig. 1a especially data of Kent (1985)).

Most natural grain-size distributions are crudely log-normal because aspect ratios must be >1 and normal geological environments prevent the survival of high aspect ratios for mineral habits (e.g. >5). Thus, Walton (1980, 1983) modified equation (3), and that permitted the revision of the nomograms:

$$\frac{T_d[\log(Ct_d)]^2}{J_s^2(T_d)} = \frac{T_a[\log(Ct_a)]^2}{J_s^2(T_a)} \quad (4)$$

The geological community was optimistic for the initial prospects of this approach because it appeared to support higher unblocking temperatures (e.g. Middleton & Schmidt 1982). However, this is now regarded as empirical serendipity because, as Dunlop (1986a) and Jackson & Van der Voo (1986) noted, equation (4) assumes that the full gamut of grain sizes unblocks simultaneously. In reality, different portions of the grain-size spectrum unblocked at different temperatures. However, equation (3) concerns SD magnetite grains of a very restricted size range and is also unsatisfactory. Despite the 'misapplication' of nomograms based on equation (4), they may provide a better empirical explanation of actual laboratory unblocking temperatures, especially where remanences are retained in domain walls ('transdomain' remanences) in PSD grains. Despite a lack of theoretical justification, Middleton & Schmidt's nomograms are recommended without better specific information (e.g. Jackson & Van der Voo 1986).

Dunlop & Özdemir (1993) carefully prepared true SD magnetite samples with dispersed, noninteracting grains. These showed the complete absence of unusually high unblocking temperatures. Their experimentally determined unblocking temperatures agreed with those derived from Néel's theory to within a few Celsius degrees. Dunlop & Özdemir concluded that unusually high T_{UB}s of other studies were due to PSD magnetite and thus confirm the work reviewed above. Other experiments have shown that the degree to which unblocking temperatures disagree with Néel theory is very sensitive to the previous thermal history and domain structure of magnetite (Haigedahl 1993). The most recent investigation of anomalously high unblocking temperatures is a re-examination of the Sydney Basin, where that phenomenon was originally recorded (Schmidt & Embleton 1981; Middleton & Schmidt 1982). Dunlop et al. (1997a, b) now recognize a complex sequence of two thermal and two chemical magnetization components in the Basin. One of the thermal remanences is associated with shallow burial but it unblocks at 250°C. Suspecting that a contribution from multidomain (MD) magnetite spoiled the interpretation of ages based on SD thermomagnetic theory, Dunlop et al. (1997a) cleaned the MD contribution by cyclical bathing of the samples in liquid nitrogen. Low-temperature demagnetization (LTD) selectively demagnetizes MD magnetite, leaving SD and PSD grains largely unaffected. For most samples this reduced the laboratory temperature required to demagnetize the samples and better estimated the geological ages (Dunlop et al. 1997b). Unfortunately, this procedure was not so successful with the rocks used in this study, probably because the magnetite is not truly MD but rather PSD magnetite (see Figs 4 and 5, below).

High thermal stability VRM is of great concern in palaeomagnetism and could be a much more serious problem than is generally realized (e.g. Moon & Merrill 1986). This is of special concern for carbonates, in which particularly suitable PSD magnetite is ubiquitous (Borradaile et al. 1993), probably often of bacterial origin (Kirschvink & Chang 1984; Kirschvink & Walker 1985). This may be because it was formerly difficult to obtain appropriate rock magnetic information or electron-microscopy observations, which, in turn, led to an underestimation of how commonly VRM-sensitive conditions are found in nature. With more ready access to modern rock magnetic equipment such as the Micromag gradient force magnetometer the rarity of truly SD magnetite will be recognized. It is possible that unusually high thermal stability of VRM may often go unnoticed in limestones and may be misinterpreted as a primary, or at least characteristic, remanence component. Moon & Merrill (1986) went as far as to suggest that the thermal stability of sluggishly acquired VRMs may exceed the thermal stability of a primary remanence. In such extreme cases, identifying a primary remanence would be impossible and, still worse, one would not know that the remanence was not primary.

Theoretical work suggests that transdomain ('domain wall') remanence with high thermal stability may be carried in small magnetite grains just above the SD size (Dunlop et al. 1994). Magnetite grains with this domain structure are common in limestones. The following examples from English limestones with PSD magnetite (and some with hematite) will illustrate VRMs that require temperatures >400°C to unblock.

Cretaceous and Jurassic limestones of eastern England

The following examples show that limestones with high thermal stability remanence probably acquired most, if not all, their remanence in the Brunhes epoch, since the last reversal of the Earth's field. The limestones are a soft Cretaceous Chalk both white and red in colour from northern Lincolnshire and southern Yorkshire. A coarser, shelly or oolitic Jurassic limestone was sampled also from the area around and to the southeast of Lincoln. Both limestones are incorporated in historically dated buildings that were also sampled with permission of the appropriate authorities. Both limestones have been studied under the scanning electron microscope (Borradaile 1991, 1994a). Energy dispersive analysis revealed iron oxides, usually pure

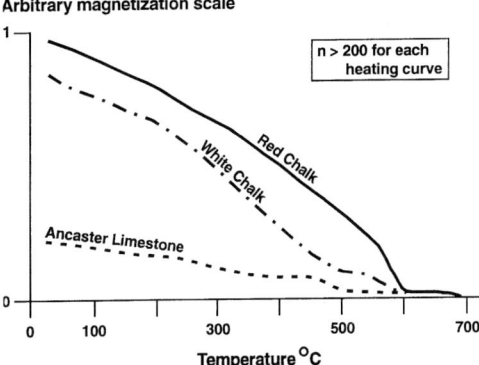

Fig. 3. The Curie temperatures determinations. These were measured in a Sapphire Instruments translational balance and reveal sharp reductions in magnetization near 580°C for all samples, indicating the importance of their magnetite contribution. However, a signal from hematite is also clear near 670°C, especially for the Red Chalk. Number of measurements is given by n.

magnetite, in PSD size grains. This was rarely of detrital origin and bacterial sources are suspected (e.g. Kirschvink & Chang 1984; Kirschvink & Walker 1985). Some titanomagnetite was found, perhaps of detrital origin, and secondary hematite was abundant in the Red Chalk as an alteration product after detrital ferruginous material. All the limestones have remanences that are easily measurable with the JR5a spinner magnetometer and most were suitable for incremental demagnetization by AF or thermal methods. Curie balance tests using a Sapphire Instruments S16 translation-style balance show clear magnetite Curie points ($c.\,580°C$) in all cases. Hematite ($c.\,700°C$) was also present in the Red Chalk, and perhaps other Curie points were caused by traces of titanium-bearing oxides or goethite at lower temperatures in the White Chalk and Ancaster Limestone (Fig. 3). The acquisition of isothermal remanent magnetization (IRM) was studied using a Sapphire Instruments S14 pulse magnetizer up to peak values of 1.2 T. Excepting the Red Chalk, these curves show the importance of PSD magnetite because they saturate below 200 mT (Fig. 4). The Red Chalk fails to saturate because the peak field could not completely magnetize hematite. These and other now standard rock magnetic tests (Lowrie 1990) suggest the presence of PSD magnetite in all samples.

Low-temperature demagnetization may also be used as a simple means of selectively removing MD remanence and therefore isolating remaining PSD or SD remanence (Merrill 1970; Dunlop & Argyle 1991). The procedure uses liquid nitrogen (77 K) to cool magnetite in a shielded space. While the liquid nitrogen boils away, the samples warm in the zero field. At 130 K, the first constant in the expression for magnetocrystalline anisotropy changes sign. If the grain is

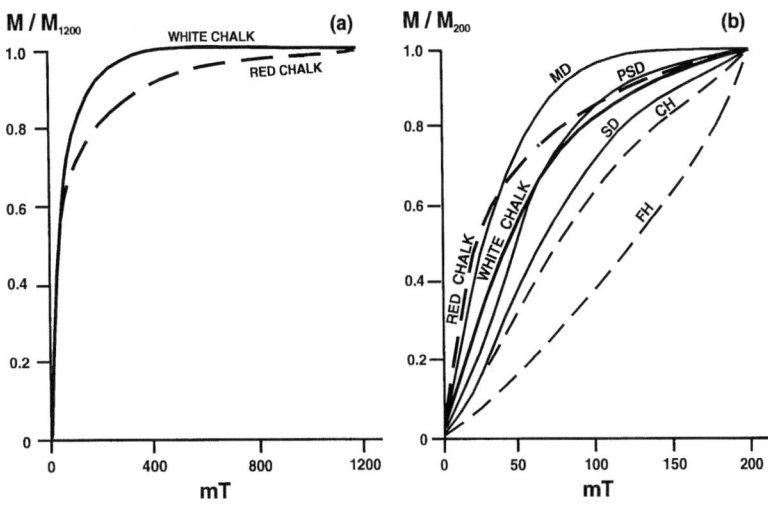

Fig. 4. Isothermal remanent behaviour. (a) Acquisition of IRM in a Sapphire Instruments pulse magnetizer permits an estimation of the roles of MD, PSD and SD magnetite in the samples according to the rate at which they approach a saturation remanence. The Red Chalk contains hematite and thus fails to reach any plateau within the range of fields applied. (b) For comparison, IRM acquisition responses of MD, PSD and SD magnetite; also fine and coarse hematite (FH, CH) from Dunlop (1971, 1981).

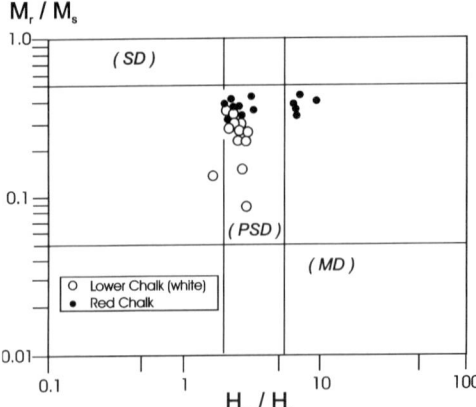

Fig. 5. The coercivity properties. (M_r, remanence; M_s, saturation remanence; H_{cr}, coercivity of remanence; H_c, coercivity) and the fields defining MD, PSD and SD behaviour of magnetite (from Dunlop 1973, 1983b, 1986a, b; Dunlop et al. 1987; Dunlop & Argyle 1991)). The limestones usually show PSD characteristics.

a large MD grain, this permits domain wall rearrangements. As the sample is in zero field the magnetic moments are randomized and the sample experiences low-temperature demagnetization. Truly SD remanences, and those in small PSD grains, survive many cycles of this treatment, which removes the MD remanences just as effectively as subjecting the samples to 200 MPa (2 kbar) hydrostatic pressure (Borradaile 1994b). Low-temperature demagnetization, applied in five cycles, was applied to the limestones to assess the relative importance of MD and SD remanence. Samples of the White Chalk reveal negligible MD or SD contribution to the remanence; the signature is mostly PSD. This technique reveals a moderate contribution from MD magnetite in Ancaster Limestone, and the Red Chalk may show this also. However, we must remember that the hematite content of the Red Chalk can give misleading results by low-temperature demagnetization, because high-coercivity hematite remanences can be cleaned out more easily than low-coercivity ones using the low-temperature technique (Borradaile 1994b). Hysteresis measurements made with an alternating gradient force magnetometer (Princeton Instruments Micromag 2900) also confirm the PSD response of most samples (Fig. 5).

Lower Chalk

The NRM of the White and Red Chalks was progressively demagnetized by either thermal or AF methods. AF demagnetization used fine increments of 2.5 mT to 20 mT, then 5 mT increments to 50 mT, and finally 10 mT increments to 200 mT. Thermal demagnetization proceeded in intervals of c. 30°C until 700°C. Figure 6 shows data from a typical sample of Cretaceous White Chalk bedrock of northern Lincolnshire. The limestone is almost horizontally bedded over the region and shows no evidence of tectonism that could disturb remanences. The stereonet and vector plot (Fig. 6a and b) show a distinct down-to-the-north seeking remanence that was stable up to a demagnetization temperature of 650°C (Fig. 6c). Inspection of the vector plot reveals that the remanence is univectorial to the origin and appears to have all the features of a primary remanence.

Stable end vectors were derived from the vector plots by principal component analysis (PCA; Kirschvink, 1980). This procedure yields the vector that best represents an ancient field direction represented by a series of magnetic components on a vector plot. The procedure used at least five data points from each vector plot in this study although usually more than ten points were used. Site-mean data reveal the estimated field directions for ten sites of Red Chalk and 26 sites of White Chalk, and are presented in Fig. 7 by the PCA method.

We are justified in considering these unblocking temperatures anomalously high because they unblock above 250°C despite their having been apparently acquired in the present geomagnetic field. The scatter of the PCA vectors about the present local geomagnetic field direction appears to support this although the local geomagnetic field direction is not very different from the palaeofield direction believed to exist at the time of deposition (Kerth & Hailwood 1988). We will propose that, in this region, it is not a 'primary' Mesozoic remanence, but one acquired in the Brunhes epoch, since the last reversal of the geomagnetic field, at c. 0.78 Ma. This is arguable on the basis of Fig. 7 alone, but will be confirmed below using a 'conglomerate' test (e.g. Tarling 1983).

Fortunately, the tectonically monotonous Chalks do show some disturbance along the coastline, north of Bridlington in Yorkshire. Documented by several cartographic surveys since 1772, cliff slumping has caused cliff-line retreat of 0.33 m per year, slumping blocks from the size of kilograms to many tonnes in a crushed limestone and clay matrix toward the sea (Borradaile 1998). The disturbed blocks provide total disorientation of the originally horizontal chalk; sedimentary structures show that the limestone blocks are tilted vertically, inverted, and investigation of the matrix leads

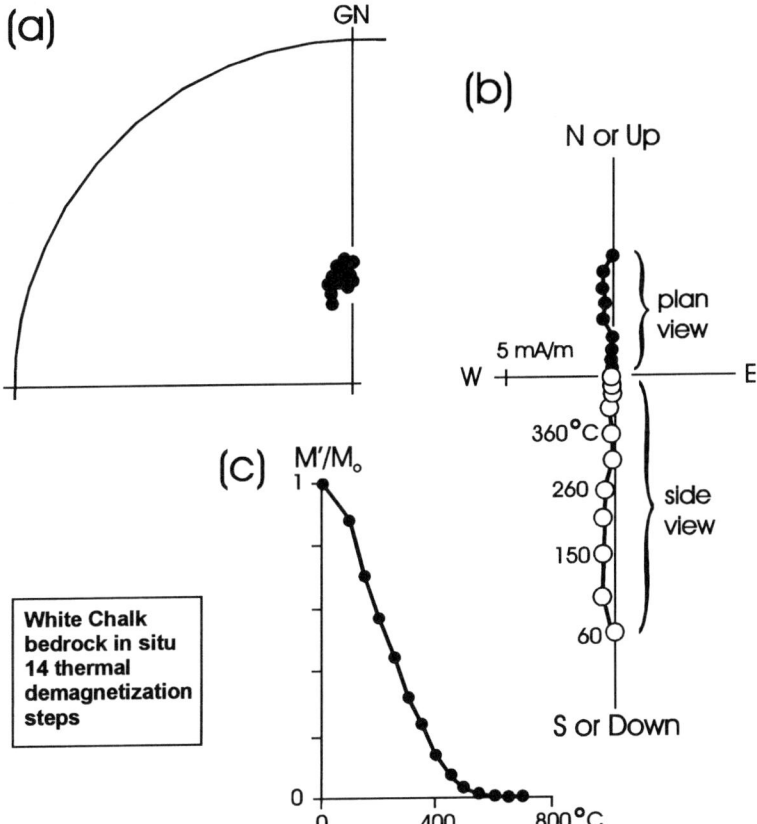

Fig. 6. Demagnetization data for a typical sample of *in situ* White Chalk. (a) Stereonet, (b) vector plot (see Zijderveld 1967; Tarling 1983) and (c) intensity decay plot. These show clear univectorial remanences, decaying to the origin at temperatures near the Curie point. Thus, they appear to represent very stable, primary remanences. Geographic North shown.

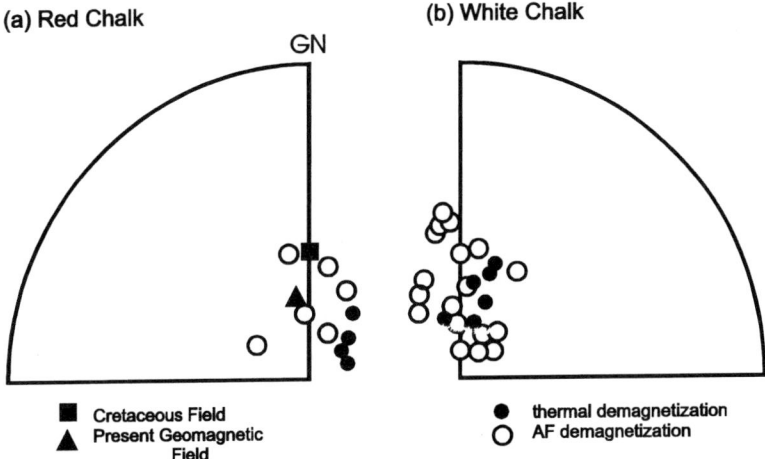

Fig. 7. Stereonets with all vectors down-seeking. The stereonets reveal the stable end-vectors derived by bestline fit to at least five points of each demagnetization plot for each sample. This 'PCA' vector (Kirschvink 1980) then represents the best estimate of palaeomagnetization of the sample. These tend to group around the present geomagnetic field rather than the appropriate palaeofield. However, this is argumentative here and superceded by superior information later in the paper.

one to suspect that some blocks may have rotated several times as they slumped seaward. The slump breccia provides abundant samples of both Red and White Chalk blocks.

Oriented samples of these blocks were taken and progressively demagnetized, noting their remanence components in both geographic and sample coordinates. Figure 8 shows typical demagnetization plots for slumped blocks at Speeton Cliff. In Fig. 8a we see an example of a recently disrupted fragment that suffered rigid body rotation more rapidly than any suspected viscous remagnetization rate. The remanence is stable to a high temperature but is unrelated to the present or Mesozoic geomagnetic field directions. Because the fragment was recently disrupted from the bedrock we find no evidence of anomalously resistant VRM in the present north direction. Figure 8b shows a similar example, but this block has been tilted in

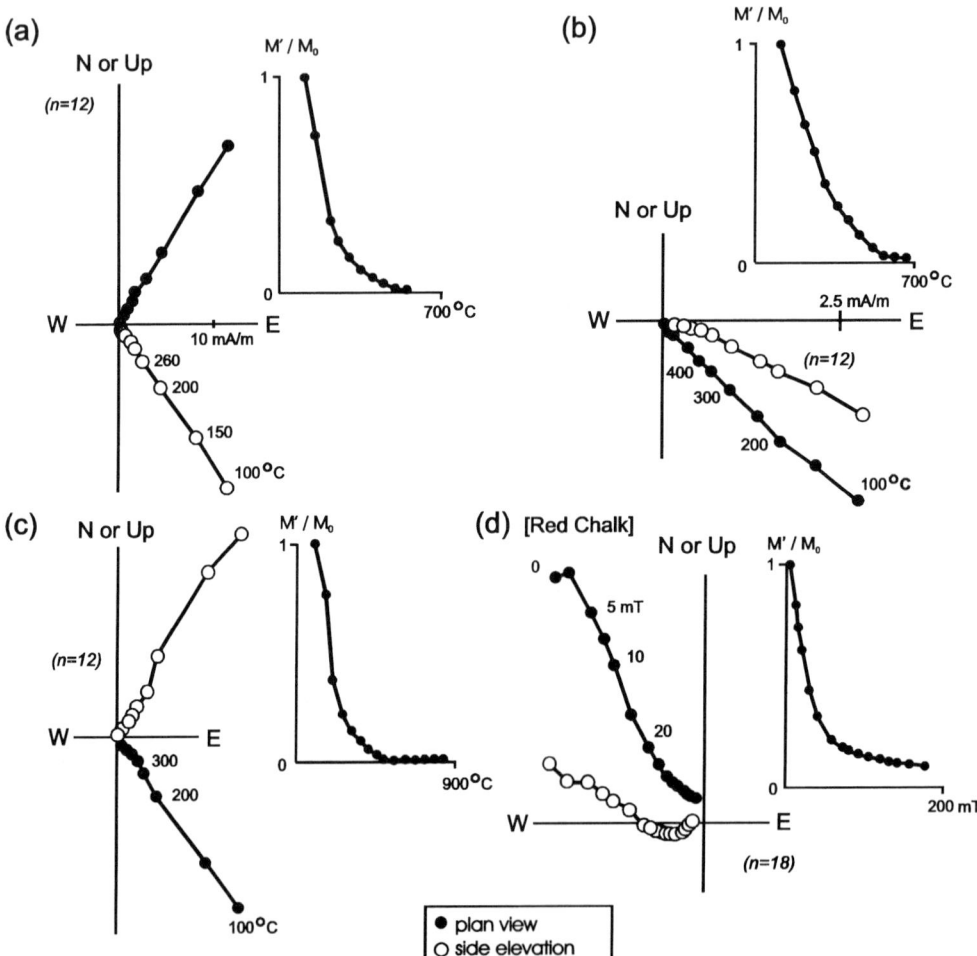

Fig. 8. The White and Red Chalks disorient in debris flows caused by long-term failure of sea cliffs north of Bridlington, Yorkshire. Inspection of the bedding orientation and way-up indicators reveals that blocks weighing up to many tonnes have been rotated and even inverted. The vector plots (**a, b, c**) show cases where univectorial remanences are tilted to yield bedding orientations that are incompatible with the palaeofield (or present geomagnetic field). All the remanences show the same high thermal stability as the *in situ* country rock (Fig. 6). One suspects a long history of slow rotation and movement for some blocks, especially near the base of the flow. Such samples reveal smeared, multicomponent remanences (**d**) caused by progressive remagnetization as the slump blocks rotated in the geomagnetic field.

another direction because of the inconsistent tumbling action in the slump. Smeared or curving demagnetization paths appear in Fig. 8c and d. These blocks occur lower down the slump and have been rotating and moving for longer periods. Thus, they have remagnetized partially each time they stabilized in the Earth's field. Of course, eventually they will reach the shore line, become buried beneath the beach and after sufficient time they will become completely remagnetized to possess univectorial magnetizations as shown in Fig. 6, pointing to the present north. The only difference is that their stratification will be tilted in some random fashion so that any future remanence will be meaningless in sample or bedding coordinates.

The oldest components of remanence in recently slumped blocks have been calculated using Kirschvink's technique of PCA. This defines the north-seeking vectors, in our case for stable, high unblocking temperature ($T > 300\,°C$) magnetization. As this discussion suggests, these vectors are completely scattered, pointing both up and down with random azimuths (Fig. 9) because the recent slumping of these blocks simulates the 'conglomerate test' of palaeomagnetism (Tarling 1983). However, as will be apparent from the conclusions (e.g. Fig. 14) the VRM-stability sensitivity of a limestone is sensitive to lithological variation. Therefore, it is not surprising that the 'same' Chalk horizons, far away in southern England, show a positive fold test, confirming a pre-Alpine age for remanence (Montgomery et al. 1998).

Slumped blocks that moved more slowly, and those with greater disorientation, show more complex, chaotic demagnetization paths that are most easily seen in stereograms. These limestone fragments moved slowly enough that they could remagnetize progressively during rotation (Fig. 10). Occasionally, the newest component of magnetization clearly has attempted to remagnetize towards the present geomagnetic field direction (Fig. 10a and b). The remagnetization path is an inverse record of the actual particle rotation. Thus for a VRM acquired over a time scale similar to the slumping, the 'conglomerate test' is not decisively positive for all fragments. Nevertheless, for the fragments that were simply tilted more rapidly than the viscous remagnetization rate (Fig. 8a–c) the 'conglomerate test' is positive, confirming that the remanences in the Cretaceous rocks are not primary but Bruhnes epoch VRM remagnetizations.

To advance rock magnetism theory but also for practical palaeomagnetic and geomorphological purposes we need to know the rate at which the remagnetization occurs. Traditionally, palaeomagnetism has utilized the conglomerate test to discriminate between primary and secondary remanences in clasts (Fig. 11). This is inappropriate where the remanence developed during the reorientation of slowly spinning slump fragments in the Speeton debris flow. The smeared, multicomponent remanences in the slumped blocks (Fig. 10) provide a positive 'debris flow test' that confirms the syn-slumping acquisition of VRM (Fig. 12).

Further proof of the recent, viscous acquisition of remanence is provided by a 'masonry test' (Fig. 13). This area has been the site of continuous occupation since the Pleistocene glaciation and, since Roman times (50 BC), Chalk has been used in construction. Permission was granted to sample numerous buildings of historically documented age to perform a masonry test. When a sample of bedrock suitable for viscous remagnetization is incorporated in a building, it commences remagnetization at the date of construction. The extent of remagnetization is directly related to the time that elapsed since construction and thus we calibrate the remagnetization rate without any assumptions or estimates about acquisition rates (cf. Heller & Markert 1973). Of course, precautions have to be taken to exclude the effects of chemical

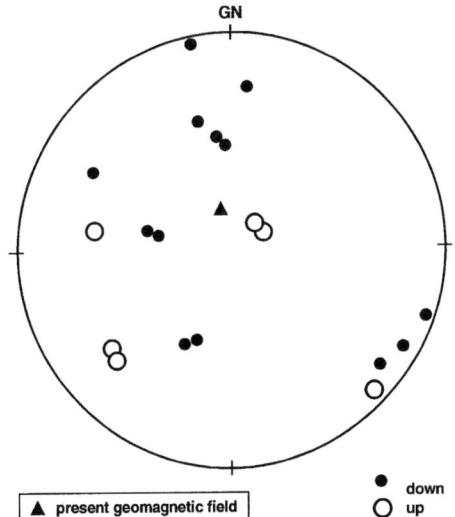

Fig. 9. The stable end-vectors of the slumped blocks. Each point shows the stable end-vector derived from at least five data points, for demagnetization temperatures >250°C. Superficially, the wide scatter would appear to provide a positive 'conglomerate test'; i.e., the remanences should be 'primary'. However, blocks with smeared curving vector plots (e.g. Fig. 8d) show that a high-stability VRM is more probable, especially in Red Chalk.

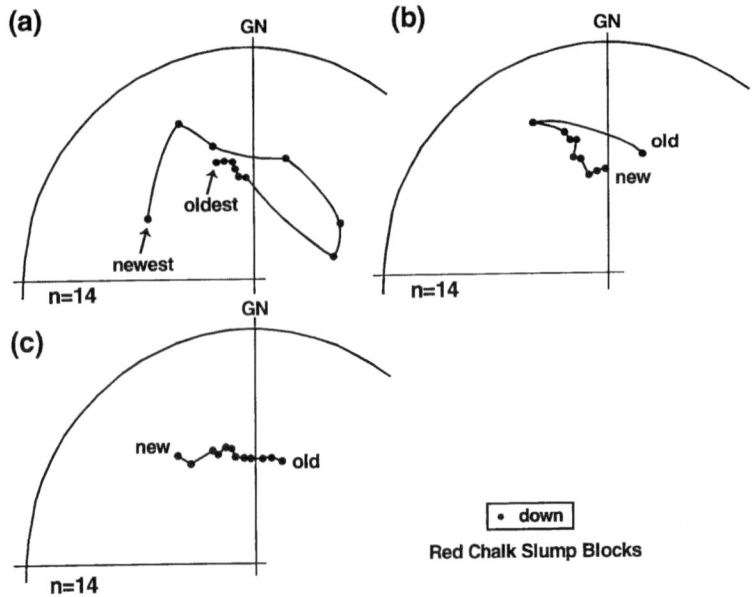

Fig. 10. Lower hemisphere stereograms showing the multi-component remanences found in slowly moving Chalk blocks in the debris flow. The 14 demagnetization steps represent increments from 60°C to 650°C with good stability to the final demagnetization stage.

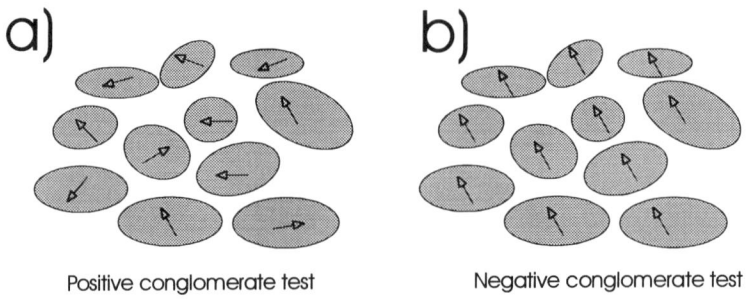

Fig. 11. The classical conglomerate test. This sharply distinguishes between (**a**) primary and (**b**) post-depositional remanences but is unsuitable for the detection of VRM that is younger than the conglomerate. Physical tests that incorporate more than one age of reorientation are a prerequisite for the detection of VRM (e.g. Figs 12 and 13).

weathering, e.g. by sampling indoors, in dry dungeons, in core-walls, etc.

Jurassic limestone, Lincoln, England

The Lincoln Limestone and the Ancaster Limestone are two Jurassic limestones of similar age that from part of the limestone ridge occupied by Lincoln city. Many fine historical buildings, most importantly Lincoln Cathedral, were built of these limestones since the Roman occupation in the early centuries after Christ. The bedrock dips very gently eastward but shows no evidence of deformation that could disturb remanence. Outcrop remanences are stable to >360°C and align with the present geomagnetic field.

However, large disoriented blocks in an abandoned mediaeval quarry (which is now protected inside a mediaeval underground chamber) show the disorientation of the highest unblocking temperature components of remanence. Moreover, the lower unblocking temperature components are partly remagnetized toward the present Earth's field. Masonry was sampled from the Roman (c. AD 300), and from many building

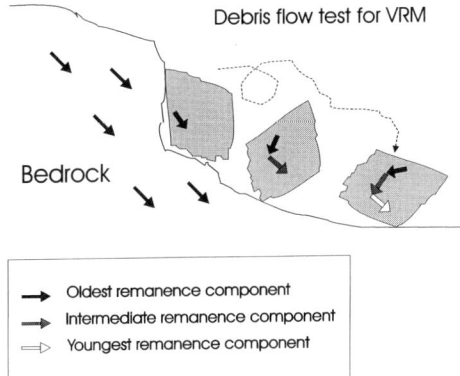

Fig. 12. The debris flow test. This provides slowly reoriented slump blocks that may record components of the geomagnetic field in different orientations within the sample, if the orientation changes rate at a similar pace to viscous remagnetization. This is superficially comparable with the conglomerate test of traditional palaeomagnetism (Tarling 1983), but an important difference exists. The conglomerate test represents one discrete reorientation of material that may, or may not, remagnetize after reorientation (Fig. 11). Here, the slump block has the potential for multiple reorientations. The presence of multiple magnetization components proves that the remanence is acquired viscously.

phases and repairs in the Mediaeval (c. 1160–1450), post Civil War (1640–1720), and Victorian periods. These were all dated historically, more accurately than would be possible by any physical technique such as radiocarbon or dendrochronology. Progressive thermal demagnetization and analysis of vector plots yielded clear evidence of progressive remagnetization with age (Borradaile & Brann 1997). Sharp changes in vector direction occur at the time of incorporation of the stone into the building or during remodelling (Fig. 15). These samples document well the acquisition rate of VRM (Fig. 14), which differs from that of the Cretaceous Chalks.

To the southeast of Lincoln and the Ancaster quarries the Witham chain of Norman Abbeys is found as a series of ruins. These are also constructed of limestones barged in along canals in this formerly partially flooded region. Tupholme Abbey is one of this chain and shows numerous building phases since its establishment between AD 1155 and 1166. A similar study of the unblocking temperatures of masonry fragments from different building phases of this abbey (Borradaile et al. 1999) yielded yet another empirical acquisition-rate curve (Fig. 14) for a different source of the same geological formation.

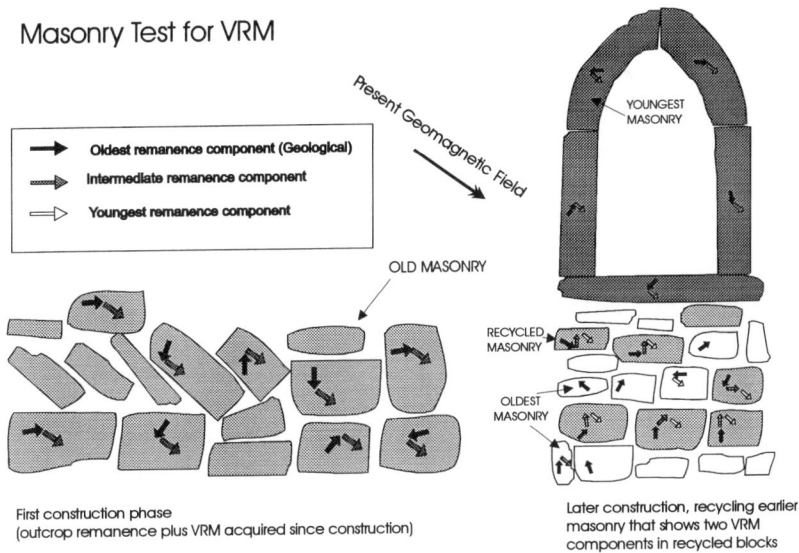

Fig. 13. The 'masonry' test. Like the debris flow test of Fig. 12, the masonry test may also expose the rock to multiple magnetizations during cultural reorientation of the original quarried material. This permits us to determine the rate of acquisition of viscous magnetization over historical or archaeological time scales. Conversely, it also provides us with method of dating the age of stabilization of a rock sample. Relative ages can be determined easily, but if some historical calibration is available this permits high-resolution archaeological dating (Borradaile & Brann 1997).

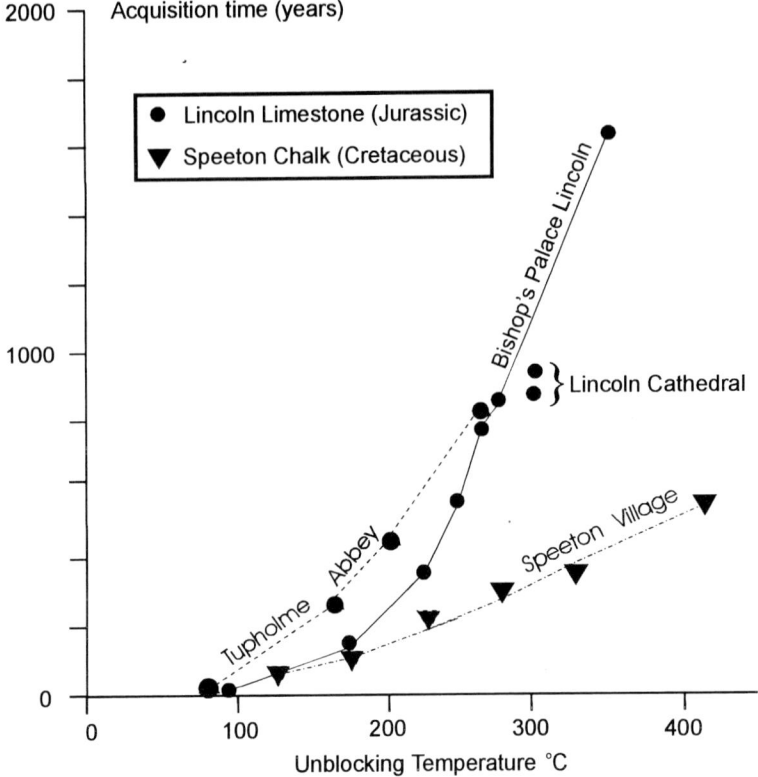

Fig. 14. Chronometric curves showing the rate of acquisition of remanence for masonry of well-documented ages in eastern England. These buildings were constructed of the limestones described earlier in the paper. The curves may be used for dating poorly documented buildings or geomorphological features involving ground disturbance. They should not be extrapolated to other lithologies nor to much earlier times.

Use and nuisance of anomalously stable VRM

An understanding of VRM is important for several reasons. Here we concentrated on the importance to rock magnetic studies and the relationship of VRM to grain size (Tivey & Johnson 1984; Yamazaki & Katsura 1990), domain structure (Dunlop & Stirling 1977; Dunlop 1983a; Sporer 1984; Dunlop & Özdemir 1990, 1993) and time–temperature relationships (Heller & Markert 1973; Pullaiah et al. 1975; Middleton & Schmidt 1982; Kent 1985; Jackson & Van der Voo 1986; Kent & Miller 1987) and perhaps even some chemical effects (Özdemir & Dunlop 1989). These studies all allow deeper insight into the acquisition of remanence by thermal or viscous mechanisms.

For palaeomagnetists, these studies provide comfort in the knowledge of the stability and primary character of a remanence. We saw here that VRM is not easily distinguished from a primary remanence and may even be more stable under some circumstances (Moon & Merrill 1986). Thermal demagnetization, low-temperature demagnetization and analysis of vector plots may fail to isolate the primary remanence. Thus, careful field tests, such as the conglomerate test or a masonry test, should be performed to validate the primary character, as with these English limestones. It must be appreciated that it may be very difficult to distinguish a primary remanence from a VRM of anomalously high thermal stability.

Whereas VRM is generally a nuisance to most palaeomagnetists, it has been shown here that it may be useful for dating processes that proceed at a similar rate to VRM acquisition. For example, the kinematics and rate of geomorphological processes such as historically slow debris flows may be assessed. Also, archaeological events may be dated from masonry of different building phases, and, occasionally, recycling of building materials can been proven

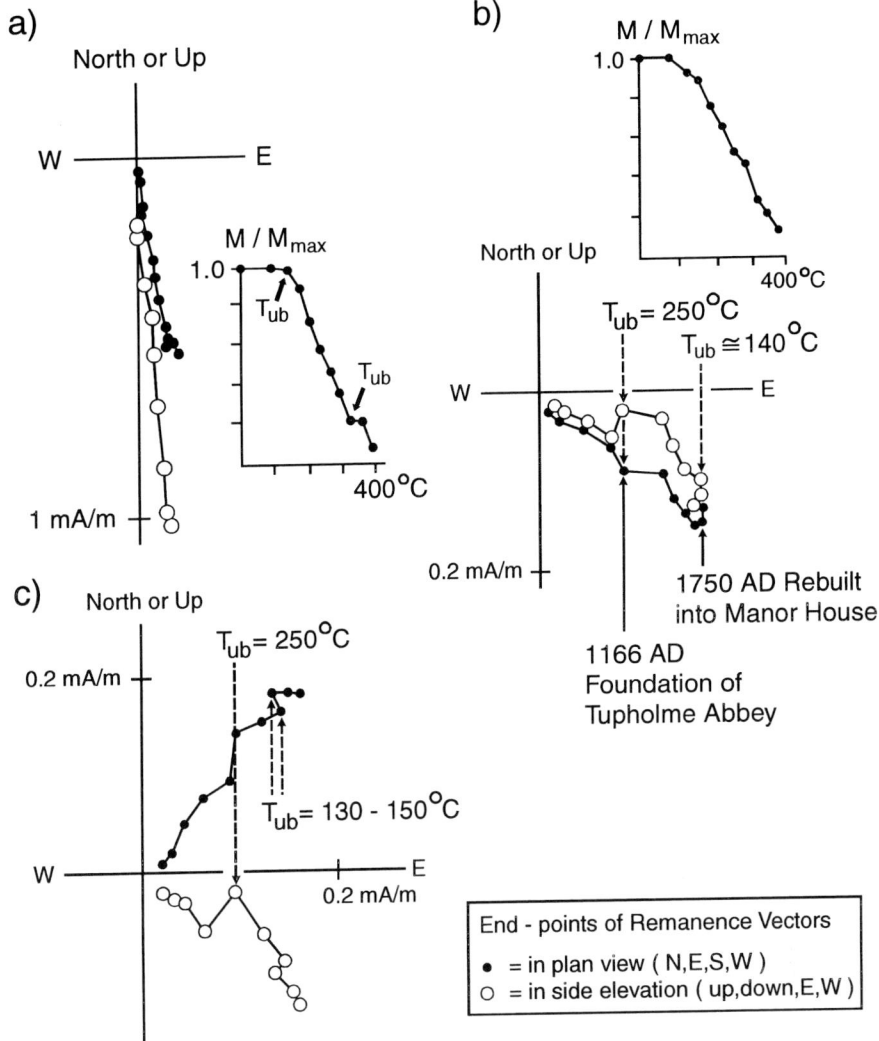

Fig. 15. Vector plots from limestone masonry in Tupholme Abbey, Lincolnshire. The masonry has documented ages of construction and reconstruction (Borradaile *et al.* 1998). (**a**) A few samples show single remanence components that do not relate to the present geomagnetic field and thus have no evidence of VRM overprint. Nevertheless, slight breaks in the intensity decay curve correspond to historically dated construction and remodelling. (Note that in (**a**) some points near the origin of the horizontal projection have been omitted for clarity.) However, most samples show multicomponent remanences (**b, c**), most commonly with reorientation of the sample corresponding to the foundation of the abbey (AD 1166) or the rebuilding into the manor house, c. AD 1750. (The resolution of the remanence measurements with the JR5a spinner magnetometer is about 0.01 mA/m, thus the negligible error bars justify the turning points identified.)

by using multiple VRM components (Borradaile & Brann 1997).

For regional magnetic studies, one now realizes that regional VRM contributions can considerably modify aeromagnetic signatures (e.g. Johnson 1985; Treloar *et al.* 1986; Arkani-Hamed, 1988, 1989, 1990, 1991). Thus, a fuller appreciation of the role and significance of VRM is essential for a complete interpretation of magnetic field anomalies, especially where the intensity of remanent magnetization is significant in comparison with induced magnetization.

The studies in this paper were restricted to three limestones from two different formations in eastern England that show two essentially different acquisition rates. A high rate

Table 1. *Curve-fitting data to chronometric curves for VRM acquisition in English limestones*

	A	B	R^2	N (sites)	Cores	
Speeton White Chalk (Cretaceous)	4298E−6	1.973	0.962	5	82	power law
Lincoln Limestone (Jurassic)	1.298E−6	3.598	0.992	7	109	power law
Tupholme Limestone (Jurassic)	1245	−6085	0.968	4	47	logarithmic

Power law: $t = (A\,T_{UB})^B$; logarithmic: $t = A\log(T_{UB}) + B$.

of remagnetization occurs for the Lincoln and Ancaster Limestones in contrast to the White Chalk (Fig. 14). These graphs provide VRM experiments lasting up to 1600 years, much longer than the previously reported laboratory experiments. Lasting days or weeks, laboratory experiments consistently reveal that the intensity of magnetization depends on log(time) (Dunlop 1989).

Of course, in this study it was not possible to know what the intensity of magnetization was at previous times. Indeed, it would have been a combination of earlier and later VRMs, not a single remanence component as in the laboratory experiments. Thus, these archaeologically derived results cannot easily confirm or reject the dependence of intensity on log(time). However, the relationship between the age or, more precisely, duration (t in years), of acquisition can be empirically correlated with the unblocking temperature (T_{UB}, in °C). Despite the few archaeologically derived data points of Fig. 14, the long time span and high precision of the historical dates justify curve-fitting, to improve the use of the curves as a chronometric tool in archaeology or geomorphology. The Tupholme Abbey data correlate as $t = A\log(T_{UB}) + B$ at an acceptable level with $R^2 = 0.968$, where A and B are arbitrary coefficients. On the other hand, the same formation quarried and used nearer Lincoln is satisfied better by a power-law fit with $t = A\log T_{UB}^B$ with $R^2 = 0.99$. Another power law satisfies the Speeton Chalk data with $R^2 = 0.96$. The empirical curve-fitting data are summarized in Table 1.

These empirical fits should not be taken to represent an underlying theoretical relationship, especially in view of the paucity of sites. Nevertheless, the length of these archaeological 'experiments', the convincing viscous nature of the remanence from the debris-flow test and masonry test, and the high accuracy of the ages may recommend the curves for future use. It should be noted that the two limestones show different acquisition rates (Fig. 14); even the Jurassic Lincoln Limestones from different sites shows some difference. This is to be expected because the magnetic mineralogy of clastic carbonate rocks with bacterial and detrital ferromagnetic minerals would be expected to show some variability. Therefore, such chronometric curves should not be extrapolated to other lithologies. Nor should they be extrapolated too far back in time in view of the small number of sites available at present.

We are very grateful and deeply conscious of the privilege of being permitted to sample sites of great religious, architectural and historical significance. For this, we thank the Bishop of York, English Heritage, Heritage Lincolnshire (D. Start), Lincoln Archaeology Unit (M. Jones, M. Brann), and landowners Mr Coleman and Mr Rogerson of Speeton, Yorkshire. M. Brann collaborated with archaeological advice at the Bishop's Palace, Lincoln, as did D. Start with Tupholme Abbey. C. Goulder helped with laboratory work and J. Borradaile and C. Borradaile helped with fieldwork. The Natural Science and Engineering Research Council of Canada funded the laboratory research. D. Dunlop provided helpful correspondence during an earlier part of the project.

References

ARKANI-HAMED, J. 1988. Remanent magnetization of the oceanic upper mantle. *Geophysical Research Letters*, **15**, 48–51.
—— 1989. Thermoviscous remanent magnetization of oceanic lithosphere inferred from its thermal evolution. *Journal of Geophysical Research*, **94**, 17421–17436.
—— 1990. Magnetization of the oceanic crust beneath the Labrador Sea. *Journal of Geophysical Research*, **95**, 7101–7110.
—— 1991. Thermoremanent magnetization of the oceanic lithosphere inferred from a thermal evolution model: implications for the source of marine magnetic anomalies. *Tectonophysics*, **192**, 81–96.
BIQUAND, D. & PRÉVOT, M. 1971. A.F. demagnetization of viscous remanent magnetization in rocks. *Zeitschrift fur Physik*, **37**, 471–485.
BORRADAILE, G. J. 1991. Remanent magnetism and ductile deformation in an experimentally deformed magnetite-bearing limestone. *Physics of the Earth and Planetary Interiors*, **67**, 363–373.

—— 1994a. The magnetic remanence of the Chalk of Eastern England: an unusually resistant VRM? *Geological Magazine*, **131**, 593–608.

—— 1994b. Low temperature demagnetization and ice-pressure demagnetization. *Geophysical Journal International*, **16**, 571–584.

—— 1996. An 1800-year archeological experiment in remagnetization. *Geophysical Research Letters*, **23**, 1585–1588.

—— 1998. Rock magnetic constraints on long-term cliff-slump rates and coastal erosion. *Géotechnique*, **48**, 271–279.

—— & BRANN, M. 1997. Remagnetization-dating of Roman and mediaeval masonry. *Journal of Archaeological Science*, **24**, 813–824.

——, GOULDER, C., STEWART, D. & START, D. 1999. Dating of a medieval Abbey by Viscous Renanent Magnetization. *Archaeometry*, in press.

——, CHOW, N. & WERNER, T. 1993. Magnetic hysteresis of limestones: facies control? *Physics of the Earth and Planetary Interiors*, **76**, 241–252.

BUTLER, R. F. 1992. *Paleomagnetism.* Blackwell Scientific, Oxford.

DUNLOP, D. J. 1971. Magnetic properties of fine particle hematite. *Annales Geophysicae*, **27**, 269–293.

—— 1973. Superparamagnetic and single-domain threshold sizes in magnetite, *Journal of Geophysical Research*, **78**, 1780–1793.

—— 1981. The rock magnetism of fine particles. *Physics of the Earth and Planetary Interiors*, **26**, 1–26.

—— 1983a. Viscous magnetization of 0.04–100 pm magnetites, *Journal of Geophysical Research*, **741**, 667–687.

—— 1983b. Determination of domain structure in igneous rocks by alternating field and other methods. *Earth and Planetary Science Letters*, **63**, 353–367.

—— 1986a. Coercive forces and coercivity spectra of submicron magnetites. *Earth and Planetary Science Letters*, **78**, 288–295.

—— 1986b. Hysteresis properties of magnetite and their dependence on particle size: a test of pseudo-single-domain remanence models. *Journal of Geophysical Research*, **91**, 9569–9584.

—— 1989. Viscous remanent magnetization (VRM) and viscous remagnetization. In: JAMES, D. E. (ed.) *Encyclopedia of Solid Earth Geophysics.* Van Nostrand Reinhold, Stroudsburg, PA, 1297–1303.

—— & ARGYLE, K. S. 1991. Separating multidomain and single-domain-like remanences in pseudo-single domain magnetites (215–540 nm) by low-temperature demagnetization, *Journal of Geophysical Research*, **96**, 2007–2017.

—— & ÖZDEMIR, Ö. 1990. Alternating field stability of high-temperature viscous remanent magnetization. *Physics of the Earth and Planetary Interiors*, **65**, 188–196.

—— & —— 1993. Thermal demagnetization of VRM and PTRM of single domain magnetite: no evidence for anomalously high unblocking temperatures. *Geophysical Research Letters*, **20**, 1939–1942.

—— & —— 1997. *Rock Magnetism: Fundamentals and Frontiers.* Cambridge Studies in Magnetism. Cambridge University Press, Cambridge.

—— & STIRLING, J. M. 1977. 'Hard' viscous remanent magnetization (VRM) in fine-grained hematite. *Geophysical Research Letters*, **4**, 163–166.

—— & WEST, G. F. 1969. An experimental evaluation of single domain theories. *Reviews in Geophysics*, **7**, 709–757.

——, NEWELL, A. J. & ENKIN, R. J. 1994. Trans-domain thermoremanent magnetization. *Journal of Geophysical Research*, **99**, 19741–19755.

——, ÖZDEMIR, Ö. & ENKIN, R. J. 1987. Multidomain and single-domain relations between susceptibility and coercive force, *Physics of the Earth and Planetary Interiors*, **49**, 181–191.

——, —— & SCHMIDT, P. W. 1997a. Paleomagnetism and paleothermometry of the Sydney Basin: (2) Origin of anomalously high unblocking temperatures. *Journal of Geophysical Research*, **102**, 27 851–27 295.

——, SCHMIDT, P. W., OZDEMIR, Ö. & CLARK, D. A. 1997b. Paleomagnetism and paleothermometry of the Sydney Basin: (1) Thermoviscous and chemical overprinting of the Milton Monzonite. *Journal of Geophysical Research*, **102**, 27 271–27 283.

ENKIN, R. J., & DUNLOP, D. J. 1988. The demagnetization temperature necessary to remove viscous remanent magnetization. *Geophysical Research Letters*, **15**, 514–517.

HAIGEDAHL, S. 1993. Experiments to investigate the origin of anomalously elevated unblocking temperatures. *Journal of Geophysical Research*, **98B**, 22 443–22 460.

HELLER, F. & MARKERT, H. 1973. The age of the viscous remanent magnetisation of Hadrian's Wall (Northern England). *Geophysical Journal of the Royal Astronomical Society*, **31**, 395–406.

JACKSON, M. J. & VAN DER VOO, R. 1986. Thermally activated viscous remanence in some magnetite- and hematite-bearing dolomites. *Geophysical Research Letters*, **13**, 1434–1437.

JOHNSON, B. D. 1985. Viscous remanent magnetization model for the Broken Ridge satellite anomaly, *Journal of Geophysical Research*, **90B**, 2640–2646.

KENT, D. V. 1985. Thermoviscous remagnetization in some Appalachian limestones. *Geophysical Research Letters*, **12**, 805–808.

—— & MILLER, J. D. 1987. Redbeds and thermoviscous magnetization theory. *Geophysical Research Letters*, **14**, 327–330.

KERTH, M. & HAILWOOD, E. A. 1988. Magnetostratigraphy of the Lower Cretaceous Vectis Formation (Wealden Group) on the Isle of Wight, Southern England. *Journal of the Geological Society, London*, **145**, 351–360.

KIRSCHVINK, J. L 1980. The least-squares line and plane and the analysis of paleomagnetic data, *Geophysical Journal of the Royal Astronomical Society*, **62**, 699–718.

—— & CHANG, S.-B. R. 1984. Ultrafine-grained magnetite in deep-sea sediments, possible bacterial magnetofossils. *Geology*, **12**, 559–562.

—— & WALKER, M. M. 1985. Particle-size considerations for magnetite-based magnetoreceptors. *In*: KIRSCHVINK, J. L., JONES, D. S. &, MACFADDEN, B. J. (eds) *Magnetite Biomineralization and Magnetoreception in Organisms*. Plenum, New York, 243–254.

LOWRIE, W. 1990. Identification of ferromagnetic minerals in a rock by coercivity and unblocking temperature properties. *Geophysical Research Letters*, **17**, 159–162.

MERRILL, R. T. 1970. Low temperature treatments of magnetite and magnetite-bearing rocks. *Journal of Geophysical Research*, **75**, 3343–3349.

MIDDLETON, M. F. & SCHMIDT, P. W. 1982. Paleothermometry of the Sydney basin, *Journal of Geophysical Research*, **87**, 5351–5359.

MONTGOMERY, P., HAILWOOD, E. A., GALE, A. S. & BURNETT, J. A. 1998. The magnetostratigraphy of Coniacian – Late Campanian chalk sequences in Southern England. *Earth and Planetary Science Letters*, **156**, 209–224.

MOON, T. & MERRILL, R. T. 1986. A new mechanism for stable viscous remanent magnetization and overprinting during long magnetic polarity intervals. *Geophysical Research Letters*, **13**, 737–740.

NÉEL, L. 1955. Some theoretical aspects of rock magnetism. *Advances in Physics*, **4**, 191–243.

ÖZDEMIR, Ö. & DUNLOP, D. J. 1989. Chemico-viscous remanent magnetization in the Fe_3O_4–gFe_2O_3 system. *Science*, **243**, 1043–1047.

PULLAIAH, G., IRVING, E., BUCHAN, K. L. & DUNLOP, D. J. 1975. Magnetization changes caused by burial and uplift. *Earth and Planetary Science Letters*, **28**, 133–143.

SCHMIDT, P. W. & EMBLETON, B. J. J. 1981. Magnetic overprinting in southeastern Australia and the thermal history of its rifted margin. *Journal of Geophysical Research*, **86**, 3998–4008.

SPORER, H. 1984. On viscous remanent magnetization of synthetic multidomain magnetite. *Geophysical Research Letters*, **11**, 209–212.

TARLING, D. H. 1983 *Palaeomagnetism*. Chapman and Hall, London.

TIVEY, M. & JOHNSON, H. P. 1984. The characterization of viscous remanent magnetization in large and small magnetite particles. *Journal of Geophysical Research*, **89B**, 543–552.

TRELOAR, N. A., SHIEVE, P. N. & FOUNTAIN, D. M. 1986. Viscous remanence acquisition in deep crustal rocks. *Eos, Transactions of the American Geophysical Union*, **67**, 266.

WALTON, D. 1980. Time–temperature relations in the magnetization of assemblies of single-domain grains. *Nature*, **286**, 245–247.

—— 1983. Viscous magnetization, *Nature*, **305**, 616–619.

—— & DUNLOP, D. J. 1985. The magnetization of a random assembly of interacting moments. *Solid State Communications*, **53**, 359–362.

WILLIAMS, W. & WALTON, D. 1988. Thermal cleaning of viscous magnetic moments. *Geophysical Research Letters*, **15**, 1089–1092.

YAMAZAKI, T. & KATSURA, I. 1990. Magnetic grain size and viscous remanent magnetization of pelagic clay. *Journal of Geophysical Research*, **95B**, 4373–4382.

ZIJDERVELD, J. D. A. 1967. A.C. demagnetization of rocks: analysis of results. *In*: COLLINSON, D. W., CREER, K. M. & RUNCORN, S. K. (eds) *Methods in Paleomagnetism*. Elsevier, New York, 254–256.

The significance of magnetotactic bacteria for the palaeomagnetic and rock magnetic record of Quaternary sediments and soils

BARBARA A. MAHER & MARK W. HOUNSLOW

School of Environmental Sciences, University of East Anglia, Norwich NR4 7TJ, UK

Magnetotactic bacteria are micro-organisms that form crystals of ferrimagnetic minerals intracellularly (Blakemore 1975). Many species of magnetotactic bacteria produce crystals of magnetite, which may 'age' and oxidize towards maghemite; some species, however, have been found to precipitate the ferrimagnetic iron sulphide greigite (Bazylinski *et al.* 1995). The ferrimagnetic particles are precipitated within an organic envelope and thus their size and shape are determined by this biological structure. The crystals dominantly fall within the single-domain (SD) grain size range (c. 0.03–0.05 μm for magnetite), are euhedral, often adopting unique and distinctive morphologies (including cubes, octahedra, 'bullet' and 'boot' shapes), and are aligned in chains. Some of these distinctive crystal morphologies are illustrated in the electron micrographs (Figs 1–3). The close linear arrangement of the magnetic particles results in positive interactions between them, which align the individual moments parallel to each other along the chain direction. Thus, the entire chain acts as an SD magnetic dipole. Microscopic investigation of *Magnetobacterium bavaricum*, a species found in fresh-water lakes of Bavaria, and some magnetic cocci shows that they possess as many as five discrete magnetosome chains (Hanzlik *et al.* 1996). The spatial arrangement of the chains is such that they are separated by the maximum possible distance, forcing them to be in direct contact with the cell envelope. As a result, the magnetic torque acting on the chains from the Earth's magnetic field is transferred very effectively to the whole bacterial cell. Magnetotactic bacteria are highly motile

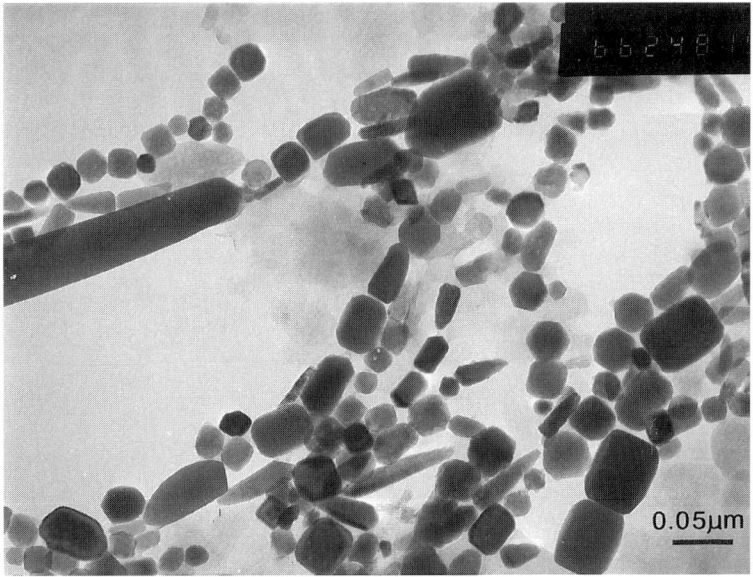

Fig. 1. Transmission electron micrograph of bacterial ferrimagnets magnetically extracted from Quaternary hemi-pelagic ooze, the Bahamas. A range of crystal sizes and morphologies are present, including four-sided, eight-sided and bullet-shaped grains, together with rather unusual, elongate blade-shaped grains.

MAHER, B. A. & HOUNSLOW, M. W. 1999. The significance of magnetotactic bacteria for the palaeomagnetic and rock magnetic record of Quaternary sediments and soils. *In*: TARLING, D. H. & TURNER, P. (eds) *Palaeomagnetism and Diagenesis in Sediments*, Geological Society, London, Special Publications, **151**, 43–46.

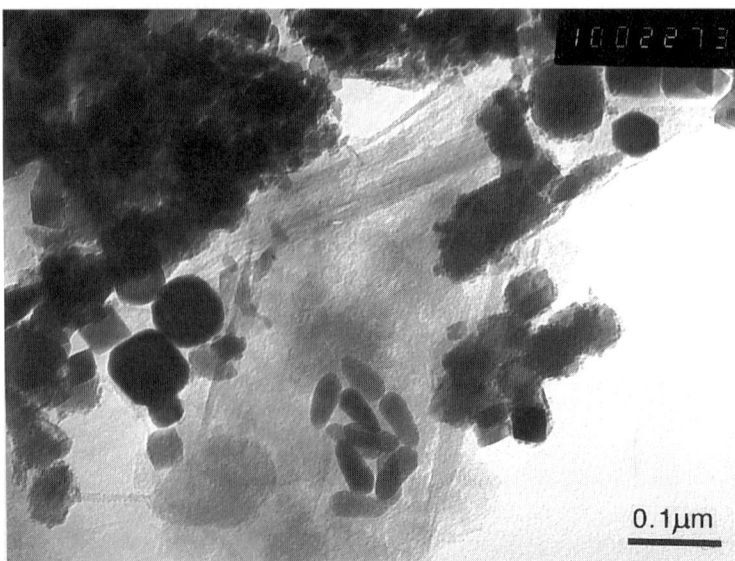

Fig. 2. Cluster of bullet-shaped bacterial ferrimagnets, from top $c.$ 5 m of sediment, ODP Hole 711.

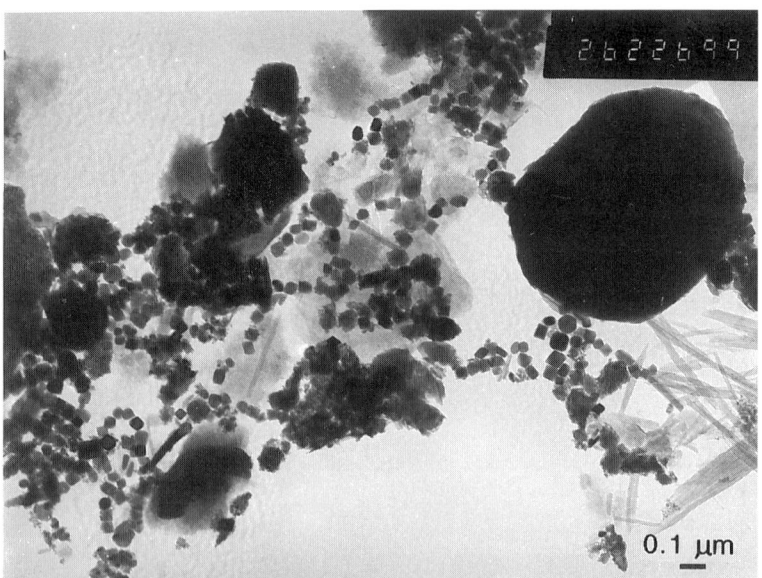

Fig. 3. Abundant bacterial ferrimagnets, together with some detrital iron oxides, from sediment depth of $c.$ 0–15 cm, ODP Hole 709.

and equipped with a propulsive flagella. The net effect of the intracellular magnets on the motility of the bacteria, and whether or not it confers any ecological advantage, remains as yet undefined. Living populations of magnetotactic bacteria have been found in diverse modern environments, including soils, fresh- and brackish-water lakes, sewage ponds, salt-marshes, rivers, estuaries and the deep sea (both within the water column and in the top few centimetres of deep-sea sediments). Fossil bacterial ferrimagnets, displaying diagnostically distinctive crystal sizes, shapes and concatenation, have been identified in palaeosols in the Quaternary loess sequences of north–central China, and in those sedimentary records of lakes and the deep sea that have been unaffected by post-depositional diagenesis.

The significance of bacterial magnetofossils in geological settings is that they may act as recorders of (a) the Earth's geomagnetic field at, or after, the time of sediment deposition, and (b) changes in the chemistry of the environment of deposition, i.e. they may represent a possible proxy of changes in climate, environment or oceanography. Magnetic criteria for detecting the presence of magnetofossils are desirable, as magnetic measurements can be rapid and non-destructive of samples. Room-temperature measurements of the acquisition and demagnetization of artificial isothermal remanences can identify the presence of the rather narrow grain-size range of bacterial magnetites, but are not uniquely diagnostic. When the magnetofossils are preserved as intact chains of particles, they will be characterized by high values of anhysteretic remanent magnetization (ARM), because of the strong interactions between the particles. Such interactions will also affect their low-temperature remanence behaviour; Moskowitz et al. (1993) found distinct magnetic behaviour for bacterial chains of magnetite, depending upon whether the samples were cooled (to $c.20\,K$) in an applied field or in zero field. The field-cooled remanence displayed a marked decrease at the Verwey transition whereas the zero field cooled remanence displayed only a very minor decline. However, such non-destructive magnetic measurements may provide ambiguous results where the bacterial chains have not been preserved intact, where grains have been maghemitized, and where grain interactions vary because of the presence of varying grain morphologies. Thus, 'calibration' of the magnetic analyses by independent techniques is required. Magnetic extraction, for direct observation, analysis and quantification of the magnetic carriers, provides a basis for improved understanding of the sources of magnetic minerals, their causal links with changing climate and/or environment, and enhanced interpretation and evolution of magnetic analyses. We have applied quantitative magnetic extraction methods to a range of Quaternary sediments and soils, including deep-sea sediments from the Indian Ocean (Ocean Drilling Program (ODP) Holes 722, 709 and 711), and modern soils in the UK and loessic palaeosols of north–central China. To identify the effectiveness of the extraction procedures, measurements of susceptibility, ARM and saturation isothermal remanent magnetization (SIRM) are made both before and after the extraction (Hounslow & Maher 1996). As bacterial ferrimagnets contribute preferentially to ARM, the ARM extraction efficiencies of a range of samples will be particularly noted here. For the modern soils and palaeosols, these range from $c.50$ to 80%. For the Indian Ocean sediments, ARM extraction efficiencies vary from 33 to 94%, with a mean of 61%. Lowest extraction efficiencies result from magnetically weak samples, in which a significant contribution to the ARM is made by particles occurring as inclusions within host diamagnetic silicates (Hounslow & Maher 1996).

Magnetic grains formed by magnetotactic bacteria were found in magnetic extracts from both the modern soil and the palaeosol samples and also within the top $c.1.5\,m$ of sediment (dominantly foraminiferal ooze) from ODP Hole 722, the Owen Ridge, Indian Ocean. However, compared with other magnetic components present in these samples, they are of minor significance, forming <1% of the extracted grain population. Below $c.1.5\,m$ in the Owen Ridge sediments, diagenetic dissolution has removed all the fine-grained ferrimagnets, whether of biogenic or detrital orgin (Maher & Hounslow 1999). Conversely, biogenic ferrimagnets are dominant and prolific within the upper $c.25\,m$ of (Pleistocene age) sediment from ODP Hole 709, the Mascarene Plateau, Indian Ocean. At this location, they may provide a new Pleistocene palaeoenvironmental proxy because the number of magnetofossils demonstrates a marked correlation with climate stage. High populations of bacterial ferrimagnets (Fig. 3)

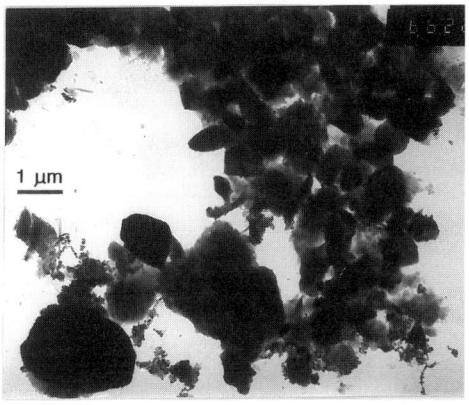

Fig. 4. This magnetic extract, from sediment depth of 10–40 cm, ODP Hole 709, is dominated by detrital ferrimagnets, with only a minor number of bacterial grains.

correlate strongly with interglacial and interstadial climate stages (age model information from S. Robinson, pers. comm.). Glacial climate stages are associated with low numbers of bacterial ferrimagnets (Fig. 4). Paradoxically, despite the abundant presence of these bacterial, SD ferrimagnets within the upper part of Hole 709, the quality of much of the palaeomagnetic record (c. 10–20 m below sea floor (bsf)) is poor, rendering impossible any reliable interpretation of polarity. Between c. 25 and 120 mbsf, the sediments (mostly nannofossil ooze) display a zone of marked magnetic depletion (S. Robinson, pers. comm.) and bacterial magnetofossils are notably absent within this Pliocene–middle Miocene interval. However, below c.120 mbsf, the magnetic content of the sediment is as high as or higher than in the upper c. 20 m and the numbers of preserved magnetofossils are again notably high.

More detailed discussion of these data will be made elsewhere. Some preliminary conclusions can, however, be drawn at this point. First, unambiguous detection of magnetofossils in sediments by magnetic analysis is difficult, because of the possibility of varied grain interactions, fragmentation of the particle chains and maghemitization upon oxidation. Second, magnetic extractions can be highly efficient in concentrating bacterial ferrimagnets and hence providing representative extracts for quantitative analysis. Third, analysis of magnetic extracts from a range of sediments and soils shows that bacterial ferrimagnets are magnetically significant only in rather restricted Quaternary sedimentary sequences. For deep-sea sediments, only those unaffected by reductive diagenesis can preserve any palaeomagnetic and/or palaeoenvironmental signal from bacterial magnetofossils. Bacterial ferrimagnets appear insignificant as magnetic carriers in many modern soils and in the palaeosols of the Chinese loess sequences (Fig. 5). Finally, the contribution of magnetotactic bacteria to the magnetic record of sediments and soils needs to be assessed quantitatively on both a site-specific and a time-specific basis.

References.

BAZYLINSKI, D. A., FRANKEL, R. B., HEYWOOD, B. R., MANN, S., KING, J. W., DONAGHAY, P. L. & HANSON, A. K. 1995. Controlled biomineralization of magnetite (Fe_3O_4) and greigite (Fe_3S_4) in a magnetotactic bacterium. *Applied Environmental Microbiology*, **61**, 3232–3239.

BLAKEMORE, R. P. 1975. Magnetotactic bacteria. *Science*, **190**, 377–379.

HANZLIK, M., WINKLHOFER, M. & PETERSEN, N. 1996. Spatial arrangement of chains of magnetosomes in magnetotactic bacteria. *Earth and Planetary Science Letters*, **145**, 125–134.

HOUNSLOW, M. W. & MAHER, B. A. 1996. Quantitative extraction and analysis of carriers of magnetisation in sediments. *Geophysical Journal International*, **124**, 57–74.

MAHER, B. A. 1991. Inorganic formation of ultrafine grained magnetite. *In*: FRANKEL, R. B. & BLAKEMORE, R. P. (eds) *Iron Biominerals*. Plenum, New York, 179–191.

—— 1998. Magnetic properties of modern soils and loessic paleosols: implications for paleoclimate. *Palaeogeography, Palaeoclimatology, Palaeoecology*, **137**, 25–54.

—— & HOUNSLOW, M. W. 1999. Palaeomonsoons II: Magnetic records of aeolian dust in Quaternary sediments of the Indian Ocean. *In*: MAHER, B. A. & THOMPSON, R. (eds) *Quaternary Climates, Environments and Magnetism*. Cambridge University Press, Cambridge, in press.

MOSKOWITZ, B. M., FRANKEL, R. B. & BAZYLINSKI, D. A. 1993. Rock magnetic criteria for the detection of biogenic magnetite. *Earth and Planetary Science Letters*, **120**, 283–300.

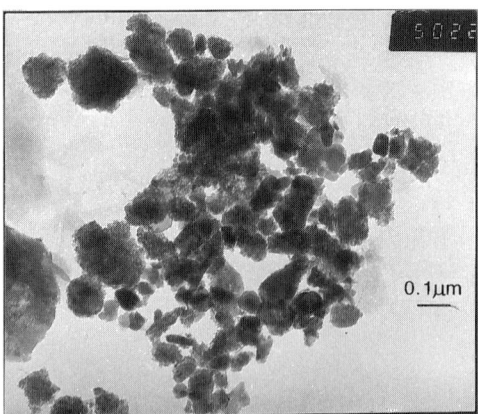

Fig. 5. Magnetic extract from a palaeosol sample from the Quaternary loess–soil sequences of north–central China. Bacterial ferrimagnets appear an insignificant component of these samples; ferrimagnets with a range of (uncontrolled) grain sizes, including many superparamagnetic grains (seen as clusters around larger particles) are dominant, inferred to be of non-biogenic or extracellular origin (Maher 1991, 1998).

The locking-in of remanence in upper Pleistocene sediments of Lake Lisan (palaeo Dead Sea)

SHMUEL MARCO,[1] HAGAI RON,[2] MICHAEL O. McWILLIAMS[3] & MORDECHAI STEIN[4]

[1] *Geological Survey of Israel, 30 Malkhe Israel Street, Jerusalem 95501, Israel (e-mail: shmulik@mail.gsi.gov.il)*
[2] *Geophysical Institute of Israel, Holon, Israel*
[3] *Department of Geophysics, Stanford University, CA, USA*
[4] *Institute of Earth Sciences, Hebrew University, Jerusalem, Israel*

More than 1500 oriented samples were collected from an outcrop of the Lisan Formation ranging in age between 67 and 32 ka. The samples consist of alternating aragonite and detritus laminae deposited in Lake Lisan, the ancestor of the present Dead Sea. The mean sedimentation rate was calculated from U series ages as 0.86 mm/a (Schramm 1997), which indicates that each 2 cm thick sample provides a magnetic snapshot of about 23 years duration. Hence even short-lived geomagnetic events are potentially recorded in the Lisan Formation. Demagnetization experiments show that the natural remanent magnetization (NRM) is commonly very stable; 878 horizons yield a mean direction Dec. $= 5°$; Inc. $= 45°$ ($\alpha_{95} = 1°$; $k = 22$). Stability tests were performed on layers that exhibit a variety of intraformational soft-sediment subaqueous deformations. The slumped layer tests show post-deformation acquisition of stable remanence (Fig. 1), as earlier reported by F. Addison (in Tarling (1983), p. 60) on far fewer samples. Other tests were done on 'mixed layers', i.e. layers up to 50 cm thick composed of mixtures of fragmented laminae resembling sedimentary breccia. Their formation has been attributed to earthquake shaking that disrupted the unconsolidated sediment at the bottom of the lake (Marco & Agnon 1995; Marco *et al.* 1996). The mixed layers provide a natural redeposition experiment for testing whether the mixed layers record the field that prevailed immediately after the earthquake, or a later field. The two cases differ in the preservation of the secular variation (SV) record before the earthquake. In the first scenario, this record was lost, whereas in the second scenario the mixed layers recorded SV of a later time. The dispersion of magnetic directions within mixed layers was compared with the scatter outside the mixed layer and with the dispersion within a single horizon within the mixed layer (Fig. 2). The directional variation in the mixed layers suggests that the magnetization lags behind sedimentation by an interval longer than the time equivalent to the thickness of the mixed layer, i.e. >400 years.

Three modes of directional geomagnetic variation were observed in the stratigraphic sequences and, by inference, in time: (1) rapid directional fluctuations, shifting erratically up to several tens of degrees from sample to sample; (2) more gradual variation with directional changes of tens of degrees within several tens

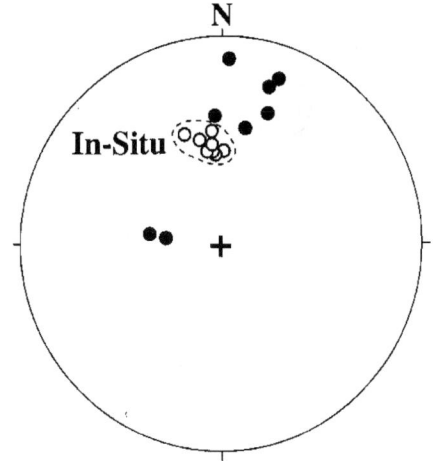

Fig. 1. A stereographic presentation of a fold test. The *in situ* population (○) is tightly clustered: $R = 0.996$, $k = 226$, $\alpha_{95} = 3.2°$. Tilt correction (●) offsets the vectors away from the expected direction and increases dispersion with $R = 0.863$, $K = 6$, $\alpha_{95} = 22°$.

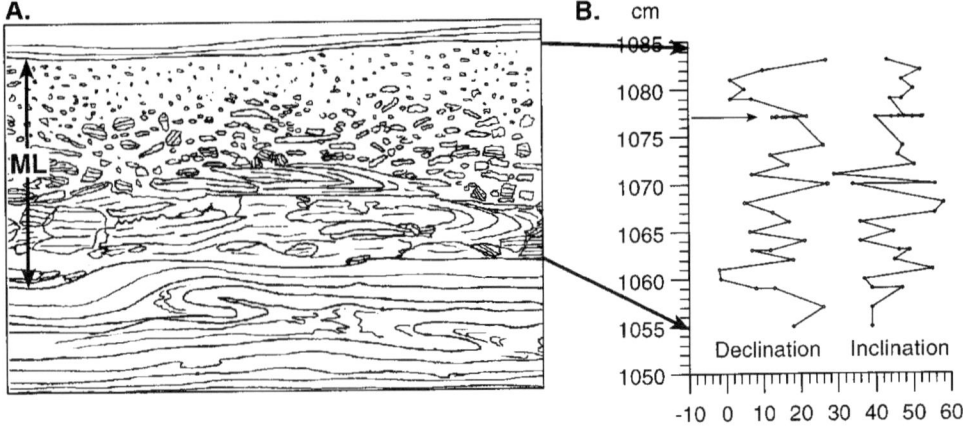

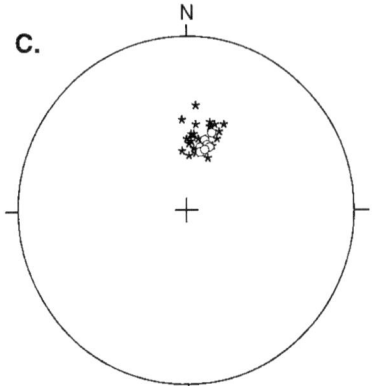

Fig. 2. A 'mixed layer' test. (a) A typical mixed layer (ML) shows a gradual upward transition from folded laminae, through fragment-supported texture, to matrix-supported texture at the top. Such mixed layers have been interpreted as earthquake deformations. The overlying undisturbed laminae are post-seismic (traced from a photograph). (b) Declinations and inclinations of samples in the 28 cm thick mixed layer, from 1055 to 1083 cm. Nine samples from level 1077 (arrow) that represent about 23 years cluster significantly tighter than the rest of the samples, which show SV. (c) Equal angle projection of the same points shows the small scatter of the horizontal set of samples (open symbols) and the significantly larger scatter of the vertical set (stars).

of centimetres to several metres; (3) a general trend in inclinations which became shallower with increasing age (Fig. 3). The rapid changes are particularly interesting because historical records suggest that such rapid changes of the geomagnetic field are probably rare. A few of the fastest apparent changes in the Lisan are caused by a single sample direction that deviates markedly from the samples above and below it, but most of the swings span several samples and are thus serially correlated, suggesting that they may record (SVs) of the geomagnetic field.

To understand how the NRM varies in space and in time, a comparison was made of the scatter of 15 horizontally and vertically sampled sites. At each site 12 samples were collected along a 1 m line. The small scatter in the horizontal sites corresponds to a precision parameter of $k = 269 \pm 150$ and angular standard deviation of $\alpha_{63} = 6 \pm 2°$, which are comparable with those of rapidly cooled Holocene basalt flows from Hawaii (McWilliams et al. 1982). In contrast, in the vertically sampled sites the precision and circular standard deviations are larger, $k = 69 \pm 65$ and $\alpha_{63} = 12 \pm 4°$, indicating that they probably reflect geomagnetic SV (Fig. 4). To examine the reproducibility, a duplicate set of samples was collected from a section between 1045 and 1250 cm where a distinct eastward deviation of declination had been observed. The second set, collected at the same site, reproduced the same general shape as the first set, with differences of the order of 15° (Fig. 5). The largest signal was of the order of 30°.

The rates of change in the Lisan record were estimated by dividing the angle between successive samples by the time difference (assuming constant sediment accumulation rate of 0.86 mm/a). The fastest angular change is 4.6°/a, with an average of $0.57° \pm 0.57°$/a. During 87% of the time the angular changes in the Lisan Formation are less than 1°/a, i.e. within the range of the historical record, but episodes with high rates are common and have peak rates of change up to ten times faster than in historical records. All the sources for scatter, except the

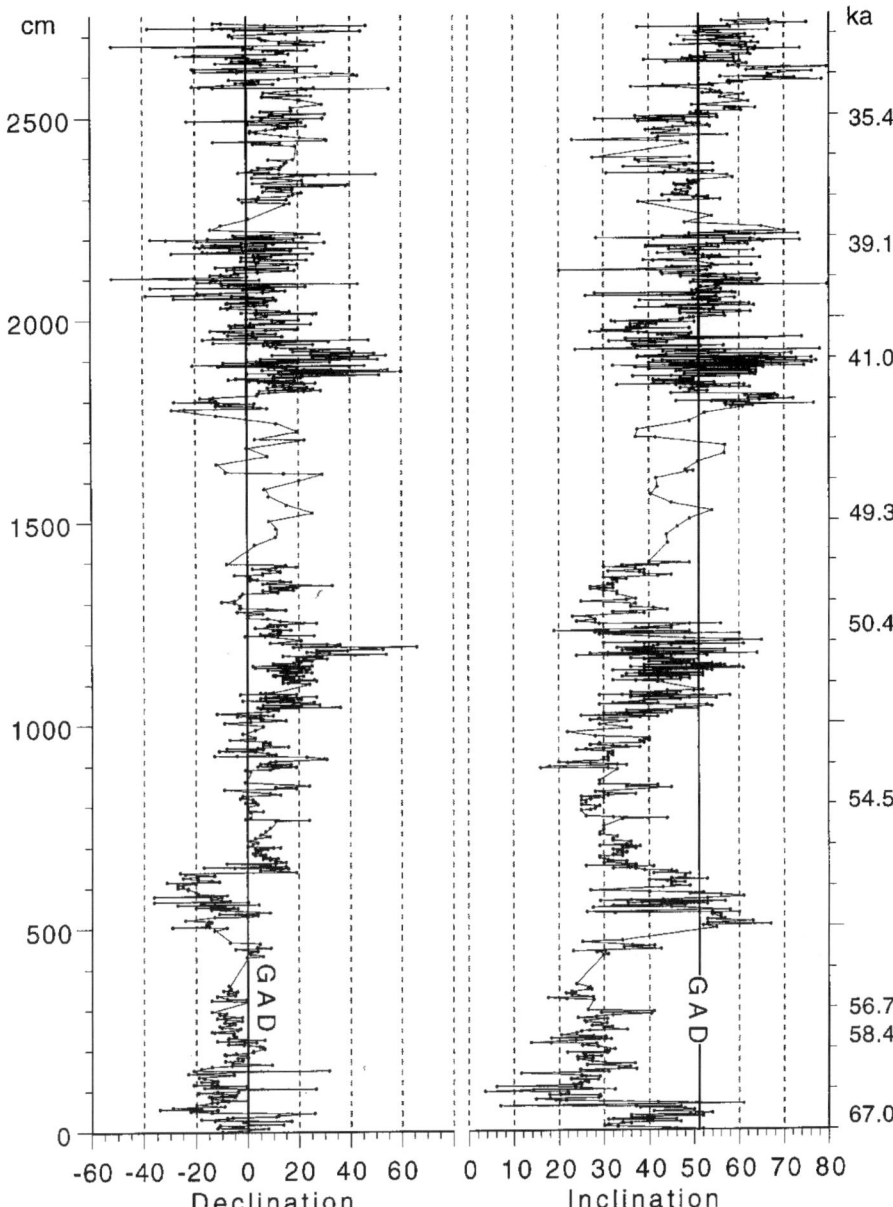

Fig. 3. Inclinations (right) and declinations (left) of 878 samples from the Lisan Formation. The samples were collected continuously along the section. Each point represents one sample, except for 15 horizons that are represented by the Fisher means of 12 samples. U-series ages by Schramm (1997) are shown at right.

geomagnetic field, probably cause random noise. The errors in sample orientation and measurement are estimated to total to less than 5°, and disturbance of the sample in the plastic boxes may also contribute up to 3 but such errors should characterize samples equally, whether they are from levels of slow or rapidly fluctuating directional change. It is considered that the small scatter of the directions within single horizons and the reproducible pattern (Fig. 5) are evidence that the high-frequency fluctuations mostly reflect the behaviour of the field, with only a small amount of scatter caused by geological processes and measurement errors.

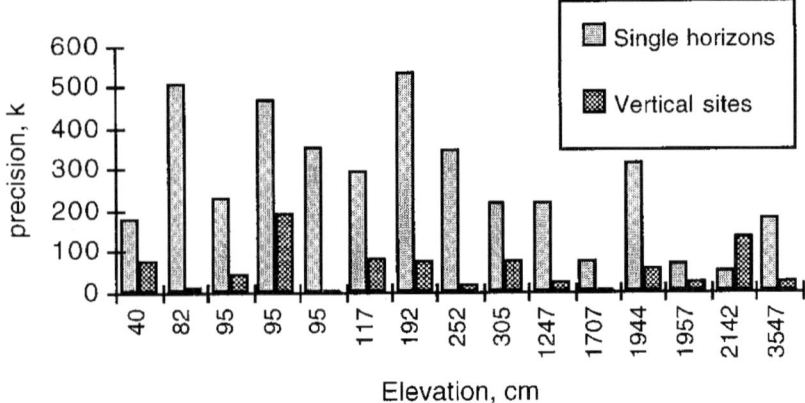

Fig. 4. Horizontal versus vertical dispersion. Precision parameters (k) of 12 sample sites show significantly larger dispersion where samples were collected vertically along 30–40 cm of the section than from single horizons.

To reduce noise, remove outliers, and emphasize the geomagnetic field behaviour, the SV curve has been smoothed. Each point in the smoothed curve is the Fisher mean of six consecutive samples. This window was then moved in single-sample steps (Fig. 6). Recalculating the angular change rate using the smoothed curve yielded a maximum of $0.66°/a$ with mean $0.10° \pm 0.10°/a$, and 68% of the time the apparent change rate is below $0.1°/a$. Although there is no 'objective' procedure for choosing the 'correct' smoothing method it is felt that the prominent features of the smoothed curve reflect the behaviour of the geomagnetic field. No reversed NRMs were observed, but geomagnetic field excursions may be present where the virtual geomagnetic poles (VGPs) deviate by more than 40 from geographic North. Such events are observed at 12 m and 19 m (dated as 52 ka, and 41 ka, respectively); the latter may represent the Laschamp excursion event. No geomagnetic event is known at 52 ka time elsewhere, save perhaps a 49–52 ka inclination anomaly in the Gulf of California interpreted as the Laschamp event (Levi & Karlin 1989), so this is tentatively termed the Lisan geomagnetic event, although its existence needs to be confirmed by future studies.

It is concluded that the characteristic remanent directions isolated in these samples are largely attributable to geomagnetic field variations. However, such remanent vectors can be traced systematically through both disturbed and undisturbed layers. The brecciated seismite layers have a sharp upper boundary with undisturbed overlying layers, but their lower boundaries are commonly gradational into the underlying undisturbed layers. Such seismic fluidization of the uppermost part of the sediment during earthquake activity provides a natural redeposition experiment. The magnetization in these disturbed layers clearly postdates the disturbance but is consistent within

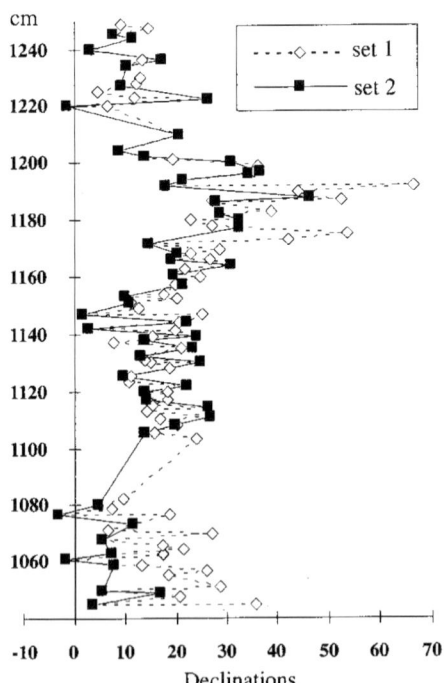

Fig. 5. A reproducibility test. Two sampling sets collected by different people from the same site to test reproducibility show similar declination pattern. The sampling noise is estimated at up to 15° but the signal is larger than 30°.

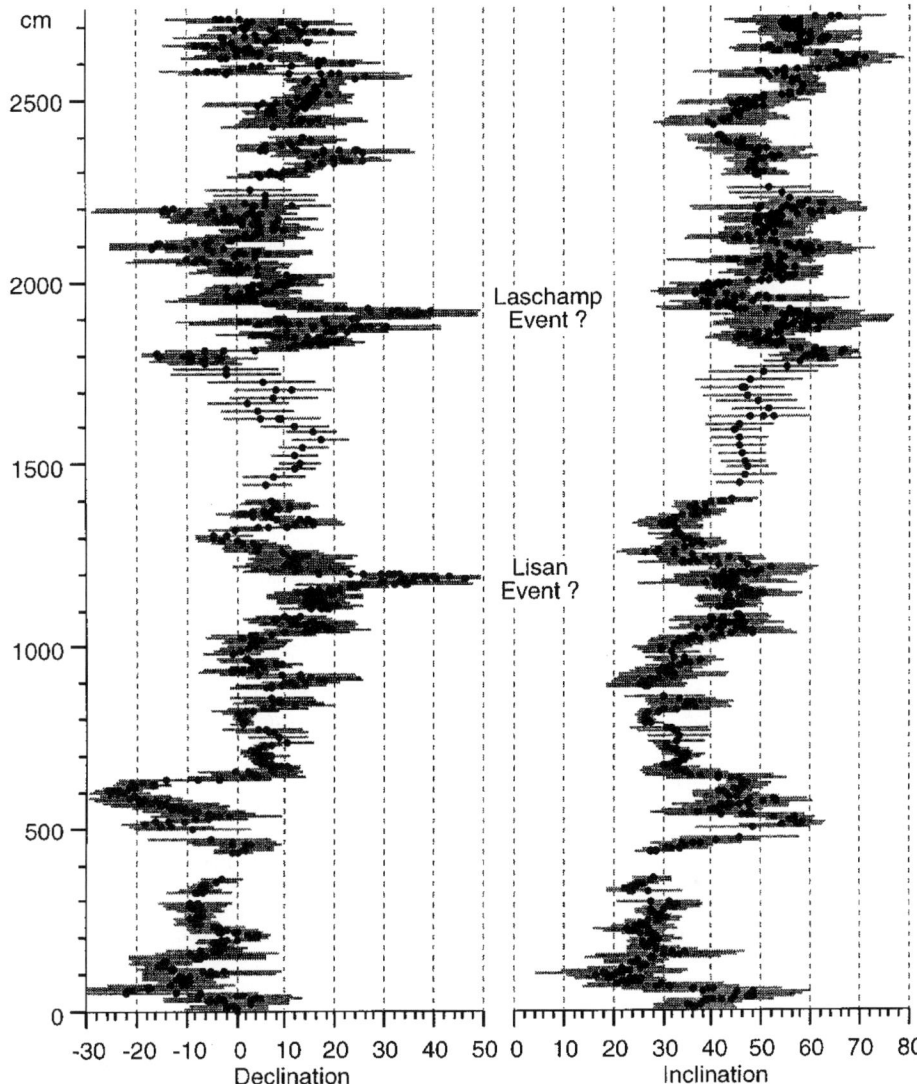

Fig. 6. The smoothed directions (filled symbols) with corresponding (α_{95} angles (grey). Each point in the smoothed curve is the Fisher mean of six consecutive samples. The averaged groups are moved by single sample steps. The declination deviation at 1900 cm was dated to 41 ka, the age of the Laschamp excursion event (Levi et al. 1990; Nowaczyk et al. 1994).

each layer and forms part of a regular pattern with the underlying layers. The Lisan Formation therefore provides evidence for the time scale of the locking-in of a remanence. Sediments down to a depth of about 1 m are dewatering and can be reset by SV changes, but below c. 1 m the increase in cohesivity means that they are no longer fluidized during the passage of seismic waves and are able to retain their directions of remanence. This means that the magnetic mineral orientations were locked at a depth corresponding to a few centuries after the original deposition, in agreement with several other estimates elsewhere (e.g. Butler 1992). However, such finds contrast with the conclusions of Tauxe (1993) that, in quiet sedimentary environments, any realignments of magnetic particles by change in the geomagnetic field were negligible, and such magnetizations locked in even before consolidation were vulnerable only to strong mechanical disturbances such as earthquakes, turbidity currents, etc. Estimates of the locking-in depth in pelagic sediments are between 30 cm (Kent & Schneider 1995) and 3–4 cm (Hartl & Tauxe 1996). In the low accumulation rates of the pelagic sediments (0.01–0.001 mm/a) this depth range is equivalent to 300–9000 years. The locking-in of magnetization in the Lisan Formation is within the time range proposed by Hartl & Tauxe (1996), but its depth is closer to that proposed by Kent & Schneider (1995). The detailed mechanism of locking-in and what controls the depth and time delay of magnetization are still open questions.

This project was funded by the US–Israel Binational Science Foundation grant #9200346. We are grateful to A. Agnon, A. Schramm and S. Goldstein for constructive and fruitful discussions and help in fieldwork. Thanks to O. Gonen and R. Ken-Tor for assistance in operating the magnetometer and in fieldwork, and to R. Weinberger, Y. Bartov, O. Klein, N. Shilony, A. Sagy and M. Machlus for help in the field.

References

BUTLER, R. F. 1992. *Paleomagnetism*. Blackwell Scientific, Oxford.
HARTL, P. & TAUXE, L. 1996. A precursor to the Matuyama/Brunhes transition field instability as recorded in pelagic sediments. *Earth and Planetary Science Letters*, **138**, 121–135.
KENT, D. V., & SCHNEIDER, D. A. 1995. Correlation of paleointensity records in the Brunhes/Matuyama polarity transition interval. *Earth Planetary Science Letters*, **129**, 135–144.
LEVI, S. & KARLIN, R. 1989. A sixty thousand year paleomagnetic record from Gulf of California sediments: secular variation, late Quaternary excursions and geomagnetic implications. *Earth and Planetary Science Letters*, **92**, 219–233.
——, AUDUNSSON, H., DUNCAN, R. A., KRISTJANSSON, L., CILLOT, P. Y. & JAKOBSSON, S. P. 1990. Late Pleistocene geomagnetic excursion in Icelandic lavas: confirmation of the Laschamp excursion. *Earth and Planetary Science Letters*, **96**, 443–457.
MARCO, S. & AGNON, A. 1995. Prehistoric earthquake deformations near Masada, Dead Sea graben. *Geology*, **23**, 695–698.
——, STEIN, M., AGNON, A. & RON, H. 1996. Long term earthquake clustering: a 50,000 year paleoseismic record in the Dead Sea Graben. *Journal of Geophysical Research*, **101**, 6179–6192.
McWILLIAMS, M. O., HOLCOMB, R. T. & CHAMPION, D. E. 1982. Geomagnetic secular variation from ^{14}C-dated lava flows on Hawaii and the question of the Pacific non-dipole low. *Philosophical Transactions of the Royal Society of London, Series A*, **306**, 211–222.
NOWACZYK, N. R., FREDERICHS, T. W., EISENHAUER, A. & GARD, G. 1994. Magnetostratigraphic data from late Quaternary sediments from the Yermak Plateau, Arctic Ocean: evidence for four geomagnetic polarity events within the last 170 ka of the Brunhes Chron. *Geophysical Journal International*, **117**, 453–471.
SCHRAMM, A. 1997. *Uranium series and C14 dating of Lake Lisan (paleo Dead Sea) sediments: implications for C14 time-scale calibration and relation to global paleoclimate* (PhD thesis), Göttingen University, Germany.
TARLING, D. H. 1983. *Palaeomagnetism, Principles and Applications in Geology, Geophysics and Archaeology*. Chapman and Hall, London.
TAUXE, L. 1993. Sedimentary records of relative paleointensity of the geomagnetic field: theory and practice. *Reviews of Geophysics*, **31**, 319–354.

Diagenesis and remanence acquisition in the Lower Pliocene Trubi marls at Punta di Maiata (southern Sicily): palaeomagnetic and rock magnetic observations

J. DINARÈS-TURELL & M. J. DEKKERS

Paleomagnetic Laboratory 'Fort Hoofddijk', Utrecht University, Budapestlaan 17, 3584 CD Utrecht, Netherlands (e-mail: dinares@geo.uu.nl)

Abstract: Three new profiles (LCM, PMD and CMD) through the lower Cochiti polarity zone allow comparison of the effects of weathering and diagenesis on the magnetic properties of the Lower Pliocene Trubi marls. These marls have been well dated, by means of astronomical calibration, and are of particular interest in terms of polarity zonation and regional tectonics. The freshest exposures (LCM profile) carry a remanence associated with pyrrhotite, which is particularly enhanced in some grey layers. This is the first time that pyrrhotite has been documented as the main remanence carrier in these marls. Previously, single-domain (SD) magnetite was recognized as the fundamental ferromagnetic phase. The mean characteristic remanent magnetization (ChRM) direction from the LCM record does not show the expected 30–35° clockwise rotation found in other studies, nor does it record the normal Cochiti interval (i.e. the sampled interval is entirely reverse polarity), suggesting a relatively late origin for the natural remanent magnetization (NRM). Standard rock magnetic techniques are used to investigate the nature of the magnetization in the three laterally equivalent profiles at Punta di Maiata. The entire dataset seems to be better explained by a complex diagenetic history involving a multistage redox history that invokes a 'late' anoxia event. The previously postulated iron migration model was put forward for the 'suboxic situation' and does not take into account the presence of magnetic sulphides, which are clearly demonstrated in the zone under study. The 'extended diagenetic model' inferred here proposes a pathway that accounts for most of the new observations. The salient feature is the postulated existence of a diagenetic phase that has altered the original NRM in some areas (dissolution of magnetite) resulting in a later-stage acquisition of chemical remanence that does not date to the timing of deposition of the sediments. The extension and the detailed mechanisms by which such a process has taken place deserve further research. None the less, the present investigation accounts for previously unrecognized features in the Trubi marls from Sicily and outlines how intricate the nature of the mechanisms contributing to the blocking of remanence in sediments can be.

The natural remanent magnetization (NRM) of undeformed sedimentary rocks is a composite of detrital input, compaction, a variety of diagenetic processes (often biologically mediated) and lithification, and fluid migrations during any of these phases. Each of these processes can give rise to authigenic growth of secondary magnetic phases, or to partial or even complete dissolution of the original detrital grains or even of secondary authigenic grains formed during earlier phases. Deformation may then even further complicate the picture, and exposure to weathering may occur after any of these phases. The present investigation involves an interdisciplinary analysis of the impact of the diagenetic and weathering processes on the diagenetic record of the Lower Pliocene Trubi marls in Sicily in an attempt to better understand the NRM acquisition processes. These deposits are particularly interesting as they show rhythmic sedimentation patterns that have allowed precise dating using astronomical precession and eccentricity cycles (Hilgen 1991; Lourens *et al.* 1996). Consequently, several studies have already been undertaken to understand their stratigraphic, palaeomagnetic and tectonic significance (e.g. Hilgen & Langereis 1988; Zachariasse *et al.* 1989; Langereis & Hilgen 1991). These previous studies will be summarized and then the results of the rock-magnetic, geochemical and petrographic observation studies of the new profiles will be outlined. The implications of these

DINARÈS-TURELL, J. & DEKKERS, M. J. 1999. Diagenesis and remanence acquisition in the Lower Pliocene Trubi marls at Punta di Maiata (southern Sicily): palaeomagnetic and rock magnetic observations. *In*: TARLING, D. H. & TURNER, P. (eds) *Palaeomagnetism and Diagenesis in Sediments*, Geological Society, London, Special Publications, **151**, 53–69.

findings for the nature of remanence acquisition in these beds will then be discussed at both a local and global scale.

The Trubi marls of Sicily

The open-marine marls of the Trubi Formation in Sicily comprise a cyclically bedded marly–calcareous biogenic sequence that usually conformably overlies non-marine argillaceous sands, known as the Arrenazzolo member, which was formed at the end of Miocene time. The boundary between these units, in the Capo Rossello section, has been informally defined as the stratotype for the Miocene–Pliocene boundary and the base of the Zanclean Stage (Cita & Gartner 1973), although it has been proposed that this stratotype boundary should be defined in the nearby Eraclea Minoa section (Hilgen & Langereis 1988). A formal decision from the International Commission on Stratigraphy has not yet been taken. The Trubi Formation is overlain by the Narbone Formation, which consists of marly clays, locally interbedded with sapropelitic layers. In this area of southern Sicily, a series of exposures (Eraclea Minoa, Punta di Maiata, Punta Grande and Punta Piccola) constitute the well-known Rossello 'composite' (Hilgen 1987; Langereis & Hilgen 1991), which contains 119 small-scale sedimentary cycles averaging c. 1 m thickness (95 cycles in the Trubi marls and 24 cycles in the Narbone Formation). In the Trubi marls, each cycle consists of distinct grey–white$_1$–beige–white$_2$ (G–W$_1$–B–W$_2$) layers (Hilgen 1987). The carbonate content varies from 60 to 85%, with the lowest values in the grey and beige layers. This cyclicity continues into the overlying Narbone Formation, where sapropelitic layers are sometimes intercalated into, or replace the grey layers of the carbonate cycle. This quatripartite rhythmic bedding has been correlated with the astronomical precessional cycle, and larger-scale cyclicity in the form of recurrence of relatively thick and/or indurated intervals has been related to the eccentricity cycle, with periods of about 100 and 400 ka (Hilgen 1991; Lourens et al. 1996). The rhythmites are interpreted as representing changes in the dilution of terrigenous material and in carbonate production in the surface waters controlled by periodical fluctuations in the precipitation and runoff related to orbital forcing (Hilgen 1987; De Visser et al. 1989). The grey layers are interpreted as being deposited under anoxic conditions, whereas suboxic to oxic conditions prevailed during deposition of the white and beige layers. Although most of the exposures along the coast are dipping gently to the N–NE, the Trubi marls have been thrust and folded into a series of southwesterly directed thrust slices facilitated by the Messinian evaporites, which acted as a décollement level (Roure et al. 1990; Catalano et al. 1995). At Punta di Maiata, the flat-lying coast exposures belong to the footwall of a syncline of which the overturned limb outcrops some 200–300 m inland. The internal structure in the hinge zone appears complicated and its lateral expression is unknown. The overthrusting hanging wall of this structure is formed by a highly deformed thickened sequence of Messinian evaporites. There are also many microstructures, such as penetrative planes and small-scale faults, in this area.

Previous palaeomagnetic and rock magnetic studies

A detailed magnetostratigraphy, with a resolution averaging 5–10 ka, has been established by Hilgen & Langereis (1988), Zachariasse et al. (1989) and Langereis & Hilgen (1991). The section, the Rossello 'composite' section, ranges from below the Thvera Subchron into the Matuyama Chron (4.86–2.45 Ma) and includes the Zanclean neostratotype and the Miocene–Pliocene boundary. The original magnetostratigraphy at Punta di Maiata (Hilgen 1991; Langereis & Hilgen 1991) was established on the western side of the cape, extending from the beach up to the cliff (Fig. 1). Van Hoof (1993) studied ten polarity transitions in great detail (temporal resolution <1 ka). These were from the lower Thvera to the upper Cochiti interval (in the Gilbert Chron), the Gilbert–Gauss reversal and the upper Kaena reversal (in the Gauss Chron). Nine reversal records are located in the Trubi Formation. Three were studied in the original sections constituting the Rossello composite (the lower and upper Thvera reversals, at Eraclea Minoa, and the upper Cochiti reversal at Punta di Maiata). The six remaining polarity transitions were studied at Capo Bianco, west of Eraclea Minoa. The Punta di Maiata section comprises 65 m stratigraphically (cycles 22–80 of the Rossello composite) and ranges from the upper part of the Sidufjall Subchron to the lowermost part of the Gauss Chron. These include the Nunivak and Cochiti subchronozones, the lower limit of the latter occurring within cycle 45 (Langereis & Hilgen 1991). Van Velzen & Zijderveld (1990) found a characteristic remanent magnetization (ChRM)

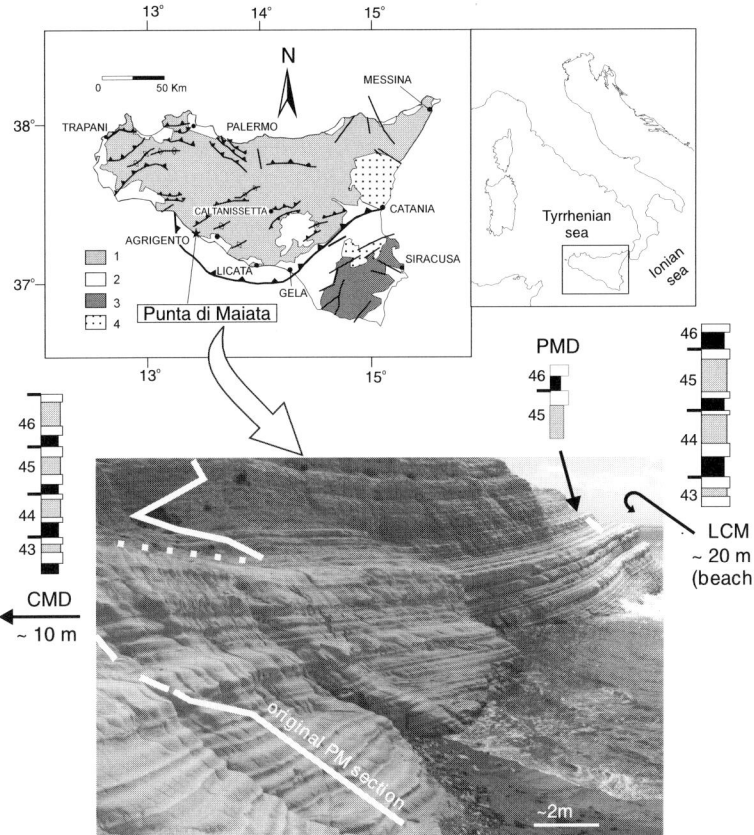

Fig. 1. Schematic geological map of Sicily with the location of the Punta di Maiata and a picture of the outcrop with the different studied sections. 1, Pre mid-Pliocene rocks; 2, mid-Pliocene to recent sediments; 3, Mesozoic volcanic rocks; 4, Pliocene and Pleistocene volcanic rocks. In the lithological columns, white denotes the white layers, black the grey layers, and shading indicates the beige layers of the individual numbered cycles.

in the Trubi marls carried by single-domain (SD) magnetite. Scheepers & Langereis (1993) reported that these ChRM directions showed a consistent clockwise rotation of $c.\ 34°$ (normal 30.7°, 50.4°, $\alpha_{95} = 2.6°$; reversed 218.3°, $-40.2°$, $\alpha_{95} = 2.2°$). They took this as support for the primary nature of the ChRM, as this rotation is widespread within the Caltanissetta basin and is related to a two-phase Pliocene rotation in the Tyrrhenian arc (Scheepers & Langereis 1993; Scheepers 1994). However, some directional behaviour along the reversal boundaries has been found to be very complex. For example, the lower Cochiti reversal at the Capo Bianco exposure is not a simple transition from reversed (R) to normal (N) polarities but involves a complex transition proceeded and followed by a R–N–R and a N–R–N 'excursion', respectively (Van Hoof 1993).

To explain such behaviour, Van Hoof (1993) postulated that magnetite could form at different times at different levels in the sediment because of early diagenetic diffusion of Fe from the anoxic grey layers into suboxic–oxic zones where secondary magnetite would then form, resulting in delayed remanence acquisition (Van Hoof & Langereis 1991). Weathering effects have also been considered to have changed the remanent properties. Van Velzen & Zijderveld (1995) showed that samples from the grey and beige layers (Eraclea Minoa) had a remanence carried by SD magnetite that exhibits unusually high coercivities thought to have originated by surface oxidation at low temperatures, i.e. weathering. These high coercivities were attributed to stresses induced by the high gradient in the degree of oxidation between the surface and core of the grains. Heating the samples to $c.\ 150°C$

was considered to release the stresses and hence the high coercivities, as well as changes in other magnetic parameters (decrease of total isothermal remanent magnetization (IRM) and anhysteretic remanent magnetization (ARM), and increase of bulk susceptibility). This model would suggest that such secondary magnetizations could be reduced or eliminated after heating to some 100–150°C, after which alternating field (AF) demagnetization could more efficiently isolate the pre-weathering component.

The new profiles

Three new sampling profiles have been made through the lower Cochiti transition interval at Punta di Maiata (Fig. 1). The zero level for each record was arbitrarily defined at a distinct boundary between two layers, and the exact stratigraphic positions of the individual standard specimens (22 mm height) were determined taking into account the drilling orientation and the attitude of the bedding plane. The mean resolution of a few millimetres implies a time resolution of $\pm <100$ years. Stepwise thermal demagnetization was performed on one specimen from each core, thus providing a resolution of about 3.5 cm (i.e. <1 ka). The additional specimens were used for rock magnetic experiments, petrographic observations and/or geochemical analysis. Profile LCM was in outcrops at sea level on the eastern side of the cape where the rocks appear fresher (i.e. bluish layers were sampled below a few centimetres of greyish surface materials) and dip at 14° towards 65°N. Sampling commenced in cycle 43 white$_1$, up to top of cycle 46 grey, totalling 310 cm of section (87 core samples). This profile was extended eight cycles upwards to cycle 54 grey, by collecting pairs of samples in almost all grey and beige layers. Profile PMD was right at the edge of the pronounced cape where the rocks dip 12° towards 90°N and had grey–white$_1$–beige–white$_2$ colour. It comprises 11 pairs of samples (each pairs some 8 cm apart) from cycle 45 white$_1$ to cycle 46 white$_1$, now recognized as typical of weathering. Additional samples were taken from the underlying cycle 45 grey and the beige and grey layers of cycles 43 and 44. As logistic problems prevented detailed sampling of this profile, the CMD profile has been taken as typical of the weathered properties. The third profile (CMD) was on the western side of the cape, in the cliff section, and about 10–15 m west of the original magnetostratigraphic profile (Fig. 1). This record spanned 295 cm of section from the middle part of cycle 43 grey up to the middle part of cycle 46 beige (139 cores). The strata dip at 20° to 95°N but included a subtle oblique-to-bedding boundary that delimits two distinct weathering profiles. The upper part of the record, above cycle 45 white$_1$, displays the characteristic grey–white$_1$–beige–white$_2$ colours, but below this level, the cycles tend to be more homogeneously greyish and the different layers are more difficult to recognize. A few metres laterally from this CMD profile, cycles 45 and above display only the homogeneous greyish surface colouring. Samples were collected at a few centimetres depth to ensure there was no substantial change in the colouring. This profile was originally chosen as representative of the most 'weathered' setting and is similar to that of the original magnetostratigraphic study and our PMD profile. (It was only after the first diagenetic analyses were available that the extent to which the subtle variations in the colour were affecting the magnetic signature along this record was realized).

Magnetic properties of the three new profiles

A variety of rock magnetic properties were investigated in the LCM and PMD profiles in addition to the standard low-field susceptibility and NRM measurements for all three profiles. These properties included the IRM at 1.5 T and the ARM imparted in a 100 mT AF with a 0.034 mT DC bias field. The back-field coercivity, B_{cr}, was measured by the stepwise increase in the DC field strength required to decrease the saturation remanence to zero. The Curie temperatures were obtained with a modified horizontal balance (Mullender et al. 1993). Unless otherwise stated, the intensity is the initial value of the intensity of NRM, i.e. before demagnetization. The ARM/IRM ratio is commonly used as an indicator of the quantity of fine (SD and pseudo-single-domain (PSD)) ferrimagnetic grains present (Hunt et al. 1995), whereas B_{cr} is independent of the presence of paramagnetic minerals and the concentration of ferromagnetic minerals. Assuming only one ferromagnetic mineral to be present, it can therefore by used as a partial discriminant between different types of ferromagnetic minerals.

The LCM profile

Thermal demagnetization (Fig. 2) showed generally two components, with a small viscous component removed at 100°C. The low-temperature component, removed by 200–400°C, had a present-day direction and was generally small or absent (Fig. 2d and h). The ChRM direction was

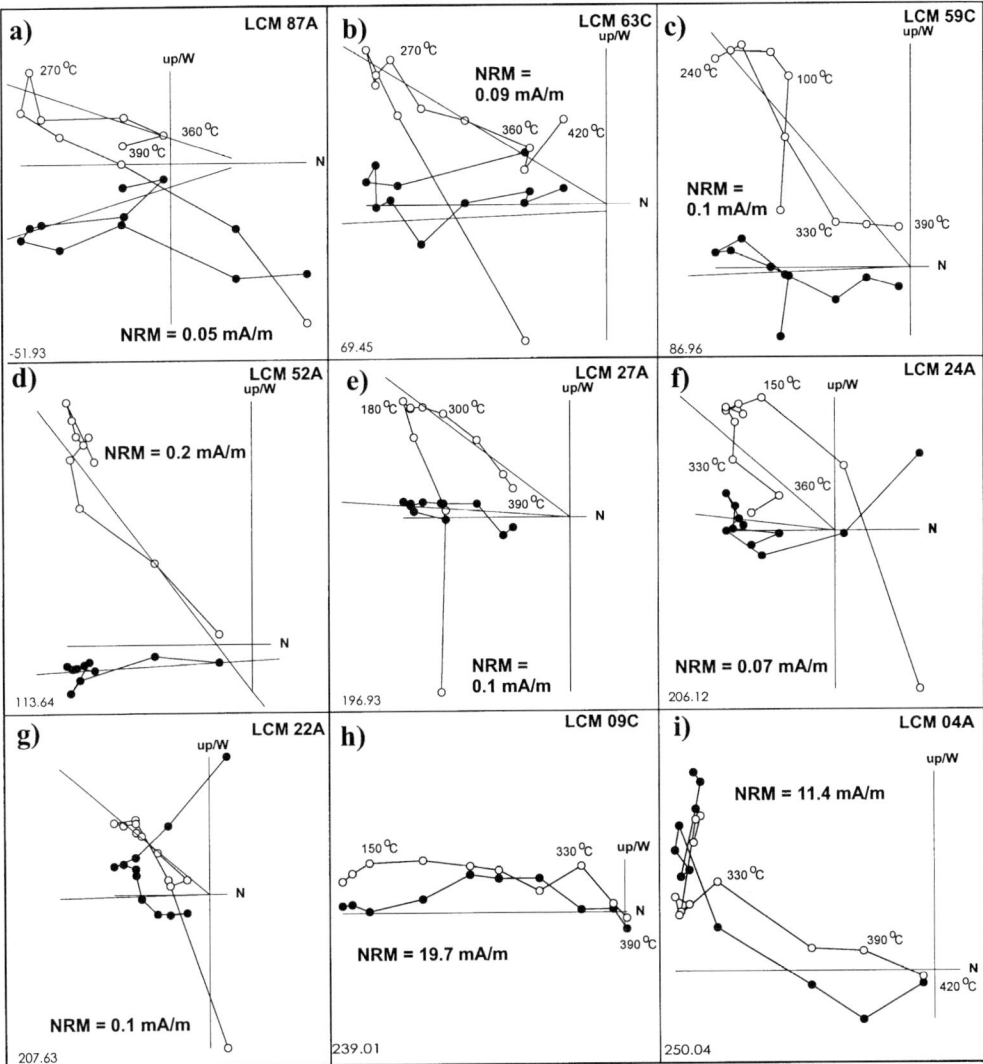

Fig. 2. Bedding-corrected orthogonal projections of stepwise thermal demagnetization of selected samples from the LCM record. Steps of 100, 150, 200, 240, 270, 300, 330, 360, 390°C (420°C) were used. Open (closed) circles denote projections on the vertical (horizontal) plane. The stratigraphic position of each sample is indicated in the lower left corner. (Note the high NRM intensity in samples LCM.04A (i) and LCM.09C (h), which belong to cycle 46 grey.)

usually removed completely below 360–390°C, although some samples remained blocked until higher temperatures (Fig. 2b and e) but also showed by an increase of bulk susceptibility and the presence of suspected spurious components. As discussed later, the ChRM appears to be predominantly carried by pyrrhotite. The linearity of the ChRM components was sometimes poor, probably reflecting the low intensity and chemical instability, but the mean direction of the better defined directions (mean angular deviation <10°) (48 of 111 samples) was statistically the same as for all of the samples (Fig. 3a) and so the total number has been take to calculate the mean ChRM direction as entirely reversed, 180.7, −37.3°, ($\alpha_{95}=5.1°$). Generally IRM was attained in a field of around 0.2 T (Fig. 4a), indicating that a low-coercivity mineral carried the IRM. The blocking temperatures, around 330°C (Fig. 4b), were identical to the Curie temperature, for which the heating and cooling curves were fully reversible up to 350°C

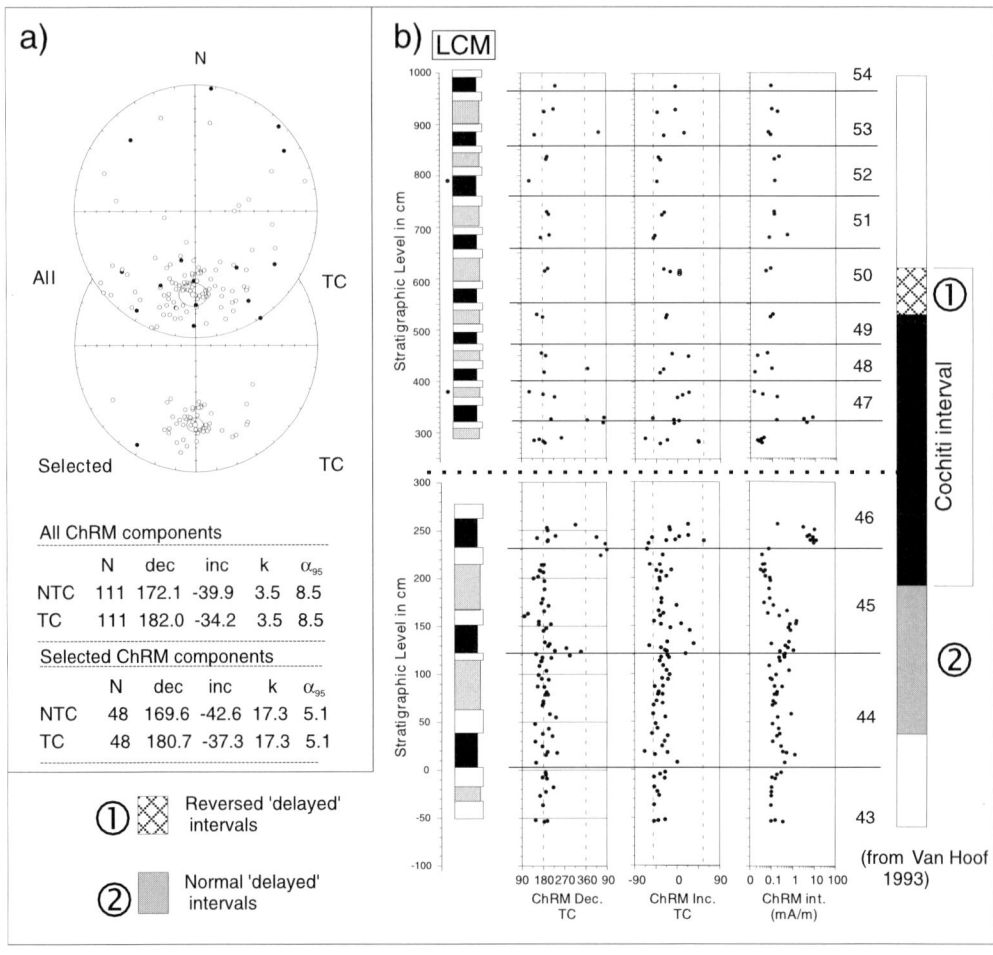

Fig. 3. Palaeomagnetic data for the LCM record. (**a**) Equal-area projection of all and a selection of the best ChRM components after bedding correction. In the table: N, number of samples; dec, declination; inc, inclination; k, precision parameter; α_{95}, radius of the 95% confidence cone about the mean direction, NTC, before bedding tilt correction; TC, after bedding tilt correction. (**b**) ChRM declination, inclination and intensity record showing that no clear record of the normal Cochiti interval exists at the LCM track (see text). The expected location of this interval and the zones where intervals holding 'delayed' magnetizations were found in previous studies (Van Hoof 1993) are shown. (Note the relatively higher ChRM intensities in the grey layers of cycles 46 and 47, which correspond also to scattered ChRM directions.)

(Fig. 4c). This indicates that pyrrhotite is the fundamental carrier of remanence, as other magnetic iron sulphides, such as greigite, usually break down before reaching the Curie point and do not have reversible Curie point curves (Roberts 1995). The intensities of remanence (NRM, IRM and ARM) were all 'enhanced' in cycle 46 grey (Fig. 5a), being one to two orders of magnitude higher at this level. The ARM/IRM ratio (Fig. 5a), commonly used as a proxy indicator of the presence of fine (SD and PSD) ferrimagnetic grains (Hunt et al. 1995), was mostly around 0.02, although the grey layers, particularly cycle 46 grey, were generally lower ($c.\,0.008$). B_{cr} was generally between 50 and 80 mT, although somewhat lower (20 mT) in the upper part (cycle 46 grey). As magnetite is not prominent, these changes in B_{cr} are likely to indicate differences in either the grain size or the type of iron sulphide. Dekkers (1988) reported coercivities for natural pyrrhotite in the <5 to 250 μm grain-size range that vary from 66 to 6 mT respectively. In the magnetically 'enhanced' cycle 46 grey level, relatively large crystals of

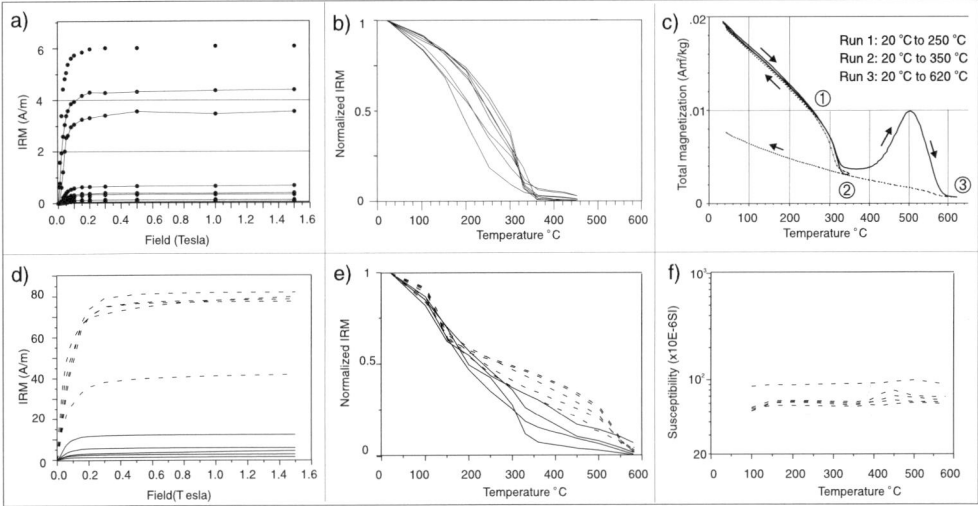

Fig. 4. Examples of isothermal remanent properties. (**a**) IRM acquisition curves from different lithologies in the LCM record. (**b**) Stepwise thermal demagnetization of the saturation IRM. (**c**) Incremental thermomagnetic runs on a sample from the grey layer of cycle 46. (**d**) IRM acquisition curves from samples from the CMD record. Dashed (continuous) curves are from the upper (lower) domain as defined in Fig. 5b. (**e**) Stepwise thermal demagnetization of the saturation IRM for the samples from the CMD record. (**f**) Low-field susceptibility measured at room temperature during thermal demagnetization of the CMD samples.

iron sulphide were observed (Fig. 6) and their selective presence might explain some of the variation in the observed B_{cr} values.

The PMD profile

Thermal demagnetization (Fig. 7a) revealed that two components were present, in addition to a small laboratory component, removed at 100–150°C. The first component was removed between 250 and 300°C and had a present-day Earth's magnetic field direction. The higher-temperature component, taken to be the ChRM, had a maximum unblocking temperature of 580°C, suggesting that it was carried by magnetite. The ChRM had both normal and reversed polarity (Fig. 7b) although it was predominantly normal. The normal polarity directions were rotated clockwise by 30° and the fewer reversed directions were rotated in the same sense, but by a greater angle, c. 51°, but this was imprecisely defined ($\alpha_{95} = 18.5°$) and appeared to have been more affected by the presence of some present-day field components, as was also observed by Scheepers & Langereis (1993) in the composite Rossello section. The transition from normal to reverse occurs between the top of cycle 45 grey and the cycle 45 white$_1$–beige boundary, and is interpreted as the lower Cochiti reversal (Fig. 7e). This reversal was placed by Van Hoof (1993) in the upper part of cycle 45 beige in the Capo Bianco section, as this layer had reverse polarities in its central part (the normal polarities above and below were interpreted as delayed remanences). No detailed rock magnetic studies were made of samples from this profile.

The CMD profile

The two zones, above and below cycle 45 white$_1$, had different characteristics. The lower zone ChRM component was generally removed at temperatures below 350°C (Fig. 8a–c) and the polarity was reversed. Somewhat higher blocking temperatures were found in the 'mixed' category (Fig. 9a). In the upper zone (above cycle 45 white$_1$), the maximum unblocking temperatures were up to 580°C, typical of magnetite (Fig. 8d–f) and the polarity was entirely normal, i.e. the boundary between the two zones coincides with the polarity transition, i.e. in cycle 45 white$_1$ (Fig. 9a). Both the normal and reversed mean ChRM directions showed clockwise rotations (Fig. 9b), 25 and 20°, respectively. The different remanences (NRM, IRM and ARM) for the CMD record (Fig. 5b) delineate the two domains with distinct characteristics. The upper domain, above cycle 45

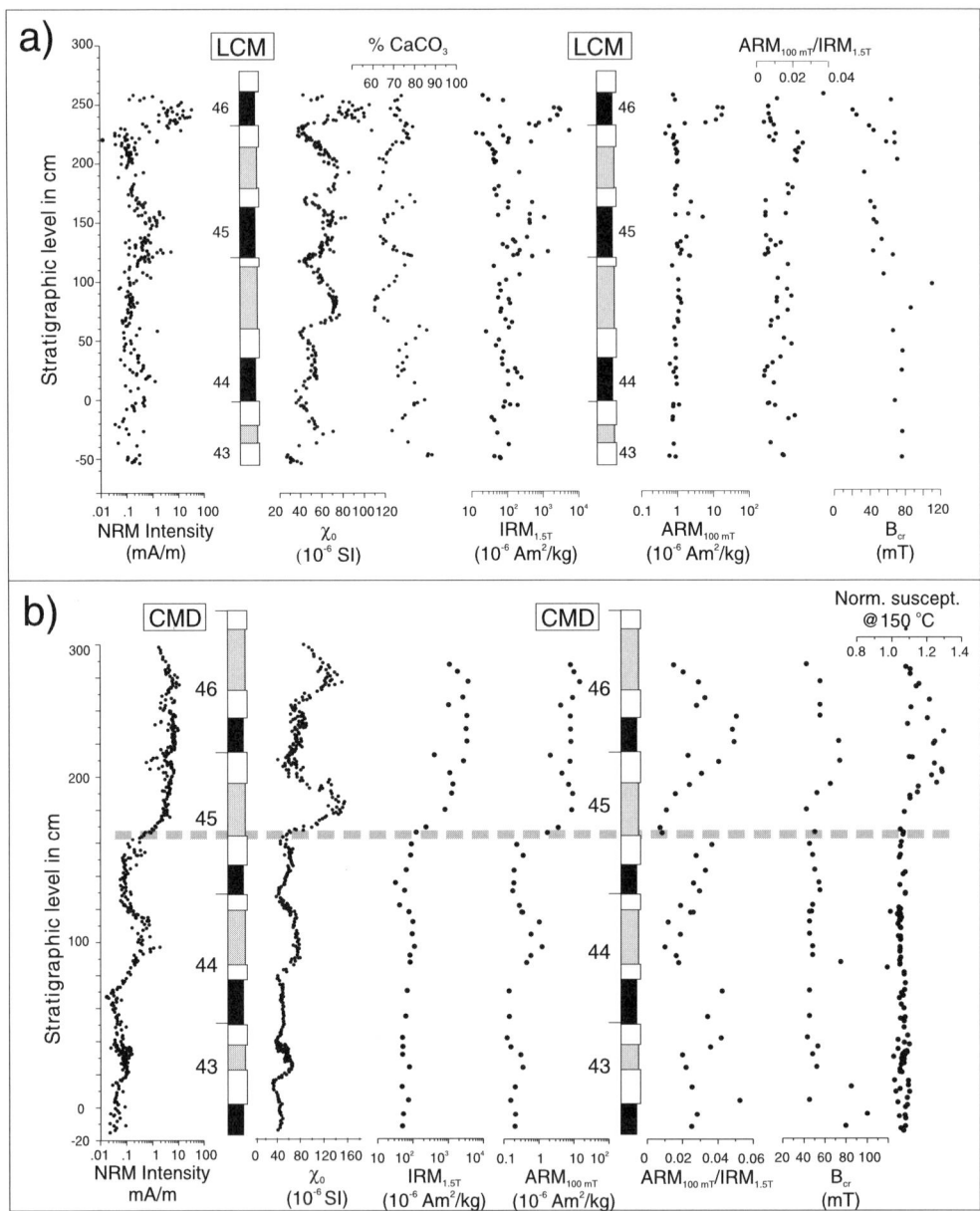

Fig. 5. Rock magnetic parameters for the LCM (**a**) and CMD (**b**) profiles. The CaCO$_3$ content for the LCM record is included. c_0, low-field susceptibility; IRM$_{1.5T}$, isothermal remanent magnetization acquired at 1.5 T; ARM$_{100mT}$, anhysteretic remanent magnetization acquired at 100 mT peak field and 34 mT DC bias field; B_{cr}, coercivity of remanence. The normalized χ_0 at 150°C as a proxy weathering parameter (see text) is shown for the CMD profile. The boundary between the two subdomains in the CMD profile, as described in the text, is shown by a horizontal dashed line.

white$_1$, presents relatively high remanences that are one to two orders of magnitude higher than those in the lower domain (Fig. 5b; note the logarithmic scale). Relatively higher intensities in cycle 44 beige, and to a lesser extent in cycle 43 beige, are observed in the lower domain. The NRM intensities from the upper domain (c. 10 mA/m) are similar to those commonly

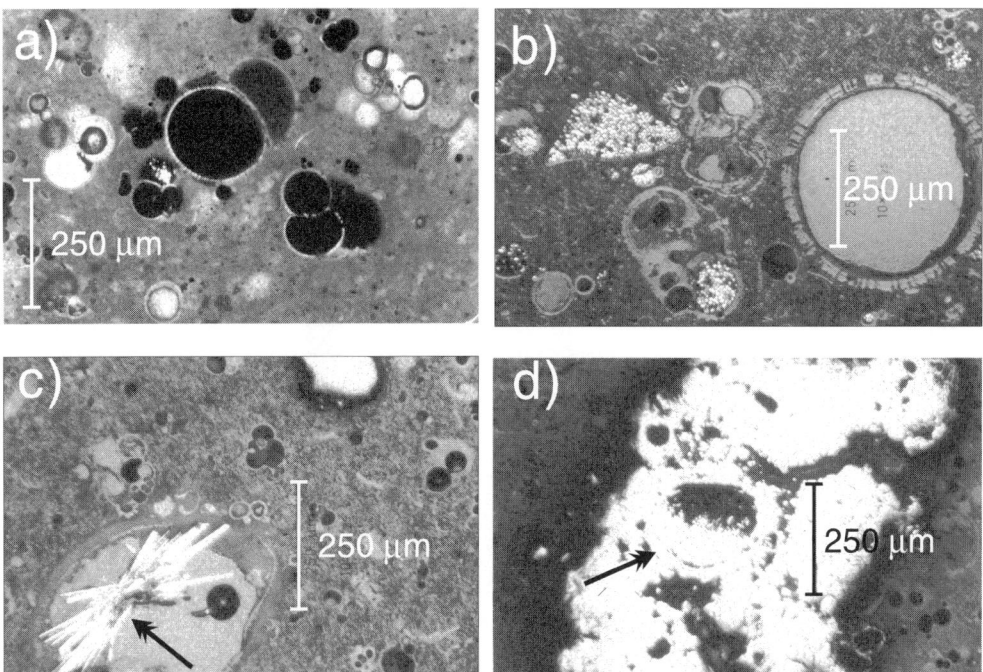

Fig. 6. Optical microscope images from LCM samples. (**a**) Refracted light micrograph illustrating the presence of microforaminifera shells that are filled by opaque minerals (framboidal pyrite). (**b**) Reflected light micrograph showing the framboidal pyrite (bright) that fills microfossil shells. (**c**) Reflected light micrograph showing (indicated by arrow) peculiar acicular aggregates (likely to be pyrrhotite) in a sample from the grey layer from cycle 46 (G46) which is characterized by 'enhanced' magnetic properties. Smaller and similar grains were observed dispersed in the rock matrix. (**d**) Reflected light micrograph of a massive irregular iron sulphide grain in another sample from G46. (Note the microforaminifera shell ghost indicated by the arrow.)

observed in previous studies in the Trubi marls. Therefore, the relatively low NRM intensities (below 1 mA/m) observed in the lower domain of the CMD profile (and in most of the LCM profile), represent a situation reported only occasionally (i.e. at the Punta Grande section located east of Punta di Maiata; Langereis & Hilgen 1991).

IRM acquisition experiments up to 1.5 T on samples from the two defined domains in the CMD record, and the subsequent thermal demagnetization of the imparted IRM, show clearly noticeable differences (Fig. 4d and e). All samples practically reach saturation below 0.2–0.3 T although in some cases a progressive small increase in intensities can be observed at higher fields. As pointed out before, samples from the upper domain acquire higher saturation IRMs (Fig. 4d). The thermal demagnetization of the IRM, for samples from the lower domain (Fig. 4e), shows the presence of unblocking temperatures typical of iron sulphides (330°C). However, in some samples, 30–40% of the remanence is lost at higher temperatures (up to 580°C) indicating that magnetite must also contribute to the IRM (these samples probably correspond to the 'mixed' category outlined before). The demagnetization behaviour of samples from the upper domain is characterized by a progressive loss in magnetization with a relatively sharp decay around 550–580°C indicating that magnetite is the predominant carrier (Fig. 4e). These samples also display a drop in intensity at $c.150°C$. This feature was also observed along the upper Cochiti record at Punta di Maiata (Van Hoof et al. 1993) and in samples from Eraclea Minoa (Van Velzen & Zijderveld 1995). The latter workers concluded that this decay of IRM intensity at 150°C, and also other magnetic changes at this temperature, are related to weathering and release of stresses in the magnetic grains as explained above.

A more noticeable difference (Fig. 4f) between the magnetic behaviour from the two domains is the variation in susceptibility upon heating (the susceptibility is measured at room temperature).

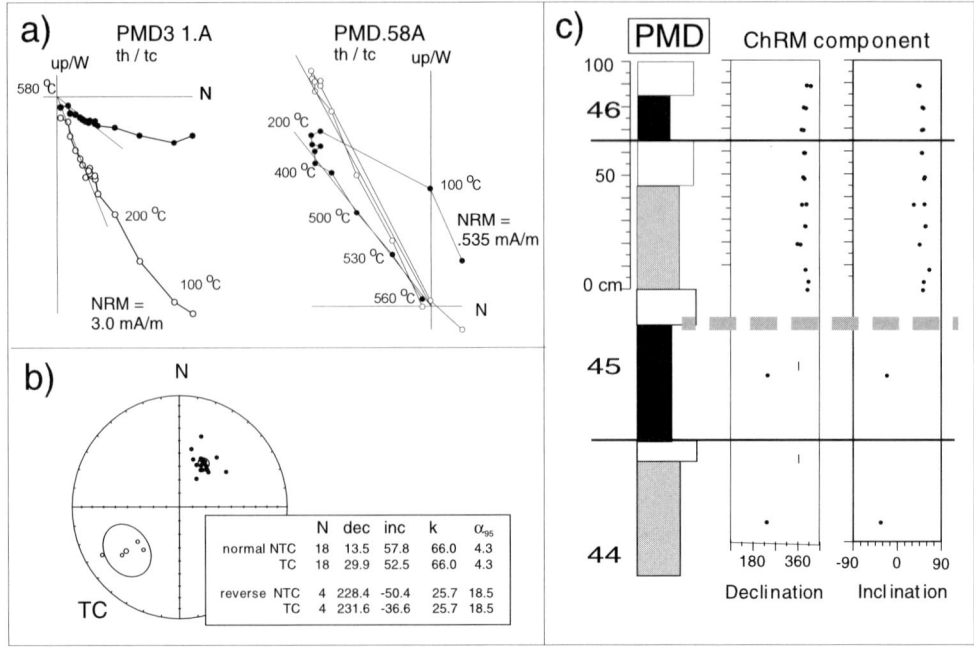

Fig. 7. Palaeomagnetic data for the PMD track. (**a**) Bedding-corrected orthogonal thermal demagnetization diagrams. (**b**) Equal-area projection of the ChRM components after bedding correction. (**c**) ChRM declination and inclination record showing that a reverse to normal transition (lower Cochiti) occurs in cycle 45 (showed by a dashed thick line between G45 and B45).

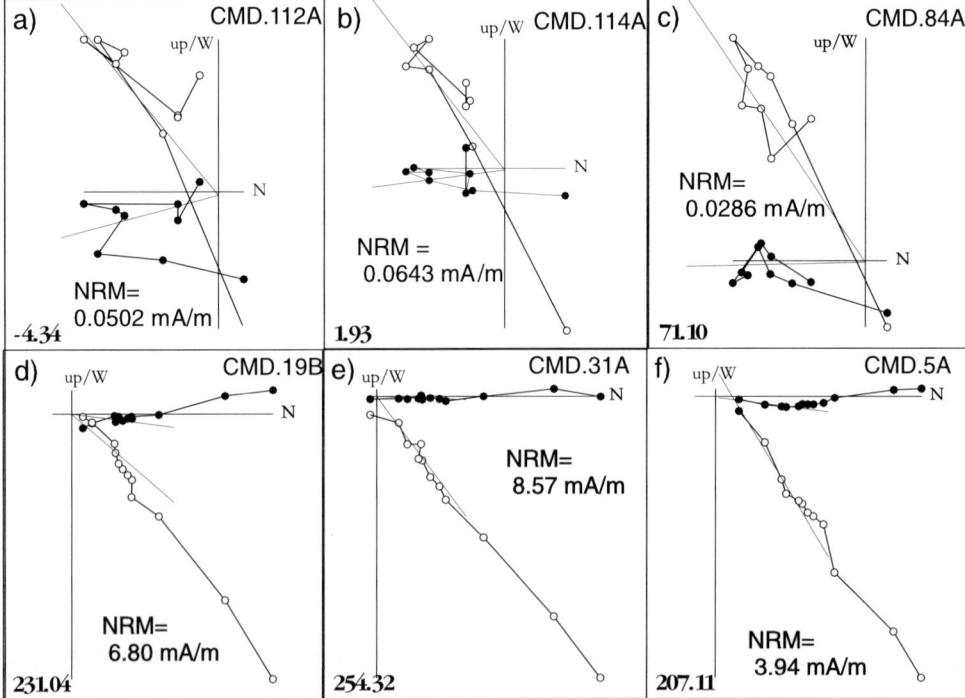

Fig. 8. Bedding-corrected orthogonal projections of stepwise thermal demagnetization of selected samples from the CMD record. Steps of 100, 150, 180, 210, 240, 270, 300, 320, 340°C were used for the top three samples. Steps of 100, 150, 200, 250, 300, 350, 400, 450, 500, 550, 585°C were used for the three lower diagrams. (Note that the low-temperature unblocking reverse samples (**a**, **b** and **c**) are located in cycles 43 and 44 whereas the high-temperature unblocking normal samples (**d**, **e** and **f**) belong to the overlying cycles 45 and 46 (see Fig. 9).

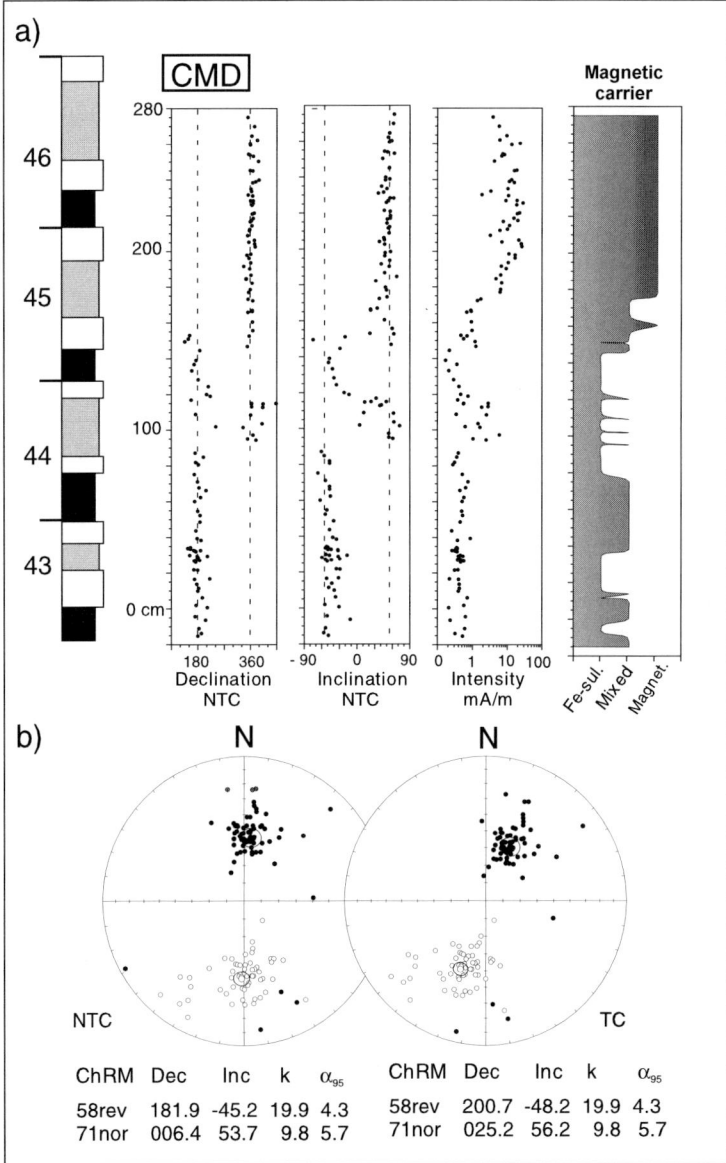

Fig. 9. Palaeomagnetic data for the CMD record. (**a**) ChRM declination, inclination and intensity record. The reversal boundary located in W1-45 also delimits a lower domain with low unblocking temperatures assigned to Fe sulphide or mixed Fe sulphide–magnetite with respect to an upper domain characterized by higher unblocking temperatures typical of magnetite. (**b**) Equal-area projections of the ChRM components before (NTC) and after bedding correction (TC).

Samples from the lower domain show a constant susceptibility after heating to temperatures of 350–400°C, where an abrupt increase occurs. This increase can be related to the breakdown of iron sulphides (pyrite) and formation of new mineral phases, including magnetite, which, after further heating, probably converts to hematite, which explains the decrease of susceptibility at high temperatures. In contrast, samples from the upper domain show a substantially different behaviour. They are characterized by a much less pronounced increase at

c. 350–400°C and by a subtle but noticeable increase of susceptibility at c. 150°C. As mentioned before, this is a peculiar feature related to weathering. The normalized susceptibility to the room temperature at 150°C is used here as an index for the weathering stage–mineralogy along the CMD record (Fig. 5b). It is evident that the upper domain displays higher values (up to 1.3) than in the lower domain, where values remain close to one. This indicates that more weathering has occurred in the upper part and coincides with the field observations of two distinct domains with different colouring profile as outlined above. The ARM/IRM ratio from the CMD record (Fig. 5b) ranges from 1×10^{-2} to 5×10^{-2} and the actual variations are difficult to interpret. In the lower domain, the lowest values are observed in cycle 44 beige. Seemingly, low values are also observed in the beige layers of cycles 45 and 46 from the upper domain. The highest values are observed in cycle 46 grey but also in several samples from the lower domain. The predominance of any of the magnetic minerals evidenced in the IRM experiments (iron sulphide, magnetite, hematite) might explain to some extent the actual observations. The B_{cr} values are around 40 mT along the lower domain with the exception of the lowermost 30 cm and in cycle 44 white$_1$, which display higher values (80–100 mT). In the upper domain, B_{cr} values are higher in the central part. The presence of a mixed mineralogy in the CMD record, including a high-coercivity mineral, precludes straightforward use of B_{cr} as a mineralogical proxy.

Comparison of the profiles

The low-field susceptibility depends on the concentration, grain size and composition of the magnetic minerals, as well as the proportion of diamagnetic and paramagnetic minerals. In the freshest LCM profile (Fig. 5a), the susceptibility minima (30–50 μSI) correlated with peaks of carbonate content, indicating that the contribution from diamagnetic carbonate might be the controlling factor. The maximum susceptibilities occurred in the grey and beige layers, typically 70 μSI. In contrast, the higher susceptibilities in the weathered PMD profile (not illustrated) and in the weathered CMD profile (Fig. 5b) were associated with the beige layers. Cycles 45 and 46 in the CMD profile had values around 150 μSI, but the beige layers from cycles 43 and 44 were only slightly higher (c. 80 μSI) than their corresponding grey layers. Previous studies of the Trubi marls (Van Hoof 1993; Van Hoof et al. 1993) showed always prominent susceptibility maxima in the beige layers (e.g. the beige layers of cycles 44 and 45 at Capo Bianco were 250 μSI). These were then interpreted as corresponding to both a higher clay content, i.e. a higher paramagnetic contribution, and a higher ferromagnetic contribution (Van Hoof et al. 1993) because the high-field (mostly paramagnetic) susceptibility did not adequately account for the maxima in the low-field susceptibility. In the profiles reported here, it can only be noted that the relative contributions of the diamagnetic, paramagnetic and ferromagnetic phases probably vary. However, it is important to note that the same beds at the different profiles show marked differences, suggesting that the rocks and hence their magnetic properties have been differentially affected by some non-depositional processes.

The main R–N transition located at cycle 45 white$_1$ in the CMD profile is consistent with the previously established lower Cochiti reversal at cycle 45. In the new PMD profile it is also located at this level. Unfortunately, this reversal boundary coincides with the boundary that separates two domains with distinct magnetic properties, which therefore casts some doubt on its genuine validity in the CMD profile (Fig. 9a). The normal 'excursion' observed at cycle 44 beige is difficult to understand as a delayed remanence because is not only held by iron sulphides but also seems to be recorded throughout the entire beige layer. Previous observations of delayed remanences of different polarity along reversal boundaries (Van Hoof et al. 1993) were accounted for by the iron-migration model (or early diagenetic model). In such a model, secondary (delayed) magnetite and not magnetic iron sulphides will form away from the grey layers and therefore the central part of the beige layers will still hold the primary polarity. Consequently, it is considered that the CMD record is a hybrid situation of the different diagenetic–weathering processes affecting the Trubi marls (see below). The coincidence of the differential complex magnetic signature with the location of the lower Cochiti reversal results in the actual puzzling observations.

The occurrence of unrotated reverse polarities throughout the LCM profile from cycle 43 to 54 is the key element in the present study. The interval sampled in the LCM profile expands both downwards and upwards from the normal Cochiti magnetozone (known to occur between cycles 45 and 49 beige layers) (Fig. 3). The expected normal polarities are not recorded and hence the ChRM directions from the LCM record must pre- or post-date the Cochiti interval. The fact that the mean direction does not imply any significant clockwise rotation (as would be expected for a Pliocene or older remanence in this region) provides a time constraint for its origin, i.e. it is post-Cochiti and likely to be of Pleistocene and pre-Brunhes date. Moreover, the presence of pyrrhotite as a carrier of such 'late' remanence in the LCM record provides a previously unrecognized feature in the Trubi marls that has important implications for the diagenetic pathway undergone by these rocks. The possible mechanism by which the later occurrence of iron sulphides has occurred in some areas is discussed in the next section and a general diagenetic model will be presented. As a general conclusion, there seems to be a direct correlation between the general magnetic properties and the expression of the colour profile of the Trubi marls. Generally, lower NRM intensities and lower unblocking temperatures are observed at locations where the typical grey–white–beige–white colouring profile is less evident or non-existent (eventually, as in the LCM profile, remanences can entirely be of 'late' origin and locally enhanced). When the colouring is more obvious, samples show higher remanence intensities, presence of magnetite and the peculiar features at low temperature ($c. 150°C$) ascribed earlier to weathering. Despite that, those magnetite remanences appear to be primary, as they record the sequence of expected Pliocene reversals derived from the oceanic magnetic anomalies in addition to local tectonic rotations (Hilgen & Langereis 1988; Zachariasse et al. 1989; Langereis & Hilgen 1991; Van Hoof, 1993; and the new PMD record).

Discussion

Palaeomagnetic and rock magnetic data for two detailed records along the lower Cochiti reversal (CMD and LCM) and a third, less detailed additional record (PMD) along the same interval at Punta di Maiata have been presented. The data clearly show the conspicuous occurrence of magnetic iron sulphides (pyrrhotite) in particular locations along these laterally equivalent profiles (notably throughout the LCM record). This feature was not known previously, and its particularities add important information that bears on the diagenetic history of the Trubi marls. The observations gathered from the Trubi marls will next be located in a general pathway framed in the diagenetic stages syndiagenesis, anadiagenesis, and epidiagenesis as defined by Fairbridge (1967) (Fig. 10).

In a normal marine environment, syndiagenesis starts at the time of deposition on the sea floor. Is during this stage that the boundary ($Eh = 0$) above which the presence of oxygen allows organic activity will delimit a lower zone where oxygen consumption exceeds the supply by diffusion (the early burial stage) and geochemical conditions may become suboxic and finally became anoxic. The cyclic nature of the Trubi marls results from sedimentary variations of this redox boundary as a consequence of climatic variations controlled by orbital forcing (chiefly influencing carbonate production and dilution of terrigenous material). The consequences of these periodic changes in sedimentation and geochemical conditions on the magnetic signature have been previously modelled under the so-called 'early diagenetic or iron-migration model' (Van Hoof et al. 1993) and described earlier in this paper. In Fig. 10, a schematic representation of this model involving three time snapshots is given. Both detrital primary magnetite formed at $t=1$ to $t=2$ and 'delayed' secondary magnetite ($t=3$) are shown. In this example, two reverse pre-transitional 'excursions' (delayed remanences) are explained by migration and later formation of magnetite away from the lowermost represented grey layer.

In the global diagenetic stage model, anadiagenesis will follow if sediments become buried to certain depths (i.e. up to 10 km). This would include compaction, dehydration and cementation and eventually metamorphism could initiate if conditions became appropriate. The Trubi marls probably did not reach the anadiagenesis stage, and probably underwent only limited burial of perhaps few kilometres at the most, as the sedimentation rate for the Trubi marls is estimated to be only 4–5 cm/ka and relatively soon after deposition ($c. 3$–4 Ma) the rocks were uplifted and incorporated into the orogenic wedge. During basin subsidence, diagenetic changes will continue, but, if subsidence is replaced by uplift, the sediments may become exposed to the influence of circulating ground water. This type of environment is that in which epidiagenesis occurs, which is characterized by the presence of a certain amount of O_2 and CO_2, and, therefore, can be aggressive towards mineral phases formed during earlier diagenesis and thus mask or destroy traces of the earlier stages. As described above, the studied outcrops in the southern part of the Caltanissetta basin have been uplifted and therefore subjected to epidiagenetic conditions before reaching their present position above the ground water table (Fig. 10). It is during this stage that an anoxic event is invoked to cause the dissolution of the former magnetite and the formation of magnetic iron sulphides (pyrrhotite). This event

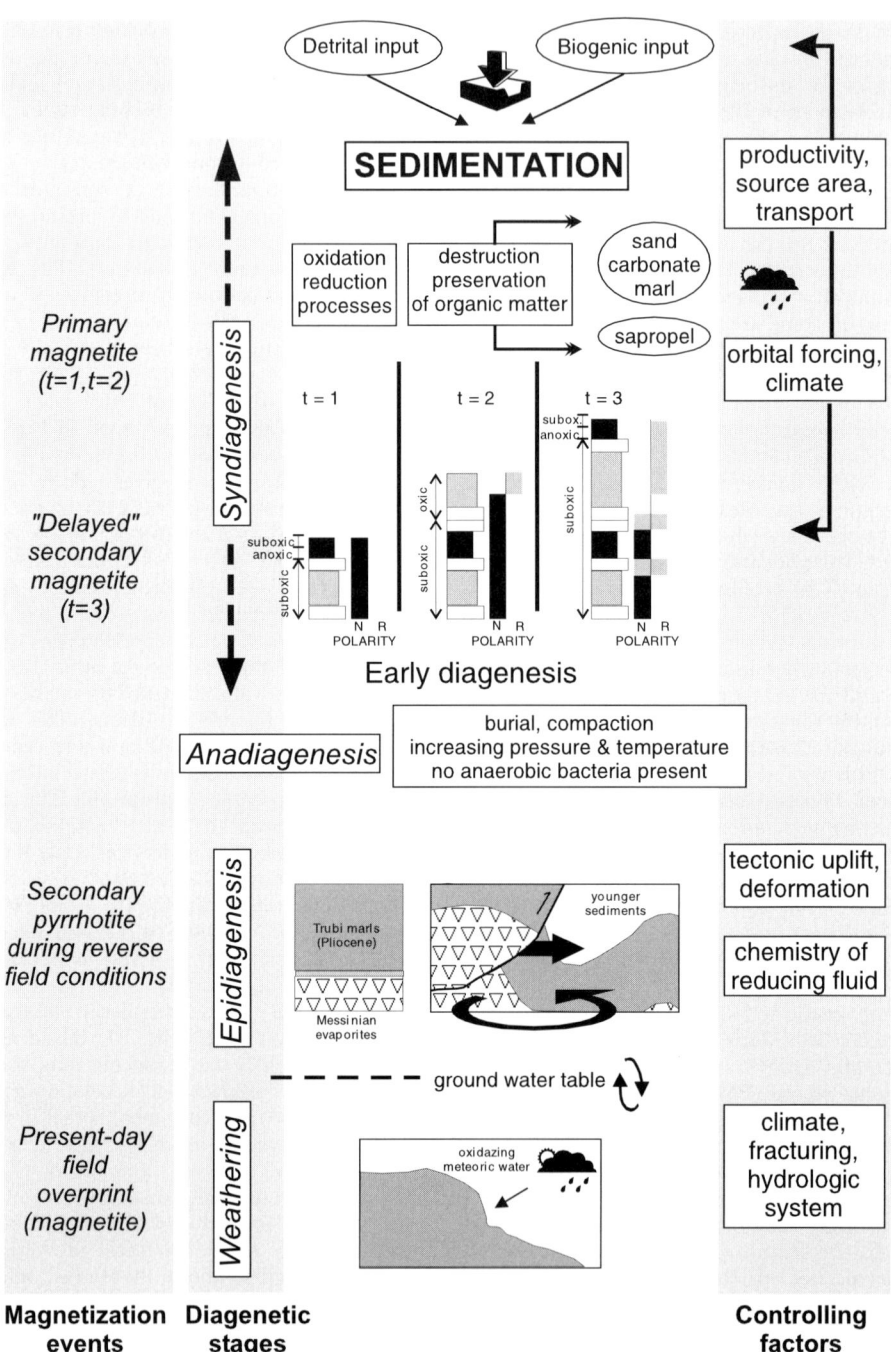

Fig. 10. The proposed diagenetic pathway for the Trubi marls together with its controlling factors. The different diagenetic stages (syndiagenesis, anadiagenesis and epidiagenesis), arranged from sedimentation to late weathering and including the 'early diagenetic model' (Van Hoof et al. 1993).

must have occurred during a reversed period of the geomagnetic field according to the polarity observed in the LCM record. No further constraints can be given on its age. It may speculatively be linked to a fluid flow related to thrust emplacement during Upper Pliocene or Pleistocene times. Such a fluid could have been enriched in sulphate after percolation and remobilization of the Messinian evaporites. The recording of this anoxic event can be rather irregular, depending on the hydrological system itself, the permeability of the rocks and the actual disposition of the outcrops in relation to the structures. The documentation of advection fluids at the front of the Sicilian subduction complex is not new but linkage to the magnetic signature of the Neogene sediments was not previously known. Larroque et al. (1996) have reported that the frontal Sicilian prism supported a localized transient fluid flow of deep fresh water from two different origins: (1) a shallow one, where aqueous fluids locally flowed along the upper décollement; (2) a deeper one where aqueous and hydrocarbon-bearing fluids have been channelled along the basal décollement from a depth as great as 10 km.

A different and unusual example of reduction in the epidiagenetic stage (otherwise generally oxidizing) occurs in the anhydrite caprock of certain salt domes (e.g. Feely & Kulp 1957; Davis & Kirkland 1979; Sassen et al. 1989). The sulphate-reducing bacteria in the sulphate-bearing caprock of a salt dome obtain their sulphate supply from the immediate vicinity. Migration and accumulation of petroleum hydrocarbons favoured by the fractured dome structure provides the nutrients for the sulphate-reducing bacteria. Another peculiar sulphate reduction setting and the formation of pyrite in the epidiagenetic phase have been described in an English Chalk aquifer (Kimblin & Johnson 1992). In this case, meteoric water, sulphate-reducing bacteria and supply of organic matter from the sediment account for the formation of pyrite in anoxic conditions located in micro-niches in a dominantly oxic environment. Champ et al. (1979) have also described sulphate reduction in ground-water aquifers where the highly oxidized state of the ground waters at recharge change progressively to a reduced state at discharge.

In the case of the Trubi marls under study, the detailed mechanisms and chemical characteristics of the remagnetizing fluid or process are not unambiguously known, although it seems reasonable to relate it to a late tectonic event perhaps connected to the previously described occurrence of fluids in the frontal Sicilian prism.

The sulphate concentration of the ground water or percolating fluid could originate from the aquifer system, i.e. from the dissolution of rock sulphate (gypsum–anhydrite) or from oxidative weathering of sulphides. When sulphate-bearing fluids penetrate strata containing organic matter, sulphate-reducing bacteria will produce H_2S. The association of dissolved sulphates and hydrocarbons is thermodynamically unstable in diagenetic environments. Hence, redox reactions occur, either with or without the mediation of bacteria. The regional structure of the organic-rich Trubi sediments in a series of thrust slices with local thick Messinian evaporites would propitiate the occurrence of a late sulphate-reducing event and dissolution of original magnetite in particular zones (in a setting comparable with that described for the salt caprock situation). The degree of dissolution is a function of the surface area of the magnetite, the concentration of dissolved sulphide and the time, together with a wide range of variables such as source rock, sulphate reduction rate, and reactivity of iron minerals contained in the sediment. It is therefore likely to conceive that magnetite dissolution and sulphitization in the Trubi marls at particular locations might have happened later in the diagenetic history when appropriate conditions were met (an anoxia event during the epidiagenetic stage). Upwards in the geological pile, epidiagenesis is replaced by weathering, with the ground-water table being the practical boundary between them (Fig. 10). It will be at this stage when the Trubi marls suffer partial oxidation inducing the magnetization of the present-day field. This late remagnetization event completes the diagenetic pathway for the Trubi marls as inferred in the present study.

Conclusions

The present research along the Cochiti interval at Punta di Maiata has demonstrated the occurrence of magnetic iron sulphides that do not record the expected polarities and tectonically related rotations. Hence, they are believed to have been formed during a late anoxic event during the epidiagenetic stage. Similar or comparable phenomena are already known related to the caprock of several salt domes. The exact geochemical conditions under which the 'anoxia' event in the Trubi marls took place is not known yet, and only the general framework for its occurrence has been put forward. Geochemical analysis, integrated with further petrographic observations and the rock magnetic data, will probably help in deciphering the details of these processes. This study also hints at the complexity of the processes by which sediments can acquire their NRM and alerts us to the necessity to use multidisciplinary approaches to unravel the exact sequence of events in the rock history.

This work has been funded by an EU Marie Curie fellowship to J.D.T. (No. ERBFMBICT961151). We thank reviewer G. Wilson for his thoughtful comments.

References

CATALANO, R., DI STEFANO, P. & VITALE, F. P. 1995. Structural trends and palaeogeography of the central and western Sicily belt: new insights. *Terra Nova*, 7, 189–199.

CHAMP, D. R., GULENS, J. & JACKSON, R. E. 1979. Oxidation–reduction sequences in ground water flow systems. *Canadian Journal of Earth Sciences*, 16, 12–23.

CITA, M. B. & GARTNER, S. 1973. The stratotype Zanclean. Foraminiferal and nannofossil biostratigraphy. *Rivista Italiana di Paleontologia*, 79, 503–558.

DAVIS, J. B. & KIRKLAND, D. W. 1979. Bioepigenetic sulfur deposits. *Economic Geology*, 74, 462–468.

DEKKERS, M. J. 1988. Magnetic properties of natural pyrrhotite Part I: Behaviour of initial susceptibility and saturation-magnetization-related rock-magnetic parameters in a grain-size dependent framework. *Physics of the Earth and Planetary Interiors*, 52, 376–393.

DE VISSER, J. P., EBBING, J. H. J., GUDJONSSON, L., HILGEN, F. J., JORINSSEN, F. J., VERKALLEN, P. J. J. M. & ZEVENBOOM, D. 1989. The origin of rhythmic bedding in the Pliocene Trubi Formation of Sicily, Southern Italy. *Palaeogeography, Palaeoclimatology, Palaeoecology*, 69, 45–66.

FAIRBRIDGE, R. W. 1967. Phases of diagenesis and authigenesis. *In*: LARSEN, J. G. & CHILINGARIAN, G. V. (eds) *Diagenesis in Sediments*, Elsevier, Amsterdam 19–89.

FEELY, H. W. & KULP, J. L. 1957. Origin of Gulf Coast salt-dome sulfur deposits. *Bulletin of the American Association of Petroleum Geologists*, 41, 1802–1853.

HILGEN, F. J. 1987. Sedimentary rhythms and high-resolution chronostratigraphic correlations in the Mediterranean Pliocene. *Newsletters of Stratigraphy*, 17, 109–127.

—— 1991. Extension of the astronomically calibrated (polarity) time scale to the Miocene/Pliocene boundary. *Earth and Planetary Science Letters*, 107, 349–368.

—— & LANGEREIS, C. G. 1988. The age of the Miocene–Pliocene boundary in the Capo Rossello area (Sicily). *Earth and Planetary Science Letters*, 91, 214–222.

HUNT, C. P., BANERJEE, S. K., HAN, J., SOLHEID, P. A., OCHES, E., SUN, W. & LIU, T. 1995. Rock-magnetic proxies of climate change in the loess–paleosol sequences of the western Loess Plateau of China. *Geophysical Journal International*, 123, 232–244.

KIMBLIN, R. T. & JOHNSON, A. C. 1992. Recent localised sulphate reduction and pyrite formation in a fissured Chalk aquifer. *Chemical Geology*, 100, 119–127.

LANGEREIS, C. G. & HILGEN, F. J. 1991. The Rossello composite: a Mediterranean and global standard reference section for the Early to early Late Pliocene. *Earth and Planetary Science Letters*, 104, 211–225.

LARROQUE, C., GUILHAUMOU, N., STEPHAN, J. F. & ROURE, F. 1996. Advection of fluids at the front of the Sicilian Neogene subduction complex. *Tectonophysics*, 254, 41–55.

LOURENS, L. J., ANTONARAKOU, A., HILGEN, F. J., VAN HOOF, A. A. M., VERGNAUD-GRAZZINI, C. & ZACHARIASSE, W. J. 1996. Evaluation of the Plio-Pleistocene astronomical timescale. *Paleoceanography*, 11(4), 391–413.

MULLENDER, T. A. T., VAN VELZEN, A. J. & DEKKERS, M. J. 1993. Continuous drift correction and separate identification of ferrimagnetic and paramagnetic contribution in thermomagnetic runs. *Geophysical Journal International*, 114, 663–672.

ROBERTS, A. P. 1995. Magnetic properties of sedimentary greigite (Fe_3S_4). *Earth and Planetary Science Letters*, 134, 227–236.

ROURE, F., HOWELL, D. G., MULLER, C. & MORETTI, I. 1990. Late Cenozoic subduction complex of Sicily. *Journal of Structural Geology*, 12, 259–266.

SCHEEPERS, P. J. J. 1994. *Tectonic rotations in the Tyrrhenian arc system during the Quaternary and late Tertiary*. PhD thesis, University of Utrecht. Geologica Ultraiectina, 112.

—— & LANGEREIS, C. G. 1993. Analysis of NRM-directions from the Rossello composite: implications for tectonic rotations of the Caltanisetta basin (Sicily). *Earth and Planetary Science Letters*, 119, 243–258.

SASSEN, R., MCCABE, C., KYLE, J. R. & CHINN, W. 1989. Deposition of magnetic pyrrhotite during alteration of crude oil and reduction of sulphate. *Organic Geochemistry*, 14, 318–392.

VAN HOOF, A. A. M. 1993. *Geomagnetic polarity transitions on the Gilbert and Gauss Chrons recorded in marine marls from Sicily*. PhD thesis, University of Utrecht. Geologica Ultraiectina, 100.

—— & LANGEREIS, C. G. 1991. Reversal records in marine marls and delayed acquisition of remanent magnetization. *Nature*, 351, 223–224.

——, & LANGEREIS, C. G. 1993a. The upper and lower Nunivak sedimentary geomagnetic transitional records from Southern Sicily. *Physics of the Earth and Planetary Interiors*, 77, 297–313.

——, VAN OS, B. J. H., RADEMAKERS, J. G., LANGEREIS, C. G. & DE LANGE, G. J. 1993. A paleomagnetic and geochemical record of the upper Cochiti reversal and two subsequent precessional cycles from southern Sicily (Italy). *Earth and planetary Science Letters*, 117, 235–250.

VAN VELZEN, A. J. & ZIJDERVELD, J. D. A. 1995. Effects of weathering on single-domain magnetite in Early Pliocene marine marls. *Geophysical Journal International*, **121**, 267–278.

ZACHARIASSE, W. J., ZIJDERVELD, J. D. A., LANGEREIS, C. G., HILGEN, F. J. & VERHALLEN, P. J. J. M. 1989. Early Late Pliocene biochronology and surface water temperature variations in the Mediterranean. *Marine Micropaleontology*, **14**, 339–355.

Magnetic properties of sediments deposited in suboxic–anoxic environments: relationships with biological and geochemical proxies

L. VIGLIOTTI,[1] L. CAPOTONDI[1] & M. TORII[2]

[1] *Istituto di Geologia Marina, C.N.R., Via P. Gobetti 101, 40129 Bologna, Italy*
(e-mail: mefo@igm.bo.cnr.it)
[2] *Department of Biosphere-Geosphere System Science, Faculty of Infomatics,*
Okayama University of Science, Okayama 700-0005,

Abstract. Several studies of the rock magnetic properties of sediments deposited in suboxic–anoxic environments show that bacterial degradation of organic matter leads to selective dissolution of magnetite with a decrease in magnetic mineral concentration and an increase in magnetic grain size. This study of upper Quaternary sediments from both marine and lacustrine environments shows many similarities, although diagenetic processes appear to be enhanced in the marine sediments. When the remanence is controlled by fine-grained materials, the magnetic susceptibility can be considered a leading indicator of such changes, and the coercivity parameters, such as S ratio and $B_{0\,cr}$, are also useful discriminants suggesting when antiferromagnetic minerals become more important in intervals characterized by reductive diagenesis. Generally, magnetic properties show close correlations with other environmental indicators: diatoms and foraminiferal concentrations and compositions and geochemical properties. This suggests that the magnetic analyses are very sensitive to environmental variations and provide powerful tools for monitoring past depositional and diagenetic environments.

Palaeomagnetic and rock magnetic investigations of marine and lacustrine sediments are being used increasingly to study decadal to millennial fluctuations in the geomagnetic field and as proxy indicators of the palaeoclimatic records represented by these sediments (Bloemendal & DeMenocal 1989; Thouveny *et al.* 1994; Vlag *et al.* 1997; Liu *et al.* 1998). The best results, for both purposes, are obtained where the down-core variations in the magnetic properties have not been affected by diagenetic processes. In such circumstances, changes in the composition, grain size and abundance of magnetic minerals can be used as signals of environmental changes in the catchment area (Thompson & Oldfield 1986). However, several studies during the last decade indicate that the magnetic properties may be drastically affected by ambient geochemical activity in the accumulating sediments (Canfield & Berner 1987; Karlin 1990; Leslie *et al.* 1990; Vigliotti 1997). These diagenetic changes are particularly controlled by the redox reactions involved in the decomposition of organic matter (Froelich *et al.* 1979). In anoxic environments, with rapid sedimentation and high organic input, the detrital iron oxides, hydroxides and sulphates are progressively dissolved and/or transformed into authigenic iron sulphides (Karlin & Levi 1983, 1985; Canfield & Berner 1987; Karlin 1990; Leslie *et al.* 1990). In such conditions, there is generally a drastic decrease in the intensity of both the natural and laboratory induced remanences, and a coarsening of the magnetic fraction. The concentration of H_2S is of primary importance in this process as its reaction with Fe^{2+} leads to the formation of sulphide minerals such as greigite, mackinawite and pyrite (Kaplan *et al.* 1963; Berner 1984). In suboxic sediments, Mn and Fe diagenetic reduction occurs but diagenesis does not usually extend as far as sulphate reduction.

This study examines the magnetic properties of marine and lacustrine sediments, deposited in suboxic to anoxic environments, which show different stages of reduction diagenesis. The magnetic features of these sediments will be compared with biological and geochemical indicators. The biological indicators comprise diatoms and planktonic Foraminifera, and the geochemical indicators include the total organic carbon content (TOC), total sulphur (S_{tot}), and the $\delta^{13}C$ isotope. It would be expected that, within the marine environment, authigenic iron minerals would form during early diagenesis, which would normally be in the form of pyrite in organic-rich anoxic sediments. In contrast, fresh-water sediments, being of low salinity, would normally have low initial sulphate concentrations, so that sulphates would be totally reduced and any H_2S would be precipitated in the form of highly insoluble iron sulphide minerals, such as greigite and mackinawite and these would not convert to pyrite (Berner 1984). To investigate such diagenetic environments, two cores were collected from the crater lakes of Albano (PAlb-6A) and Nemi (PNemi-1B) in Central Italy. The magnetic properties of these two cores have been discussed by Alvisi & Vigliotti (1996). The results from these two cores are here compared with those of a 6.8 m core from the Adriatic Sea (Pal94-9), which represents predominantly oxygenating conditions, and a 3.85 m core from the western Mediterranean (B74-27) which includes two intervals representing anoxic, sapropelitic conditions. The object of these comparisons is to investigate which environmental factors are the most important in controlling the magnetic mineralogy and hence to evaluate the degree to which magnetic minerals can be used to monitor environmental change.

Materials, methods and age

Core PAlb94-6A was recovered from 30 m water depth on the northwestern shelf of Lake Albano (41°45'N, 12°40'E) and consists of 7.8 m of olive–grey silts, laminated carbonates and muds which are divided into three lithological units (Chondrogianni *et al.* 1996*b*); (I) massive olive–grey silts and detrital sands with moss layers; (II) massive olive–grey silts intercalated with 10–20 cm thick layers of coloured, indistinctly laminated muds; (III) massive olive–grey silts intercalated with 1–6 cm thick layers of partly well-laminated carbonates. Core PNemi-1B was 9.1 m long and was obtained close to the deepest part of Lake Nemi (41°43'N, 12°43'E) at a water depth of 30m. The core comprises massive to laminated olive–black muds with diatom beds and has also been divided into three lithological units (Chondrogianni *et al.* 1996*b*); (I) diatom beds and laminae intercalated with massive olive–grey to olive–black muds and sections with indistinctly laminated coloured muds forming the uppermost 510.5 cm; (II) distinctly laminated, coloured muds intercalated with diatom beds, laminae and massive muds between 510.5 and 847.5 cm; (III) massive olive–black to dusky brown muds, intercalated with laminated muds below 847 cm. Core Pal94-9, 6.8 m long, was from the southwestern shelf and slope of the Meso-Adriatic Depression (42.6°N, 14.5°E) at 104 m water depth. The lithology is mainly homogeneous mud with some scattered bioclasts, in the upper 600 cm. Below this, there is a 60–70 cm thick graded interval of sand to silty mud, representing a discontinuity in the record (Trincardi *et al.* 1996). Core B74-27, 3.85 m long, was retrieved from 2500 m water depth in the Valencia Trough in the western part of the Algero-Provence basin (37.4°N, 0.4°E) and has a lithology characterized by grey clay muds, with some intervals of brown–yellow colour and two dark layers between 289–294 and 353–357 cm depth. Guerra (1994) correlated these dark intervals with sapropels S_4 and S_5 of the Eastern Mediterranean.

All of the cores were sampled by inserting $2 \text{ cm} \times 2 \text{ cm} \times 2 \text{ cm}$ plastic boxes, oriented with respect to the top of the core. A total of 319 samples were obtained from PAlb-6A, 378 sample from PNemi-1B, 172 samples from Pal94-9 and 48 samples from Core B74-27. All sample remanences were measured with a Minispin fluxgate magnetometer, other than the Pal94-9 core, for which a Schonstedt SSM1 spinner magnetometer was used, and a few weak PAlb-6A samples for which a Jelinek JR-4 spinner magnetometer was used. The magnetic variables for the lake sediments were all normalized per unit volume, whereas the marine core variables were normalized per unit weight. The low-field magnetic susceptibility of each sample was measured at least twice at two frequencies, 0.47 kHz (K_{lf}) and 4.7 kHz (K_{hf}), using a Bartington MS2 susceptibility meter and dual frequency sensor. (The frequency-dependent susceptibility ($K_{fd}\%$) is the percentage ratio of $K_{lf} - K_{hf}$ to K_{lf}.) The initial natural remanences were measured and then demagnetized in three steps up to a peak field of 60 mT. After that, an anhysteretic remanent magnetization (ARM in 99 mT alternating field (AF) and a 0.1 mT bias DC field) was imparted to all of the marine sediments and half of the lacustrine samples (at least one sample every 4 cm of core). An isothermal remanent magnetization (IRM; in 15 steps up to 1 T) was then induced, followed by applying a reverse field, in five steps up to 0.4 T, to the saturation IRM (SIRM) after 1 T field, allowing the measurement of the coercivity of SIRM ($B_{0\,cr}$) and the S ratios; $S_{-0.1T} = \text{IRM}_{-0.1T}/\text{SIRM}$, and $S_{-0.3T} = \text{IRM}_{-0.3T}/\text{SIRM}$ (e.g. Bloemendal 1983). (The $S_{-0.3T}$ is an indicator of the proportion of high-coercivity material (particularly the canted antiferromagnetic minerals such as hematite and goethite) to the lower-coercivity ferrimagnetic minerals (magnetite and maghemite), although, in reality, this is a discriminant only when the proportion of antiferromagnetic material is greater than 80% of the magnetic assemblage (Bloemendal et al. 1992).) The low-temperature magnetic phase transitions were determined with a Quantum Design magnetic property measurement system (MPMS-2) at the Low-Temperature Laboratory of Kyoto University. A small portion of wet sediment, c. 50 mg, was wrapped with thin plastic film before measurement. Each sample was first cooled to 5 K, with no applied DC field. An IRM was then imparted, while still at 5 K, with a DC field of 1.0 T applied for 10 s. After removal of the DC field, the IRM was measured as the samples' temperature rose from 5 K to 6 K, 8 K, 10 K, 12 K, 15 K, and then at 5 K intervals to 300 K. The increase in IRM was then calculated for each temperature increment ($\Delta IRM/\Delta T$).

The samples for microfaunal analyses were dried at 60C, then washed and sieved to produce three size fractions; >63 μm diameter, 63–125 μm, and <125 μm diameter. All three size fractions were used for qualitative analyses of the foraminiferal microfauna. More than 300 planktonic or benthic specimens were separated, using a microsplitter, and then identified and counted. The TOC for core B74-27 was measured with a Carlo Erba CHN analyser. The ages of the cores are representative of similar but not identical time spans close to the transition from late Pleistocene to Holocene time. The bottom of the Western Mediterranean core, B74-27, is considered to be a little older than Eemian time ($\delta^{18}O$ substage 5e), on the basis of the micropalaeontology (Guerra 1994), whereas the Adriatic core, Pal94-9, contains all of the Holocene sequence (Langone et al. 1996). Both lacustrine cores and have been dated by pollen, tephrochronology and ^{14}C dating. The Lake Albano core, PAlb94-6A, extended between 15.779 and 25.115 ka BP, whereas the Lake Nemi core, PNemi-1B, has an age range between the present day and 10.778 ka BP (Chondrogianni et al. 1996b).

Down-core variations in magnetic variables

The lacustrine cores (PAlb94-6A and PNemi-1B)

The main feature of the down-core plots (Fig. 1) is the clearly defined intervals of high and low magnetic content, with transition between them being abrupt, i.e. within 2 cm. The levels of high magnetic concentrations have low coercivities (30–35 mT) and higher negative S ratios ($S_{-0.3T} > -0.9$), whereas levels of low magnetic

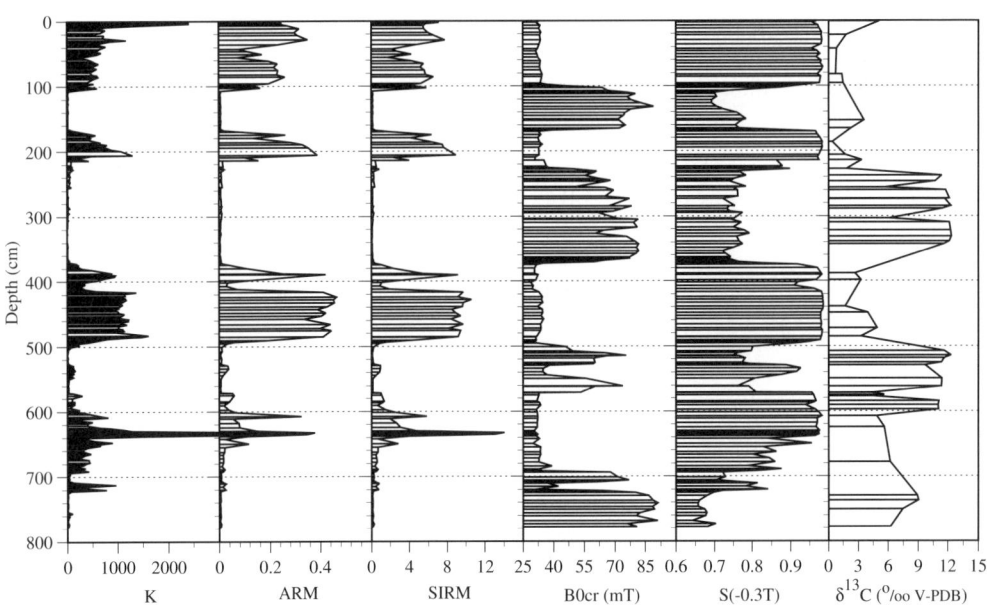

Fig. 1. Down-core magnetic properties and $\delta^{13}C$ isotopes for core PAlb94-6A. The units of susceptibility, K, are dimensionless 10^{-5} SI, and IRM and ARM are in A/m.

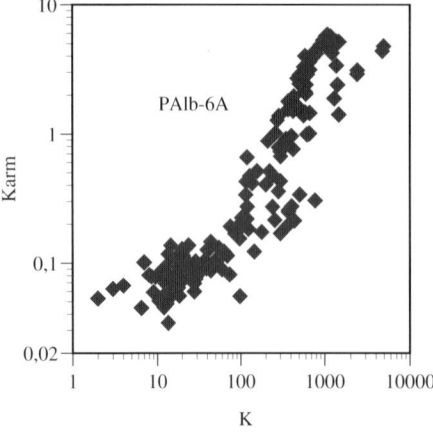
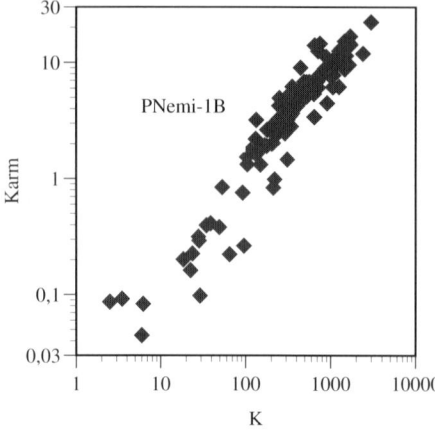

Fig. 2. Magnetic susceptibility (K) v. anhysteretic susceptibility (K_{arm}) for the samples from the Albano and Nemi Lake cores. K and K_{arm} are in 10^{-6} SI units.

concentrations show higher B_{0cr} (60–90 mT) and smaller S ratios ($S_{-0.3T} < -0.8$). The $K_{fd}\%$ values are low (<3%), suggesting that only a little of the magnetic fraction is in the superparamagnetic grain size range, but the increase in K/K_{arm} (Fig. 2) in the higher magnetization levels suggests that these changes are associated with changes in the grain sizes of the magnetic fraction. However, the main changes are probably due to the presence of titanomagnetites in the higher concentration levels, as indicated by the lower SIRM field, 0.3 T, compared with that in the low concentration levels, where the samples are not saturated even in 1 T fields (Fig. 3), suggestive of the dominance of hematite and goethite in these layers. The higher magnetic concentration levels are associated with the olive–grey silt, with intercalated macrophyte layers, whereas the low concentrations are in the laminated carbonates with intercalated organic-rich silt. Core PNemi-1B is characterized by a high ferrimagnetic mineral content, except for an interval between 570 and 820 cm depth, which has a very low magnetic concentration and higher coercivity (Fig. 4). Many samples from the lower interval exhibit diamagnetic susceptibility, testifying to the absence of ferrimagnetic minerals. IRM acquisition plots (Fig. 3) show that ferrimagnetism dominates the magnetic properties whereas canted antiferromagnetic minerals play an important role in controlling the remanence values in the higher

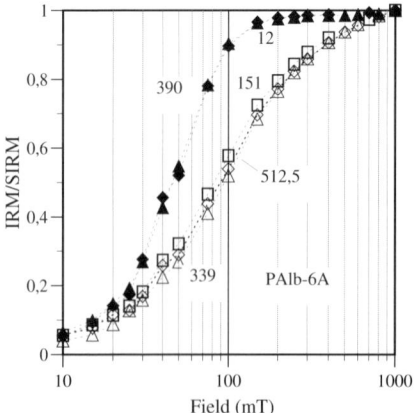
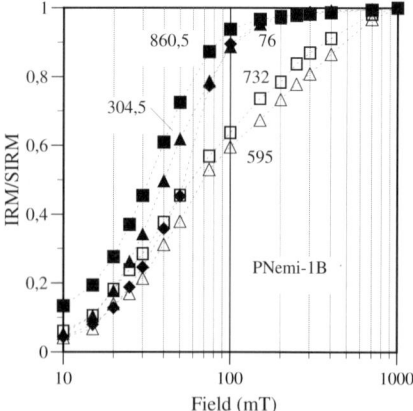

Fig. 3. Example IRM acquisition plots and normalized SIRM for selected samples of Albano and Nemi Lake cores. Open symbols refers to samples deposited in intervals of reductive diagenesis. Numbers refer to sample depth (cm).

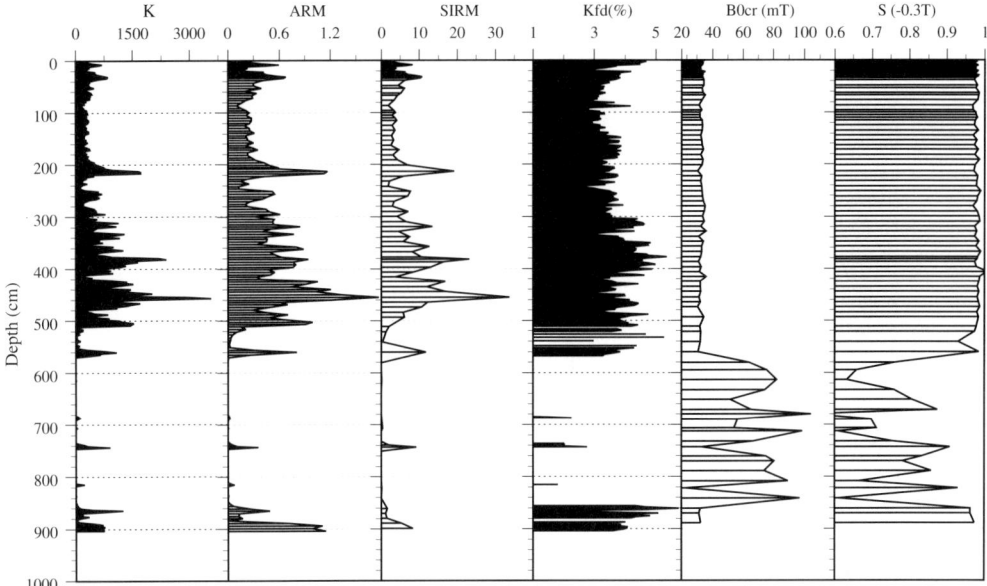

Fig. 4. Down-core magnetic properties for core PNemi-1B. The units of susceptibility, K, are dimensionless 10^{-5} SI, and IRM and ARM are in A/m.

interval between 570 and 820 cm. Despite the differences in the magnetic mineralogy, the plot of K v. K_{arm} (Fig. 2) shows a good linear correlation, suggesting that the grain size does not change significantly along the core.

Marine cores (Pal94-9 and B74-27)

Most of the Adriatic core (Pal94-9) is characterized by a fairly constant magnetic mineral content, with the susceptibility (K), ARM and

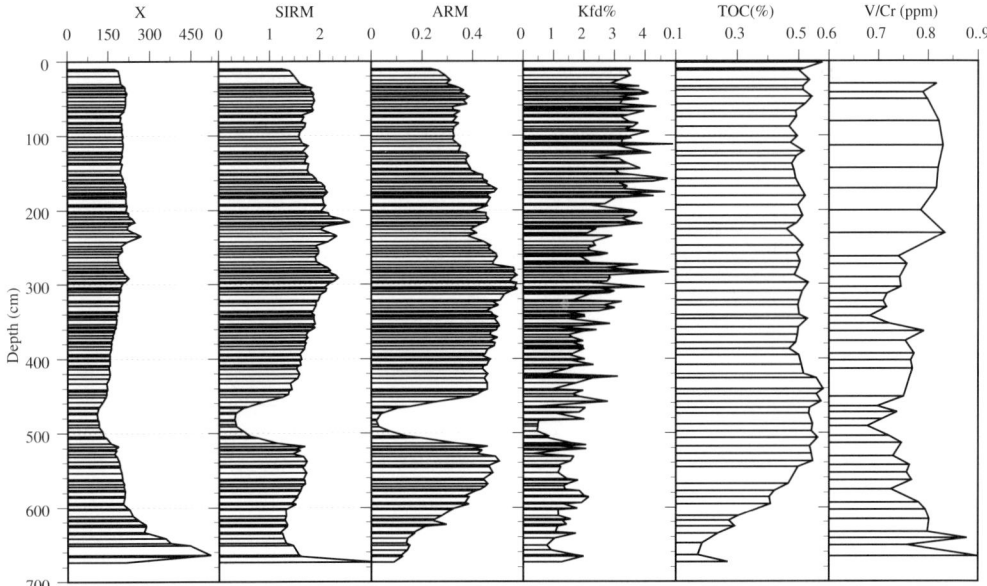

Fig. 5. Down-core magnetic properties, total organic carbon (TOC) and geochemical data (V/Cr) for core Pal94-9. X is in 10^{-6} A m^2/kg, and SIRM and ARM are in 10^{-5} A m^2/kg.

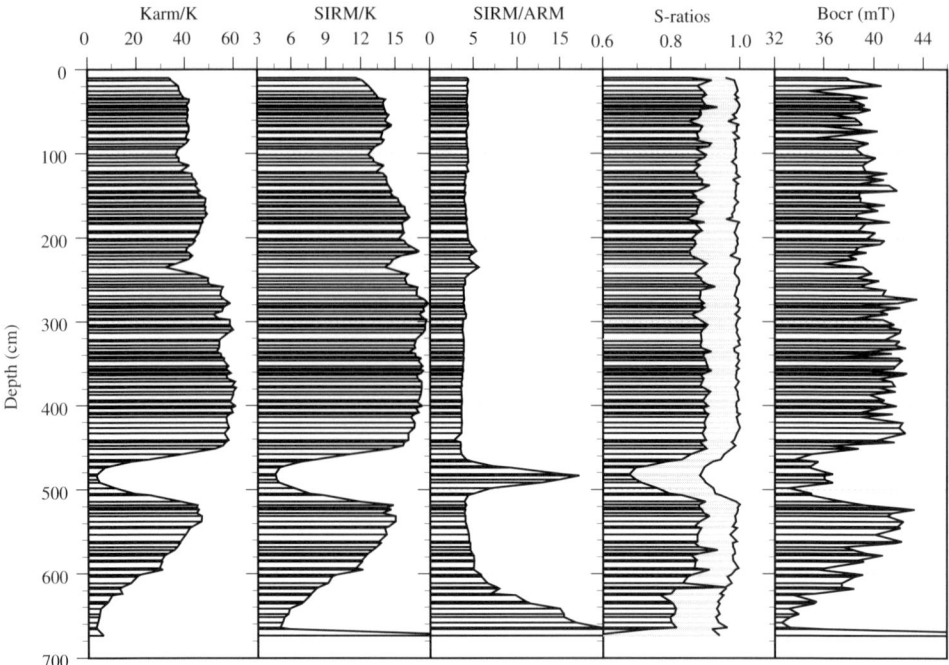

Fig. 6. Down-core interparametric ratios and coercivity of the remanence (B_{0cr}) for core Pal94-9. K_{arm}/K and SIRM/ARM are in SI units, and the S ratio and SIRM/K are in 10^3 A/m.

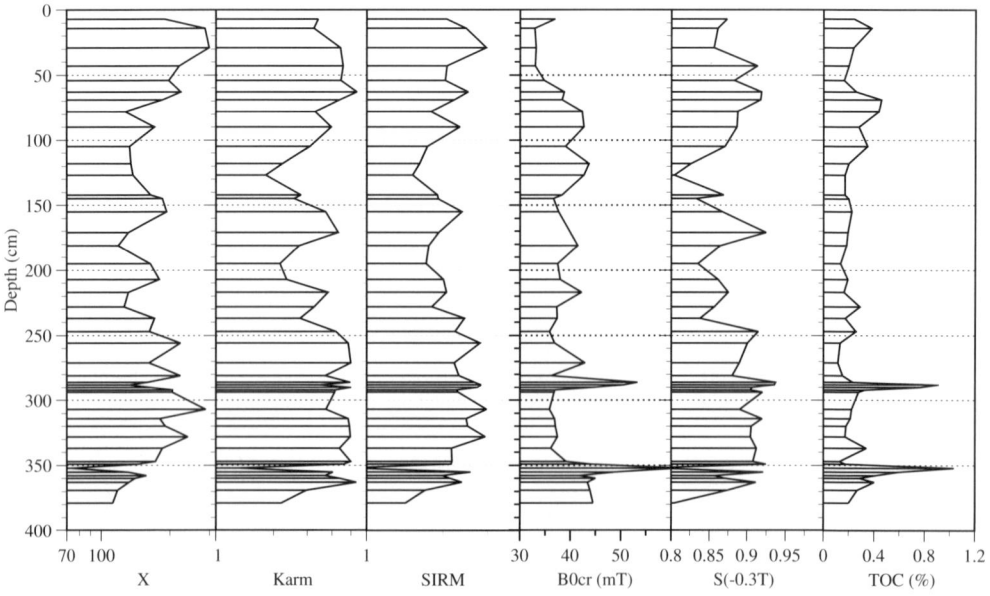

Fig. 7. Down-core magnetic properties and total organic carbon for core B74-27. X is in 10^{-6} A m^2/kg, and SIRM and K_{arm} are in 10^{-5} A m^2/kg.

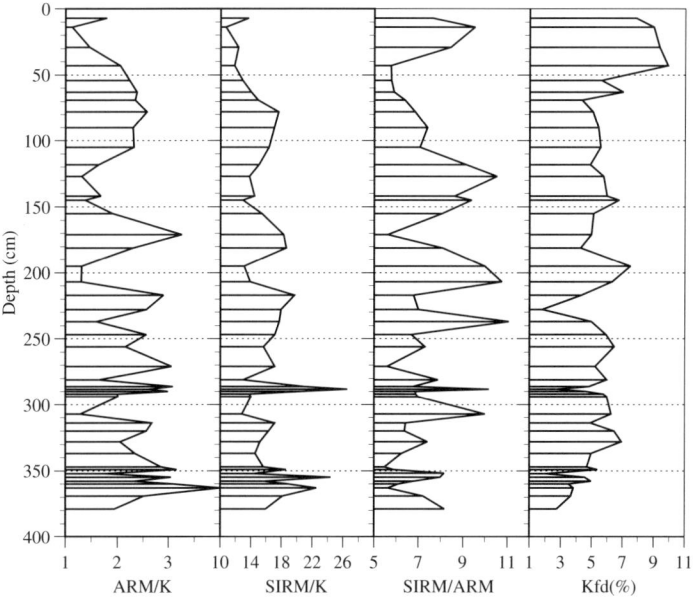

Fig. 8. Down-core interparametric ratios for core B74-27.

SIRM all showing similar trends (Fig. 5), except for an interval between 440 and 510 cm depth, which exhibits a significant drop in the values of concentration-dependent magnetic variables. This is also reflected by changes in the magnetic grain-size sensitive variables K_{arm}/K, SIRM/K, in the coercivity of remanence, the S ratios and the frequency-dependent susceptibility (Fig. 6). In the West Mediterranean core (B74-27) the magnetic variables are all fairly constant throughout, indicating a fairly uniform magnetic concentration and similar grain size (Figs 7 and 8). The main exceptions are at 288 and particularly 352 cm, where the magnetic minerals appear to be less concentrated and dominated by hematite–goethite properties, as indicated by the S ratios and higher coercivities of remanence. These two levels have a high organic carbon content and have been correlated with sapropels S_4 and S_5 on biostratigraphic evidence (Guerra 1994), corresponding to oxygen isotope substages 5c–d and 5e, respectively (Cita et al. 1977; Vergnaud-Grazzini et al. 1977).

Comparison of magnetic properties of the different environments

Lacustrine reduction environments

In the Lake Albano core, most of the concentration-related magnetic variables are inversely related to the biogenic carbonate concentration, as measured by $\delta^{13}C$ (Fig. 1). The correlation coefficients for K, ARM, and SIRM are respectively -0.64, -0.64, and -0.67. S ratios are similarly negatively correlated, $r = -0.65$, showing that the composition of the magnetic minerals is also affected by the conditions of changing carbonate production. The laminated carbonates, with high $\delta^{13}C$ values and low magnetic mineral concentrations, occur in two units (II and III) which have been interpreted as periods of high lake levels (Chondrogianni et al. 1996a). The Lake Nemi core also shows drastic decreases in magnetic content within the laminated levels. Nevertheless, the sediments exhibit a high correlation ($r = 0.96$) between susceptibility (K) and anhysteretic susceptibility (K_{arm}) implying a constant grain size (Fig. 2). In contrast, the Lake Albano core appears to be characterized by two different populations of magnetic grains with increasing grain size for samples with a lower content of magnetite (Fig. 2). Furthermore, the frequency-dependent susceptibility shows that very fine grained magnetite is present only in the sediments of the Lake Nemi core (Fig. 9). This could reflect stronger diagenetic modification of the original detrital input in the Lake Albano sediments, or indicate that the detrital grains entering this lake derived from two grain-size sources. The diatom flora of Lake Nemi assists in defining the

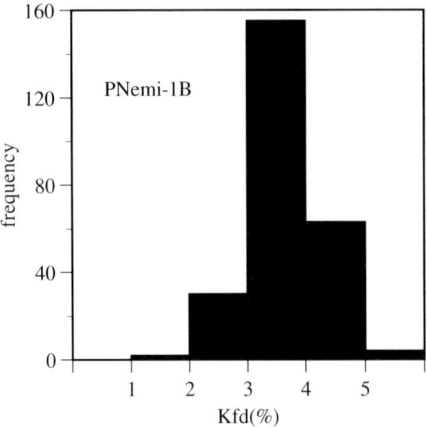
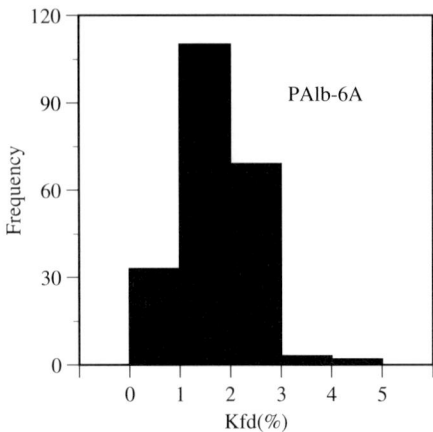

Fig. 9. Histograms of frequency-dependent susceptibility measured on the samples from Lakes Albano and Nemi. (Note that superparamagnetic grains are very poorly represented in the core from Lake Albano.)

depositional environments involved. *Cyclotella ocellata* is indicative of oligotrophic to mesotrophic conditions, whereas *Stephanodiscus minutulus* indicates increasing organic activity in the lake. Comparison of the magnetic susceptibility with these two taxa (Fig. 10) shows a clear relationship, with *S. minutulus* being dominant during intervals of reduction diagenesis, suggesting that this occurs at the same time as high organic productivity in the lake. In contrast, *Cyclotella ocellata* follows the changes in magnetic concentration. In spite of the magnetic and biological data being difficult to compare directly (the two signals may have different, non-linear responses to the environmental conditions) the linear correlation coefficient, r, between these taxa and concentration-related rock magnetic variables ranged between -0.52, for magnetic susceptibility, and -0.54 for ARM (Fig. 11). This implies that the preservation of magnetic minerals is strictly related to the variations in the oligotrophic–mesotrophic conditions of the lake, although another possibility is that iron and/or light availability (which is somewhat related to water stratification) controlled the production of these species, as has been observed for some marine diatoms (Sunda & Huntsman 1997).

Marine reduction (sapropel) environments

The main evidence for this type of environment is from the two dark layers, corresponding to Sapropels S_4 and S_5, in the Valencia Trough core, B74–27. The TOC levels in these two layers are only about 1% (Fig. 7), whereas classical sapropels are characterized by TOC > 2% (Olausson 1960). This suggests that these levels were deposited in slightly more oxic conditions (sub-oxic to slightly anoxic) than true sapropels. In the lower layer, at 352 cm, there is a drastic decrease in magnetic concentration (low X ARM and SIRM values; Fig. 7), an increase in the magnetic grain size (low $K_{fd}\%$, ARM/K, SIRM/K and ARM/SIRM; Fig. 8) and an increasing contribution of antiferromagnetic minerals (low S ratio and high $B_{0\,cr}$ values; Fig. 8). These properties suggest that this layer has been strongly influenced by reductive diagenetic processes. The higher layer, at 288 cm, appears to have been less affected by diagenetic processes, with a slight decrease

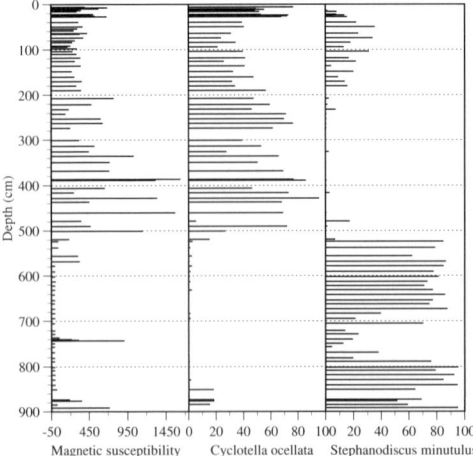

Fig. 10. Down-core magnetic susceptibility and two diatom species for core PNemi-1B. The susceptibility is in 10^{-6} SI units.

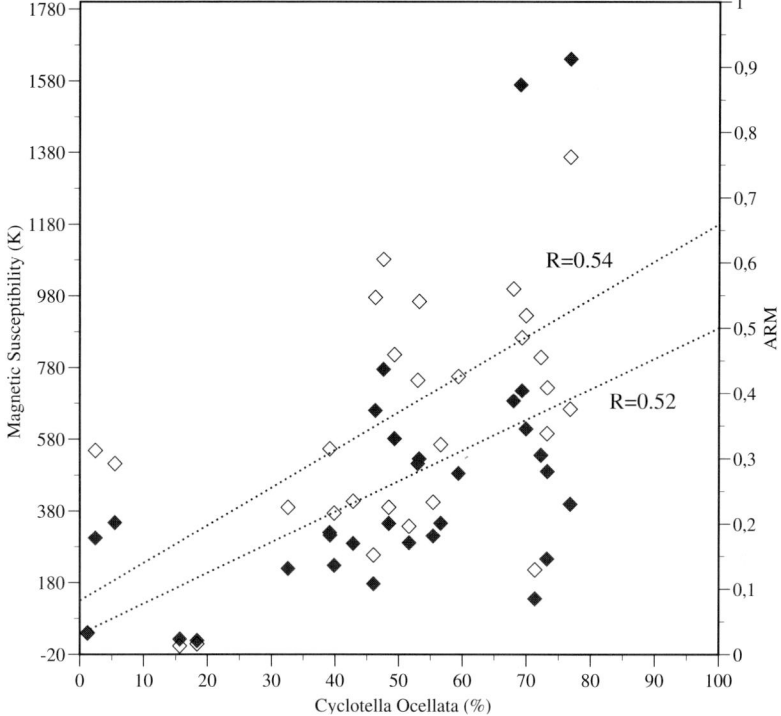

Fig. 11. Correlation between the occurrence of diatom species *Cyclotella ocellata* and susceptibility (◆) and ARM (◇) for core PNemi-1B.

in antiferromagnetic minerals and dissolution affecting only the superparamagnetic grains (lower $K_{fd}\%$ values).

Low-temperature IRM studies (Fig. 12) showed the presence of a Verwey transition in all samples but one, indicating the presence of stochiometric magnetite grains larger than the superparamagnetic grain-size range (Özdemir et al. 1993). The exception was from the most well-developed diagenetic level at 352 cm, suggesting that magnetite was completely removed under these conditions, possibly by maghemitization (Özdemir et al. 1993; Torii 1997). In the same core, the occurrence of the planktonic Foraminifera *Globigerinoides ruber* var. *rosea* and *Neogloboquadrina dutertrei* at these levels is informative. Both prefer low-salinity living conditions such as occurred during water mass stratification during the formation of sapropels (Thunell et al. 1977; Muerdter et al. 1984; Rohling & Gieskes 1989). *G. ruber* var. *rosea* prefers warmer tropical to sub-tropical waters (Van Leeuwen, 1989), whereas *N. dutertrei* is a low-salinity indicator (Cita et al. 1977; Thunell et al. 1977) and characteristic of 'cold' sapropels (Cita et al. 1982; Violanti et al. 1991). The maximum occurrence of both correlates with low susceptibility values (Fig. 13). The occurrence of *N. dutertrei* also correlates closely with low frequency-dependent susceptibility values, suggesting the preferential loss of superparamagnetic grain-size ranges.

The Adriatic core, Pal94-9, shows a dramatic change in magnetic properties between 450 and 530 cm, but there is no difference in colour and lithology. This interval has been dated, using AMS ^{14}C dating, as extending between 8720 and 6860 a BP (Langone et al. 1996). This suggests that it correlates with the S_1 sapropel in the Eastern Mediterranean, and corresponding low oxic conditions are also implied by the V/Cr ratios (Calanchi et al. 1996) in this level of the core (Fig. 5). The low $K_{fd}\%$ values indicate that superparamagnetic-sized magnetic particles are scarce, implying that processes within this zone have preferentially affected this grain-size range. The planktonic fauna is dominated by surface-dwelling species typical of warm water masses: *Globigerinoides ruber*, *Globigerinoides tenellus* and *Orbulina universa* (Pujol & Vergnaud-Grazzini 1995). The presence of *Globigerina quinqueloba* suggests high

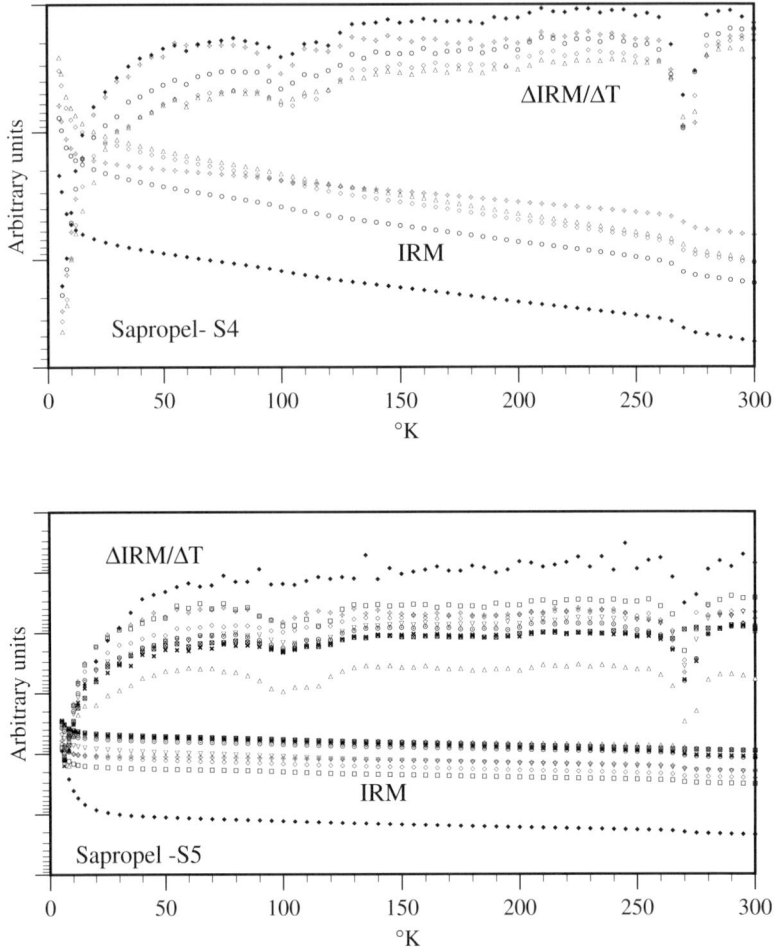

Fig. 12. Thermal demagnetization of IRM for samples from a transect in sapropels S_4 and S_5 in core B74-27. The IRM is in units of 10^{-7} A m^2. The samples with lower IRM belong to sapropel layer.

fertility in the surface waters (Rasmussen 1991) at this time. The increase in benthonic Foraminifera, such as *Uvigerina* sp., commences in the middle of the sapropel, where both the planktonic Foraminifera and magnetic mineral concentrations remain the same (Fig. 14). This later increase in the benthos presumably reflects a delay in recording by the sediments of the oceanographic changes occurring during the sapropel formation.

Conclusions

In this study magnetic techniques have been applied mainly to sediment samples from anoxic–suboxic environments and compared with geochemical and biological characteristics. Sediments from both marine and lacustrine environments show similar behaviour, even though the marine environment usually has a lower availability of organic matter. In particular, there is a decrease in the total magnetic fraction (characterized by decrease in susceptibility and ARM and SIRM magnitudes), loss of the superparamagnetic grain sizes (decreases in K_{fd}) and increases in magnetic grain size and coercivity (decreases in ARM/K, SIRM/K, ARM/SIRM, and in the S ratios). The increase in coercivity (higher $B_{0\,cr}$) further suggests that antiferromagnetic minerals (hematite and/or goethite) became more important in such zones. None the less, it is interesting to note that, at least in these anoxic–suboxic environments, the gross magnetic features appear to be more influenced by the original depositional factors than by later diagenetic process. Rock magnetic variables are

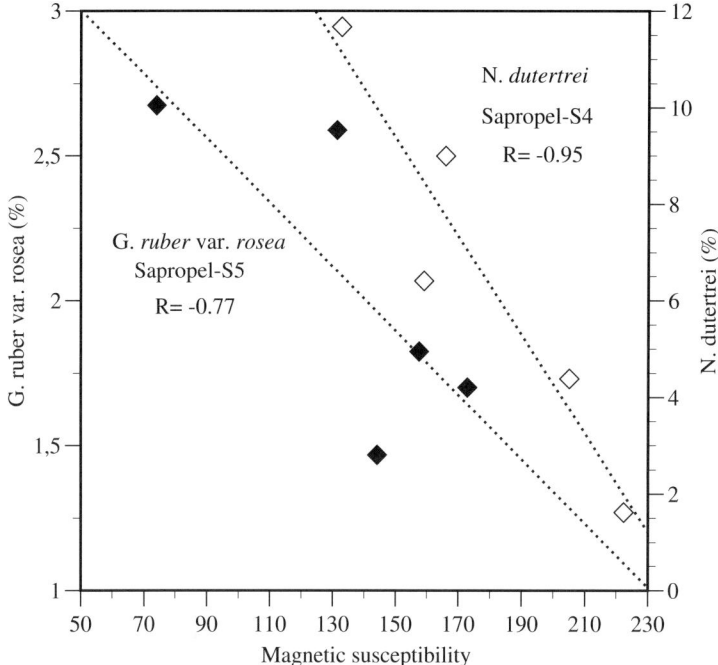

Fig. 13. Correlation between magnetic susceptibility and planktonic Foraminifera *G. ruber* and *N. dutertrei* in sapropelitic layers from core B74-27. The susceptibility is in 10^{-6} SI units.

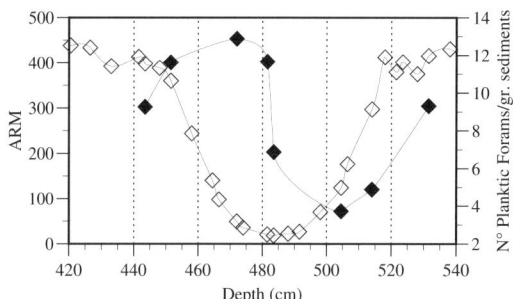

Fig. 14. Variations of ARM and organic productivity in core Pal94-9. The organic activity is expressed by the number of benthic and planktonic Foraminifera (♦), in the interval of reductive diagenesis. The anyhysteric remanence, ARM (◇) is in units of 10^{-5} A m^2/kg.

very sensitive discriminants of depositional and post-depositional processes in both marine and lacustrine sediments. It is clear, therefore, that magnetic techniques provide a rapid, effective and cheap technique for studying both depositional and diagenetic processes in sediments. Such methods are far cheaper than most geochemical analysis and also far faster than micropalaeontological techniques. However, it is also clear that their main power is when studied in conjunction with geochemical and biological proxies.

The cores from the crater lakes, Albano and Nemi, and from the Adriatic Sea, were collected under the auspices of an EU Project (PALICLAS; Contract EV5V CT93 0267).

References

ALVISI, F. & VIGLIOTTI, L. 1996. Magnetic signature of marine and lacustrine sediments from central Italy (PALICLAS Project). *In*: GUILIZZONI, P. & OLDFIELD, F. (guest eds) *Palaeoenvironmental Analysis of Italian Crater Lake and Adriatic Sediments.* Memorie Istituto Italiano Idrobiologica, **55**, 285–302.

BERNER, R. A. 1984. Sedimentary pyrite formation: an update. *Geochimica et Cosmochimica Acta*, **48**, 605–615.

BLOEMENDAL, J. 1983. Palaeoenviromental implications of the magnetic characteristics of sediments from DSDP Site 514, Southeast Argentine Basin. *In*: LUDWIG, W. J., KRASHENINNIKOV, V. A. *et al.* (eds) *Initial Reports of the Drilling Project, 71*. US Government Printing Office, Washington, DC, 1097–1108.

—— & DEMENOCAL, P. 1989. Evidence for a change in the periodicity of tropical climate cycles at 2.4 Myr from whole-core magnetic susceptibility measurements. *Nature*, **342**, 897–900.

——, KING, J. W., HALL, F. R. & DOH, S. I. 1992. Rock magnetism of Late Neogene and Pleistocene deep-sea sediments: relationship to sediment source, diagenetic processes and sediment lithology. *Journal of Geophysical Research*, **97**, 4361–4375.

CALANCHI, N., DINELLI, E., LUCCHINI, F. & MORDENTI, A. 1996. Chemostratigraphy of late Quaternary sediments from Lake Albano and central Adriatic Sea cores (PALICLAS project). *In*: GUILIZZONI, P. & OLDFIELD, F. (guest eds) *Palaeoenvironmental Analysis of Italian Crater Lake and Adriatic Sediments*. Memorie Istituto Italiano Idrobiologica, **55**, 247–263.

CANFIELD, D. E. & BERNER, R. A. 1987. Dissolution and pyritization of magnetite in anoxic marine sediments. *Geochimica et Cosmochimica Acta*, **51**, 645–659.

CITA, M. B., BROGLIA, C., MALINVERNO, A., SPEZZIBOTTIANI, G., TOMADIN, L. & VIOLANTI, D. 1982. Late Quaternary pelagic sedimentation on the Southern Calabrian Ridge and Western Mediterranean Ridge, Eastern Mediterranean. *Marine Micropalaeontology*, **7**, 135–162.

——, VERGNAUD-GRAZZINI, C., ROBERT, C., CHAMLEY, H., CIARANFI, N. & DONOFRIO, S. 1977. Palaeoclimatic record of a long deep sea core from the eastern Mediterranean. *Quaternary Research*, **8**, 205–235.

CHONDROGIANNI, C., ARIZTEGUI, D., BERNASCONI, S. M., LAFARGUE, E. & MCKENZIE, J. A. 1996*a*. Geochemical indicators tracing ecosystem response to climate change during the late Pleistocene (Lake Albano, central Italy). *In*: GUILIZZONI, P. & OLDFIELD, F. (guest eds) *Palaeoenvironmental Analysis of Italian Crater Lake and Adriatic Sediments*. Memorie Istituto Italiano Idrobiologica, **55**, 99–109.

——, ——, NIESSEN, F., OHLENDORF, C. & LISTER, G. 1996*b*. Late Pleistocene and Holocene sedimentation in Lake Albano and Lake Nemi (central Italy). *In*: GUILIZZONI, P. & OLDFIELD, F. (guest eds) *Palaeoenvironmental Analysis of Italian Crater Lake and Adriatic Sediments*. Memorie Istituto Italiano Idrobiologica, **55**, 23–38.

FROELICH, P. N., KLINKHAMMER, G. P., BENDER, M. L. *et al.* 1979. Early oxidation of organic matter in pelagic sediments of the eastern equatorial Atlantic: suboxic diagenesis. *Geochimica et Cosmochimica Acta*, **43**, 1075–1090.

GUERRA, D. 1994. *La palaeoidrologia del Mediterraneo occidentale e del Bacino di Valencia (Spagna) nel Tardo Pleistocene–Olocene, in base ai Foraminiferi planctonici e bentonici*. Thesis, University of Bologna.

KAPLAN, I. R., EMERY, K. O. & RITTENBERG, S. C. 1963. The distribution and isotopic abundance of sulphur in recent marine sediments off southern California. *Geochimica et Cosmochimica Acta*, **27**, 297–331.

KARLIN, R. 1990. Magnetic diagenesis in suboxic sediments at Bettis Site W–S, NE Pacific Ocean. *Journal of Geophysical Research*, **95**, 4421–4436.

—— & LEVI, S. 1983. Diagenesis of magnetic minerals in recent hemipelagic sediments. *Nature*, **303**, 327–330.

—— & ——1985. Geochemical and sedimentological control of the magnetic properties of hemipelagic sediments. *Journal of Geophysical Research*, **90**, 10 373–10 392.

LANGONE, L., ASIOLI, A., CORREGGIARI, A. & TRINCARDI, F. 1996. Age–depth modelling through the late Quaternary deposits of the central Adriatic basin. *In*: GUILIZZONI, P. & OLDFIELD, F. (guest eds) *Palaeoenvironmental Analysis of Italian Crater Lake and Adriatic Sediments*. Memorie Istituto Italiano Idrobiologica, **55**, 177–196.

LESLIE, B. W., HAMMOND, D. E., BURLESON, W. M. & LUND, S. P. 1990. Diagenesis in anoxic sediments from the California continental borderland and its influence on iron, sulfur, and magnetite behavior. *Journal of Geophysical Research*, **95**, 4453–4470.

LIU, X. M., HESSE, P., LIU, T. S. & BLOEMENDAL, J. 1998. High resolution climate record from the Beijng area during the last glacial–interglacial cycle. *Geophysical Research Letters*, **25**, 349–352.

MUERDTER, D. R., KENNETT, J. P. & THUNELL, R. C. 1984. Late Quaternary sapropel sediments in the Eastern Mediterranean Sea: faunal variations and chronology. *Quaternary Research*, **21**, 385–403.

OLAUSSON, E. 1960. *Description of sediment cores from the Mediterranean and the Red Sea*. Reports of the Swedish Deep-Sea Exploration 1947–1948, **VIII**, 287–334.

ÖZDEMIR, Ö., DUNLOP, D. J. & MOSKOWITZ, B. M. 1993. The effect of oxidation on the Verwey transition in magnetite. *Geophysical Research Letters*, **20**, 1671–1674.

PUJOL, C. & VERGNAUD-GRAZZINI, C. 1995. Distribution patterns of live planktic foraminifers as related to regional hydrography and productive systems of the Mediterranean Sea. *Marine Micropalaeontology*, **25**, 187–217.

RASMUSSEN, T. L. 1991. Benthonic and planktonic foraminifera in relation to the early Holocene stagnation in the Ionian Basin, Central Mediterranean. *Boreas*, **20**, 357–376.

ROHLING, E. J. & GIESKES, W. W. C. 1989. Late Quaternary changes in Mediterranean Intermediate Water density and formation rate. *Paleoceanography*, **4**, 531–545.

SUNDA, W. G. & HUNTSMAN, S. A. 1997. Interrelated influence of iron, light and cell size on marine phytoplankton growth. *Nature*, **390**, 389–392.

THOMPSON, R. & OLDFIELD, F. 1986. *Environmental Magnetism*. Allen & Unwin, New York.

THOUVENY, N., BEAULIEU, J. L. D., BONIFAY, E. *et al.* 1994. Climate variations in Europe over the past 140 kyr deduced from rock magnetism. *Nature*, **371**, 503–506.

THUNELL, R., WILLIAMS, D. F. & KENNETT, J. P. 1977. Late Quaternary palaeoclimatology, stratigraphy and sapropel history in eastern Mediterranean deep-sea sediments. *Marine Micropalaeontology*, **2**, 371–388.

TORII, M. 1997. Low temperature oxidation and subsequent down-core dissolution of magnetite in deep-sea sediments, ODP Leg 161 (Western Mediterranean). *Journal of Geomagnetism and Geoelectricity*, **49**, 1233–1245.

TRINCARDI, F., CATTANEO, A., ASIOLI, A., CORREGGIARI, A. & LANGONE, L. 1996. Stratigraphy of the late-Quaternary deposits in the central Adriatic basin and the record of short-term climatic events. *In*: GUILIZZONI, P. & OLDFIELD, F. (guest eds) *Palaeoenvironmental Analysis of Italian Crater Lake and Adriatic Sediments*. Memorie Istituto Italiano Idrobiologica, **55**, 39–70.

VAN LEEUWEN, R. J. W. 1989. Sea-floor distribution and Late Quaternary faunal patterns of planktonic and benthic foraminifers in the Angola Basin. *Utrecht Micropalaeontolical Bulletin*, **38**, 1–288.

VERGNAUD-GRAZZINI, C., RYAN, W. B. F. & CITA, M. B. 1977. Stable isotopic fractionation, climate change and episodic stagnation in the eastern Mediterranean during the late Quaternary. *Marine Micropalaeontology*, **2**, 353–370.

VIGLIOTTI, L. 1997. Magnetic properties of light and dark sediment layers from the Japan Sea: diagenetic and palaeoclimatic implications. *Quaternary Science Reviews*, **16**, 1093–1114.

VIOLANTI, D., GRECCHI, G. & CASTRADORI, C. 1991. Palaeoenvironmental interpretation of core BAN88-11GC (Eastern Mediterranean, Pleistocene–Holocene) on the grounds of Foraminifera, Thecosomata and calcareous nannofossils. *Il Quaternario*, **4**, 13–39.

VLAG, P., THOUVENY, N., WILLIAMSON, D., ANDRIEU, V., ICOLE, M. & VAN VELZEN, A. J. 1997. The rock magnetic signal of climate change in the maar lake sequence of Lac St Front (France). *Geophysical Journal International*, **131**, 724–740.

The isolation of diagenetic groups in marine sediments using fuzzy c-means cluster analyses

MICHAEL URBAT,[1,3] MARK J. DEKKERS[1] & SIMON P. VRIEND[2]

[1] *Palaeomagnetic Laboratory 'Fort Hoofddijk', Faculty of Earth Sciences, Utrecht University, Budapestlaan 17, 3584 CD Utrecht, Netherlands*
[2] *Department of Geochemistry, Faculty of Earth Sciences, Utrecht University, Budapestlaan 4 3584 CD Utrecht, Netherlands*
[3] *Present address: Department of Geology, University of Cologne, Zülpicher Strasse 499, 50674 Köln, Germany (e-mail: m.urbat@uni-koeln.de)*

Abstract: The evaluation of large rock magnetic datasets may benefit from the application of multivariate statistical methods. The combined use of fuzzy c-means clustering (FCM) and non-linear mapping (NLM) have considerable merit over conventional multivariate techniques and appear to provide a more logical choice. An example of the method of identification and characterization of early diagenetic effects on marine sediments is given using samples drilled at Ocean Drilling Program (ODP) Site 904. This confirms the clusters previously identified more laboriously by multi-variable analyses and detects clusters (diagenetic, detrital, etc.) that may only be recognized by the consideration of an array of rock magnetic data. In another example of a complex diagenetic signal the link is expanded between two papers on a long piston core (KC-01B) from the Calabrian Ridge (Central Mediterranean), where FCM and NLM provide a clue as to why the record of the lower Jaramillo Subchron is delayed in this core.

Clearly, diagenetic effects on the magnetomineralogy in marine sediments, if not detected, may severely flaw the interpretation of palaeomagnetic and rock magnetic data. Conversely, the separation of different genetic components in the magnetic signal may yield additional information on the evolution of the sediments since deposition (e.g. Urbat 1995). Untangling the extent of post-depositional processes is hampered by the fact that rock magnetic values of diagenetic and detrital iron oxides, for example, are essentially the same. Moreover, most rock magnetic techniques yield only average values of a spectrum of the magnetic minerals contained in a specimen (sediment). Several approaches have been proposed recently, which use a variety of univariate and/or bivariate techniques (e.g. Oldfield 1994) that are sensitive to specific grain sizes and/or magnetic minerals within sediment. A complete picture can only be obtained by cross-correlating numerous variables and a multi-variable approach, integrating an array of rock magnetic information, generally appears to be the best choice to interpret the different components of the magnetic signal in marine sediments (e.g. Banerjee 1997). For high-resolution studies, the size of the dataset can make this procedure tedious. The logical choice for isolating significant groups within a dataset is to use the assistance of multivariate statistical analyses. These not only facilitate the process but have the potential to extract additional information from the data that may otherwise pass unnoticed using univariate or bivariate techniques.

An assessment of available multivariate procedures (e.g. Howarth & Sinding-Larsen 1983; Davis 1986; Kaufman & Rousseeuw 1990) indicates that fuzzy c-means clustering (FCM; Bezdek *et al.* 1984) in combination with non-linear mapping (NLM; Sammon 1969) seem most appropriate techniques to accomplish significant grouping in rock magnetic data. In this paper a combination of these methods will be described, and their merits and limitations will be considered when applied to rock magnetic data. Marine sediment drilled at Ocean Drilling Program (ODP) Site 904, which has previously been successfully interpreted without the employment of cluster methods (Urbat 1995, 1996), is used to provide an established background for comparison with the results of the multivariate statistical models using the same

URBAT, M., DEKKERS, M. J. & VRIEND, S. P. 1999. The isolation of diagenetic groups in marine sediments using fuzzy c-means cluster analyses. *In*: TARLING, D. H. & TURNER, P. (eds) *Palaeomagnetism and Diagenesis in Sediments*, Geological Society, London, Special Publications, **151**, 85–93.

dataset (Urbat et al. 1997). In another example, the merits of cluster techniques are extended into other palaeomagnetic studies. Two papers on a Calabrian Ridge core, KC-01B (Dekkers et al. 1994; Langereis et al. 1997), will therefore be considered, and FCM and NLM techniques are shown to provide a clue as to the causes for the delayed record of the lower Jaramillo reversal at this locality.

Multivariate statistical methods

Clustering techniques are commonly used to discern homogeneous groups within a dataset. An advantage over more conventional techniques, such as discriminant analysis (DA), is that all available variables are employed in the analysis. Thus, the classification limits are optimal for the problem at hand and the analysis is carried out at the scale of the dataset. Yet, conventional clustering methods are rigid in that they force cases (samples) into a cluster, ignoring possible real compositional overlaps between clusters (Howarth & Sinding-Larsen 1983). This probabilistic or stochastic approach of unambiguously allocating a sample to exactly one cluster is based on a notion of uncertainty that this is the most likely configuration. Conventional (hard) c-means clustering is one example of such a multivariate statistical technique. Grouping is achieved by minimizing the distance between sample and cluster centre and concurrently maximizing distances between respective cluster centres. Samples, however, can really be intermediate between two clusters, as between detrital and diagenetic phases, yet these will be forced into one cluster and are likely to distort the true data structure. Similarly, outlying values are given the same weight as the remainder of the data and will additionally deform the model. Fuzzy c-means clustering now represents a derivative of the conventional method. Vriend et al. (1988) convincingly demonstrated the efficiency of a combination of FCM and NLM on geochemical data from Portugal, and Dekkers et al. (1994) were the first to simultaneously analyse rock magnetic and geochemical data for ascertaining early diagenetic effects in Mediterranean sediments.

Fuzzy c-means clustering

The concept of fuzziness (Zadeh 1965) takes care of the fact that in nature groupings are unlikely to be characterized by 'sharp' boundaries, as in diagenesis, which generally comprises a wide combination of transitional, gradual processes. Thus, fuzzy techniques can be expected to be a more appropriate approach. Most importantly, such procedures do not assume any a priori knowledge about classification variables. Interest is no longer in a strict assignment of one case to one cluster, but in the similarity amongst the latter and how much they are alike (Bezdek 1981). This similarity is indicated as membership (u) of a sample to a cluster, where u is a continuous function between zero (no similarity) and unity (identical). All memberships of one sample sum up to unity and intermediate cases between clusters can easily be recognized. Tracing the membership assignments of an intermediate sample to the respective clusters would indicate similar memberships to each of n clusters calculated. In this study, an initial membership threshold of $u_2 < 0.6 u_1$ was used. Two other aspects characterize the actual FCM algorithm (Bezdek et al. 1984, table 1), as follows.

(1) The algorithm involves an exponent, q which controls the extent of membership sharing between clusters and thus measures the degree of fuzziness of the model. Theoretically, q could be

Table 1. *Fuzzy c-means cluster (FCM) algorithm (Bezdek* et al. *1984)*

C clusters for N cases analysed for V variables (x) and a fuzzy exponent q.
Cluster centre CC of the ith cluster for the jth variable is given by

$$CC_{ij} = \frac{\sum_{k=1}^{N} (u_{ki})^q \cdot x_{kj}}{\sum_{k=1}^{N} (u_{ki})^q}$$

and membership (u) for the kth case to the ith cluster

$$u_{ki} = \frac{[(d_{ki})^2]^{-1/(q-1)}}{\sum_{i'=1}^{C} [(d_{ki'})^2]^{-1/(q-1)}}$$

where the standardized distance (d) of the kth case to the ith cluster (s_j as the standard deviation of the jth variable) is given by

$$(d_{ki})^2 = \sum_{j=1}^{V} [(x_{kj} - CC_{ij})/s_j]^2$$

Classification entropy H and the partition coefficient F functions and their limiting values are given by

$$H = -\sum_{k=1}^{N}\sum_{i=1}^{C} [u_{ki} \cdot \log(u_{ki})/N] \qquad 0 \leq H \leq \log(C)$$

$$F = \sum_{k=1}^{N}\sum_{i=1}^{C} [(u_{ki})^2/N] \qquad 1/C \leq F \leq 1$$

assigned any value between unity and $+\infty$ and there is no firm basis for the choice of q. For $q = 1$ the model converges towards the conventional hard c-means clustering (i.e. no 'fuzziness'), whereas for $q = +\infty$ memberships of all samples would be equal in all clusters (i.e. total 'fuzziness'). Frequently used q values range between 1.5 and three; empirically, the best results are obtained using $q = 1.5$ (Vriend et al. 1988; Dekkers et al. 1994; Urbat et al. 1997).

(2) The data need to be standardized to ensure an equal weight for all variables. Difficulties would arise when comparing different absolute scales, such as remanent magnetic intensities compared with median destructive fields. In this study, the diagonal norm (squared distance; e.g. Howarth & Sinding-Larsen 1983) over a simple Euclidean distance (no standardization of the data) has been used to provide the required scale invariance, i.e. each variable has been divided by its standard deviation. This normalization is independent of the number of clusters. (Alternative normalization methods have been discussed by Vriend et al. (1988).)

The most appropriate number of clusters that matches the true data structure must be decided. Three main approaches to the determination of the optimum number should be considered. (a) A suitable validity function, such as the classification entropy H and/or the partition coefficient F (Table 1), can be a general guide to the preferred grouping. This will, however, not provide an incontestable frame for the interpretation of the cluster model, as both functions tend to describe the properties of the cluster model rather than those of the initial raw data. A selection taking into account the likely geological, sedimentological or geochemical background of the site can be preferable, as follows. (b) The basic question is whether the grouping makes sense in the context of other information available. For example, a minimum number of clusters may be determined considering lithologies with an expected different detrital magnetic input. Available sedimentological information (e.g. composition, pathways, mode of deposition) can relatively easily be compared with the cluster model obtained. Downhole plots of magnetic parameters along with the respective cluster assignments provide valuable information about possible correlations with changes in, for example, sediment composition. Lateral changes of, for example, depositional environments can equally well be assessed plotting clusters in a regional context. Such cluster assignments can then be directly compared with geochemical indicators of sediment composition or alteration. Certainly, geochemical data sets can be clustered as well (e.g. Vriend et al. 1988) and examples of applications of FCM and NLM to combined geochemical–rock magnetic datasets will be presented subsequently (Urbat & Dekkers, in preparation). (c) An independent method is desirable, such as that provided by non-linear mapping (NLM; Sammon 1969), to judge the significance of a certain FCM cluster model. Similarities in the results of NLM and FCM analyses can then be considered to be strong evidence for the true existence of the groups obtained (Fig. 1). A major reason

Non-Linear-Mapping

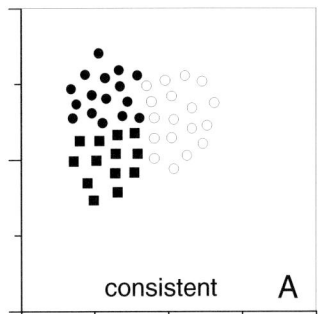
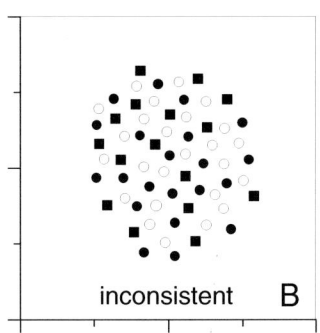

Fig. 1. Hypothetical NLM plots. Different symbols denote separate FCM clusters. The FCM clusters separate well in the NLM projection, indicating consistent results (**a**). A random distribution of the FCM cluster would in turn indicate that the results obtained by both methods are inconsistent (**b**). Increasing the number of FCM clusters might result in an increasingly less separated NLM projection. This may indicate over-interpretation of the FCM cluster model. NLM axis are in arbitrary units. (See text for explanation.)

Table 2. *Non-linear mapping (NLM) (Sammon 1969)*

$$D = \frac{1}{\sum_{k=1}^{N}\sum_{k'=1}^{k-1} d_{kk'}} \sum_{k=1}^{N}\sum_{k'=1}^{k-1} \frac{(d_{kk'} - d_{kk'}^{*})^2}{d_{kk'}}$$

where $d_{kk'}$ is the distance in the original V-dimensional variable space, $d_{kk'}^{*}$ is the distance in the two- (or three-)dimensional mapping space and D is the mapping error

to select NLM is that no assumptions about the number of clusters or even the existence of subgroups within the dataset have to be made. It is therefore considered that this should be an integral part of the analysis.

Non-linear mapping

Non-linear mapping (NLM) provides a two- (or three-) dimensional image of n-dimensional data clouds such that there is minimal distortion of the distances amongst the data (Howarth & Sinding-Larsen 1983). The algorithm (Table 2) means that the results will be closer to reality than those obtained using a simple projection plane. The data still need to be standardized in the same way as for the FCM analysis (described above), but as this technique is likely to be affected by any extreme outliers such values should preferably be removed. This is necessary to avoid an insufficient resolution of smaller distances, which results in a distortion of the scale of the NLM image. Using the so-called steepest descent method (Sammon 1969), the mapping error D is iteratively minimized over all samples until it stabilizes at a minimum value.

Selecting the variables

It is valuable to check the potential meaning of the rock magnetic variables before NLM and FCM clustering so that the most appropriate input variables for each specific dataset can be assessed. This is because it is desirable that the information on grain size, mineralogy and their relative concentration in the sediment should be well balanced amongst the variables. This can be achieved by the usual procedure of palaeomagnetic and rock magnetic data analysis (e.g. Day *et al.* 1977; Bradshaw & Thompson 1985). However, any one of the concentration-dependent variables can provide this information, but reliance on a single variable may not be adequate in detecting subtle diagenetic signals. Using only one concentration-dependent variable, such as anhysteretic remanent magnetization (ARM), preferentially for smaller ferrimagnetic grain sizes, or isothermal remanent magnetization (IRM), may lose valuable information. Similar constraints apply to the use of ratios. The ratio of IRM/χ (where χ is the magnetic susceptibility), for example, may often be used to indicate magnetite grain-size variations downhole when driven by the ferrimagnetic component. In certain parts of a profile, however, magnetic susceptibility may be dominated by the paramagnetic component and any interpretation of this ratio in terms of magnetite grain sizes may be completely erroneous. Such differences can only be detected by thorough reasoning of the remainder of the magnetic data (Urbat 1996). The choice of appropriate variables is therefore critical to the validity of any future interpretation even though the analyses may well provide clear discriminations of different groups.

Information on the population homogeneity of chosen variables is desirable, because a logarithmic transformation of log-normally distributed data provides a better scaling and may therefore enhance the significance of the cluster results. Whether the data approximate to a Gaussian, or at least symmetrical, distribution can easily be checked with the aid of histograms. Multi-modality is a common feature of natural data. Logarithmic transformation of multi-modal data will not jeopardize standardization of the data, with the proviso that extreme outliers are removed before the calculation. In view of the deductive nature of rock magnetic techniques and the environment-specific influence on the magnetic signal, it cannot be expected that one rock magnetic cluster always describes an identical detrital, diagenetic or biogenic feature. Consequently, the isolated groups must always be used as auxiliary tools in attempting to decipher the record. This may well require the use of univariate and particularly bivariate plots of the groups isolated by the cluster assignments. Samples displaying similar rock magnetic features can be traced easily through various biplots (Fig. 2) and these can then be identified according to their position in the sediment column by simply plotting that variable downhole. Ultimately, as many variables as possible must be considered when trying to resolve a complex process such as diagenesis.

Examples of the application of FCM and NLM

To illustrate the value and some of the problems of these statistical techniques in diagenetic

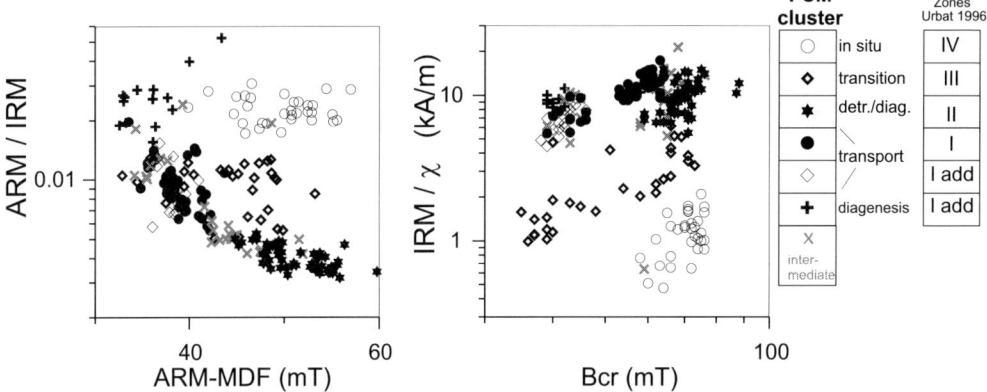

Fig. 2. Two examples of frequently used scatter-plots in rock magnetic studies, taken from Urbat et al. (1997). Different symbols denote different FCM cluster. (Note that, based on the respective bivariate scatter-plots only (without cluster tagging) the data-groups would be assigned differently or would not be recognized.) (See text for explanation.)

studies, two case studies are discussed. The two datasets are derived from an ODP drilled interval in Pleistocene marine sediments from the continental shelf of the USA and from a long piston core of Pliocene sediments from the Calabrian Ridge in the Central Mediterranean. The diagenetic processes in the Atlantic core, although complex, could be expected to be more uniform than in the even more complex lithologies of the Mediterranean core. It is also possible to illustrate, in the Mediterranean core, how cluster-tagging can be used as assistance in determining the magnetostratigraphy.

The New Jersey continental slope: ODP Hole 904A

This core was drilled roughly 141 km ESE of Barnegat Inlet, New Jersey, on the modern upper continental slope in a water depth of 1123 m (Mountain et al. 1994). The core was from 0 to 55 m depth and comprised homogeneous to heavily bioturbated clays which were sampled at approximately every 10–15 cm, yielding 235 specimens. The original study of the rock magnetism of these samples (Urbat 1996) investigated the influence of abrupt changes in sedimentation rate on the preservation of the primary magnetic signal within a hemipelagic setting. Standard rock magnetic measurements (natural remanent magnetization (NRM), anisotropy of magnetic susceptibility (AMS), ARM, IRM and hysteresis measurements) formed four main zones (I–IV) of similar magnetic characteristics but with different genetic implications. These zones occur in distinct sections of the profile and, to a varying extent, have been affected by early diagenetic alteration of the primary detrital magnetic signal. This alteration is restricted to the finer ferrimagnetic fraction; the coarser magnetite grains remain relatively unaltered. The dissolution–precipitation patterns of magnetite can be related to the interface of unconformably deposited sediment packages, i.e. abrupt sedimentation rate changes. Hence, the development of non-steady-state conditions seems to preserve markers of, for example, fluctuating redox conditions throughout the sampled sediment column (see Urbat (1996) for further details). However, these four zones were separated on a larger scale, as at least two prominent diagenetic sections occur in magnetic Zone I, where authigenic fine-grained magnetite is preserved as a function of the occurrence of sedimentary unconformities. The bimodal magnetomineralogy consisting of relatively coarse-grained detrital magnetite accompanied by fine-grained authigenic magnetite could be recognized only by thorough evaluation of the relative differences between all rock magnetic variables. The most important differences were between ARM and IRM and those variables derived from them, e.g. ARM/IRM and the alternating field (AF) demagnetization trajectories of ARM and IRM. The results of FCM analyses of the dataset were then presented by Urbat et al. (1997) and these are compared here.

On a single sample basis, a six-cluster FCM analysis is perfectly coherent with the non-statistical results of Urbat (1996) (Fig. 3). Magnetic Zones II, III, and IV are each represented by one cluster, whereas FCM divides magnetic Zone I into three clusters. The main features of Zone I are still defined, but cluster analysis also identifies the diagenetic sections in Zone I that had been determined only with great difficulty in the earlier analysis. There was also an additional cluster that had not previously been recognized. This delineates thin sections on top of and below the first diagenesis cluster, and then forms a significant part of the lowermost part of the profile, including the main Zone I

cluster, which had previously been undivided. On the basis of previous grounds of interpretation, this second additional cluster seems to represent samples where authigenic fine-grained magnetite is present, but not to the same concentration levels that are characteristic for the 'diagenesis' cluster in Zone I.

It is clear that FCM and NLM analyses tackle groups within the dataset that are recognized only with difficulty using a conventional approach by processing univariate and bivariate information simultaneously. FCM does find true, non-trivial groups within the data, the reality of which can be confirmed by detailed study of the multi-variable analyses of Urbat (1996). One limitation of clustering also becomes apparent. The rock magnetic data from Hole 904 A suggest a second diagenetic front, indicated by a continuously increasing paramagnetic/ferrimagnetic ratio extending over almost the entire sampled section. This is superimposed on a smaller-scale downhole variation, κ_{ARM}/κ, which generally decreases from the sediment–water interface to the top of magnetic Zone IV. This specific feature of the magnetic signal is not recognized by the clustering method, but this is not surprising because this would require a 'gradual cluster' that extends from the top to the bottom of the sedimentary interval under investigation. Such a cluster would include those described above and thus define a different scale of analysis. Certainly, such a cluster could be identified by clustering techniques if interest was in larger-scale features of the magnetic signal and in a longer sampled interval.

The Calabrian Ridge: KC-01B

The sediment in this 37 m long piston core was sampled at 10 cm intervals, providing a total of 337 samples (Langereis et al. 1997). The sediments appeared to have been continuously deposited from just below the Jaramillo Subchron up to the present, so that each sample had an average resolution of some 3 ka. The sediment represents various depositional conditions cycling from oxic–suboxic (beige and white layers) through to suboxic–anoxic (grey layers) and anoxic (sapropelitic layers) conditions. These cycles were attributed to climatic variations in the supply of both terrigenous (mainly clay minerals) and biogenic (mainly $CaCO_3$) materials. The rock magnetic (ARM, χ) and geochemical data ($CaCO_3$, S, Ba, Mn) were subjected to FCM and NLM analysis by Dekkers et al. (1994), providing the first example of such analyses in such a context. This also

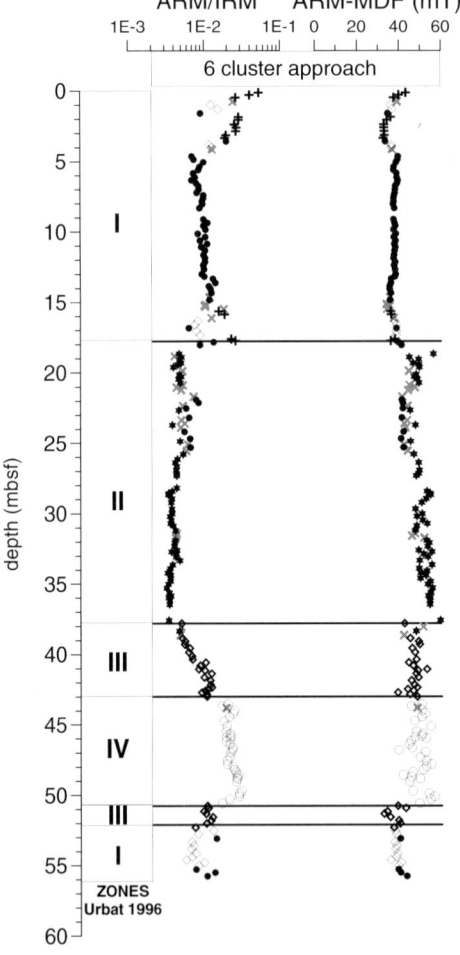

Fig. 3. Interpretation of the magnetic signal at ODP Hole 904 A (redrawn from Urbat et al. 1997). Selected variables. Left-hand column: magnetic Zones I–IV as result of the non-statistical approach of Urbat (1996). Central plot: different symbols denote separate FCM clusters (see Fig. 2 for legend). Light grey crosses denote intermediate cases. (See text for explanation.)

proved to lead to a straightforward interpretation of early diagenetic alteration of the sediment, as it discriminated discrete zones of magnetite dissolution and precipitation as functions of the sapropelitic layers in the sediment. Thus the eight-cluster model of Dekkers et al. (1994) constitutes two main categories: one expressing mainly diagenetic phenomena, and the other predominantly expressing primary depositional conditions (Fig. 4). This diagenetic phenomenon would clearly result in delayed acquisition of the NRM and hence would obscure any attempted magnetostratigraphic assessment (e.g. Van Hoof & Langereis, 1991; Van Hoof et al. 1993). In this case, the determination of the boundary of the lower Jaramillo reversal was not straightforward (Langereis et al. 1997) because this appeared to occur in an interval dominantly characterized by diagenetically altered samples (Dekkers et al. 1994). This association is easily demonstrated and further clarified when the cluster affiliations are considered in stratigraphic order (Fig. 5). Cluster 6 (Dekkers et al. 1994) encompasses all samples from the sapropelitic layers within which magnetite, which is the main remanence carrier in KC-01B, is being dissolved under anoxic conditions (e.g. Passier et al. 1996). (Sulphate-reducing bacteria may, however, subsequently cause the precipitation of authigenic remanence-carrying minerals in cluster 6). Cluster 4 is closely related to cluster 6 but here, the absence of sulphate-reducing bacteria (reduced organic matter contents) results in merely dissolving magnetite under still anoxic conditions. Cluster 3 is occasionally present and is characterized by mild dissolution associated with a gradual increase in suboxic conditions. Cluster 2 is associated with suboxic conditions where Fe^{2+}, which has diffused out of the dissolution intervals, has been reprecipitated as magnetite. Consequently, this cluster is associated with the level immediate below and above the sapropels. These interrelationships can be illustrated diagrammatically (Fig. 5) as a result of the dissolution–precipitation of magnetite as triggered by the sapropelitic layers. If, in steady state at a certain time, T_1, the geomagnetic polarity changes, the reversal boundary will be recorded in the sediment column only as a function of the lock-in depth of the remanence carriers (Fig. 5 column a). If, in contrast, a polarity change at T_1 occurs during the precipitation of magnetite in cluster 2 (Fig. 5, column b), this will result in a delayed NRM acquisition. In cluster 4 and cluster 3 the primary NRM will coincidentally be destroyed, with the actual extent of destruction being a function of the duration of continuing magnetite dissolution. NRM destruction in the same way also occurs in cluster 6; however, subsequently the sapropels may favour precipitation (particularly of remanence-carrying sulphides; Roberts et al. 1998).

Conclusions

Generally, groups in nature are characterized by a certain overlap and are prone to outliers. The approach presented here is most suitable to handle these matters appropriately. All available information can be employed and thus, the scale of interpretation is optimal for the problem

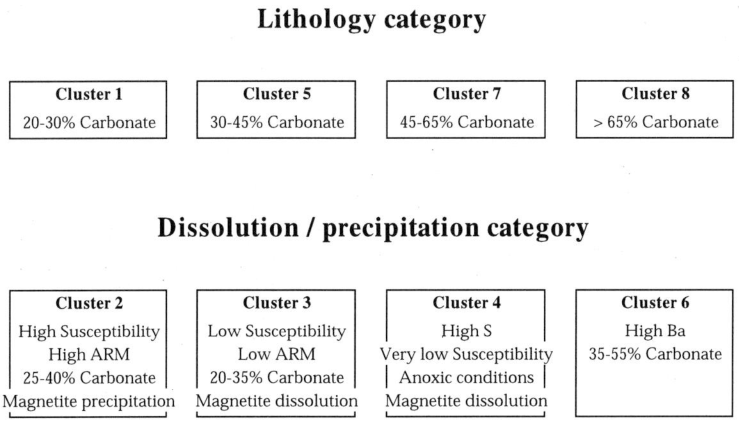

Lithology category

Cluster 1	Cluster 5	Cluster 7	Cluster 8
20-30% Carbonate	30-45% Carbonate	45-65% Carbonate	> 65% Carbonate

Dissolution / precipitation category

Cluster 2	Cluster 3	Cluster 4	Cluster 6
High Susceptibility	Low Susceptibility	High S	High Ba
High ARM	Low ARM	Very low Susceptibility	35-55% Carbonate
25-40% Carbonate	20-35% Carbonate	Anoxic conditions	
Magnetite precipitation	Magnetite dissolution	Magnetite dissolution	

Fig. 4. Cluster interpretation of core KC-01B, redrawn from Dekkers et al. (1994).

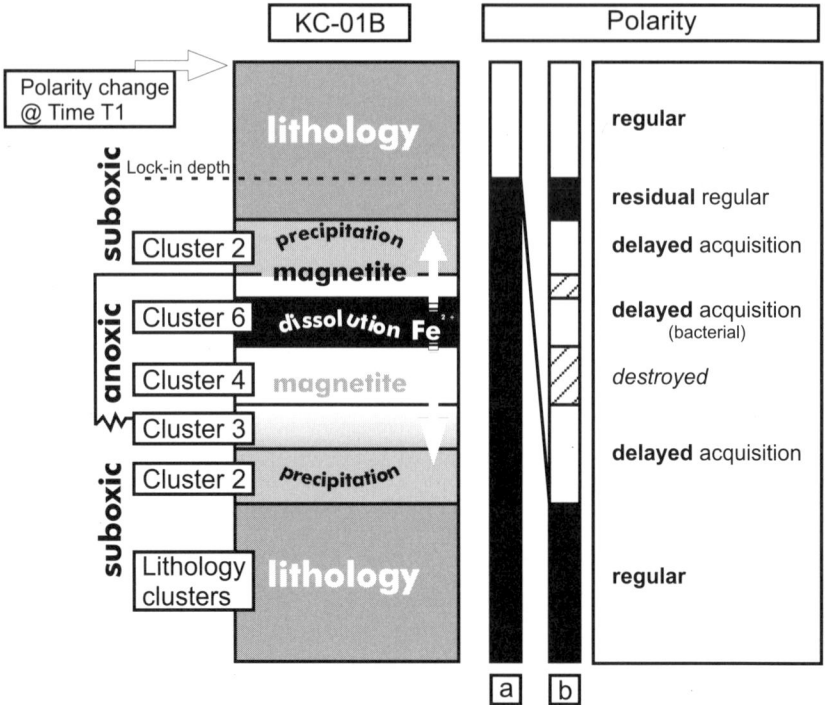

Fig. 5. Cluster assignments in the vicinity of a sapropel layer in core KC-01B. Schematized interval, not to scale. (See text for explanation.)

in hand. A combination of NLM and FCM techniques clearly provides a useful tool to aid in the interpretation of post-depositional alterations of the magnetic signal in marine sediments. Importantly, the multivariate analysis detects groups in the dataset which require a consideration of complex rock magnetic information simultaneously. A simple grouping of the data with respect to, for example, more or less obvious changes in relative concentration would not be a reason to promote the use of cluster techniques in magnetic studies. In particular, FCM accounts for gradual diagenetic effects on the magnetic signal and allows intermediate cases between clusters to be recognized, and thus yields extra information on the relationship between groups, because boundaries (crisp or gradual) can be analysed. Multivariate analysis, however, requires a thorough consideration of the magnetic variables used, their specific meaning for a specific site and their uniqueness in terms of information about characteristics of the magnetic signal. Both redundancy (similar information inherent in several variables) and scarcity (missing information on features of the magnetic signal) will also significantly influence the derived cluster models, and their interpretations will always remain the investigator's challenge.

This work was funded by European Community Marie Curie Grant FMBICT 960799 to M.U.

References

BANERJEE, S. K. 1997. Low temperature sediment magnetism – a novel approach in environmental reconstruction. UKGA – 21, University of Southampton.

BEZDEK, C. J. 1981. *Pattern recognition with fuzzy objective function algorithms*. Plenum Press.

——, EHRLICH, R. & FULL, W. 1984. FCM: the fuzzy c-means cluster algorithm. *Computers and Geosciences*, **10**, 191–203.

BRADSHAW, R. H. W. & THOMPSON, R. 1985. The use of magnetic measurements to investigate the mineralogy of some Icelandic lake sediments and to study catchment processes. *Boreas*, **14**, 203–215.

DAVIS, J. C. 1986. *Statistics and Data Analysis in Geology*. John Wiley, New York.

DAY, R., FULLER, M. & SCHMIDT, V. A. 1977. Hysteresis properties of titanomagnetites: grain-size and compositional dependence. *Physics of the Earth and Planetary Interiors*, **13**, 260–267.

DEKKERS, M. J., LANGEREIS, C. G., VRIEND, S. P., VAN SANTVOORT, P. J. M. & DE LANGE, G. J. 1994. Fuzzy c-means cluster analysis of early diagenetic effects on natural remanent magnetisation acquisition in a 1.1 Myr piston core from the Central Mediterranean. *Physics of the Earth and Planetary Interiors*, **85**, 155–171.

HOWARTH, R. J. & SINDING-LARSEN, R. 1983. Multivariate analysis. *In*: HOWARTH, R. J. (ed.) *Statistics and Data Analysis in Geochemical Prospecting, Vol. 2*. Elsevier, 207–283.

KAUFMAN, L. & ROUSSEEUW, P. J. 1990. *Finding Groups in Data. An Introduction to Cluster Analysis*. John Wiley, New York.

LANGEREIS, C. G., DEKKERS, M. J., DE LANGE, G. J., PATERNE, M. & VAN SANTVOORT, P. J. M. 1997. Magnetostratigraphy and astronomical calibration of the last 1.1 Myr from an eastern Mediterranean piston core and dating of short events in the Brunhes. *Geophysical Journal International*, **129**, 75–94.

MOUNTAIN, G. S., MILLER, K. G., BLUM, P., *et al.* (eds) 1994. *Proceedings of the Ocean Drilling Program, Initial Reports, 150*. Ocean Drilling Program, College Station, TX.

OLDFIELD, F. 1994. Towards the discrimination of fine-grained ferrimagnets by magnetic measurements in lake and near-shore marine sediments. *Journal of Geophysical Research*, **99**, 9045–9050.

PASSIER, H. F., MIDDELBURG, J. J., VAN OS, B. J. H. & DE LANGE, G. J. 1996. Diagenetic pyritisation under eastern Mediterranean sapropels caused by downward sulphide diffusion. *Geochimica et Cosmochimica Acta*, **60**, 751–763.

ROBERTS, A. P., STONER, J. S. & RICHTER, C. 1998. Diagenetic magnetic enhancement of sapropels from the eastern Mediterranean Sea. *Marine Geology*, in press.

SAMMON, J. W. 1969. A non-linear mapping for data structure analysis. *IEEE Transactions Computers*, **C18**, 401–409.

URBAT, M. 1995. *Rock magnetic properties of Pleistocene marine sediments of the, U.S. middle Atlantic margin and the Japan Sea. Case studies on the diagenesis and environmental changes (Ocean Drilling Program Sites 903, 904 and 798)*. PhD thesis, University of Cologne.

—— 1996. Rock-magnetic properties of Pleistocene passive margin sediments: environmental change and diagenesis offshore New Jersey. *In*: MOUNTAIN, G. S., MILLER, K. G., BLUM, P., POAG, C. W. & TWICHELL, D. C. (eds) *Proceedings of the Ocean Drilling Program, Scientific Results*, 150. Ocean Drilling Program, College Station, TX, 347–359.

——, DEKKERS, M. J. & VRIEND, S. P. 1997. Fuzzy c-means clustering as an aid in the interpretation of the natural remanent magnetization in various types of sediment. *In*: PAWLOWSKY-GLAHN, V. (ed.) *Proceedings of IAMG '97, The Third Annual Conference of the International Association for Mathematical Geology*. CIMNE, Spain, Part I, 395–400.

VAN HOOF, A. A. M. & LANGEREIS, C. G. 1991. Reversal records in marine marls and delayed acquisition of remanent magnetization. *Nature*, **351**(6323), 223–225.

——, VAN OS, B. J. H., RADEMAKERS, J. P., LANGEREIS, C. G. & DE LANGE, G. J. 1993. A paleomagnetic and geochemical record of the upper Cochiti reversal and two subsequent precessional cycles from southern Sicily (Italy). *Earth and Planetary Science Letters*, **117**, 246–255.

VRIEND, S. P., VAN GAANS, P. F. MIDDELBURG, J. & DE NIJS, A. 1988. The application of fuzzy c-means cluster analysis and non-linear mapping to geochemical data sets: examples from Portugal. *Applied Geochemistry*, **3**, 213–224.

ZADEH, L. A. 1965. Fuzzy sets. *Information and Control*, **8**, 338–353.

Diagenesis of magnetic mineral assemblages in multiply redeposited siliciclastic marine sediments, Wanganui basin, New Zealand

GARY S. WILSON[1,2] & ANDREW P. ROBERTS[3]

[1] *Research School of Earth Sciences, Victoria University of Wellington, P.O. Box 600, Wellington, New Zealand*
[2] *Present address: Department of Earth Sciences, University of Oxford, Parks Road, Oxford OX1 3PR, UK (e-mail: gary.wilson@earth.ox.ac.uk)*
[3] *School of Ocean and Earth Science, University of Southampton, Southampton Oceanography Centre, European Way, Southampton SO14 3ZH, UK*

Abstract: In many sedimentary environments, early diagenetic sulphidization reactions cause extensive dissolution of detrital magnetic minerals which can alter or destroy the palaeomagnetic signal. Pliocene sediments from the Wanganui basin, North Island, New Zealand, have been subjected to multiple cycles of erosion and redeposition in sulphate-reducing environments before reaching their present setting. Despite this, the sediments retain a measurable and stable remanent magnetization. The only Fe–Ti oxide recognized from detailed sedimentary petrographic characterization of magnetic extracts and bulk sediments is ilmenite, with compositions that lie within the range from which paramagnetic behaviour is expected. There are no Fe–Ti oxide grains with compositions from which ferrimagnetic behaviour is expected (except for rare chromite). A direct link was made between magnetic behaviour and mineralogy by conducting electron probe microanalysis and back-scattered electron imaging of the same grains that were subjected to magnetic analysis. These studies demonstrate that stable magnetic behaviour is displayed by grains that appear to be homogeneously composed of ilmenite. This magnetic behaviour is attributed to the presence of ferrimagnetic Fe-enriched (hemo-ilmenite) microstructural domains that have been previously reported from ilmenite grains. These hemo-ilmenite domains appear to be dominantly responsible for the remanence of the Wanganui basin Pliocene sediments. Calculations of the reactivity of detrital iron-bearing minerals to sulphidization indicate that ilmenite is much less reactive than magnetite and other detrital magnetic minerals. Given the resistance of hemo-ilmenite, to dissolution, it is suggested that microstructural ferrimagnetic domains within ilmenite grains may be a source of palaeomagnetic information in sediments that have undergone diagenetic magnetic mineral dissolution during multiple episodes of erosion and redeposition.

Thick, generally rapidly deposited, Neogene and Quaternary marine sedimentary sequences have been uplifted above sea level and now crop out over large parts of New Zealand. Extensive studies of such sequences over at least the last half-century have allowed the development of a biostratigraphic framework that now serves as a standard reference for temperate southern latitudes (Hornibrook *et al.* 1989). Outcrop-based studies of rapidly deposited sequences can provide an unequalled perspective for sequence stratigraphic studies (e.g. Carter *et al.* 1991; Abbott & Carter 1994), for high-resolution studies of palaeoclimate and eustasy (e.g. Wright & Vella 1988; Kamp & Turner 1990; Turner & Kamp 1990; Pillans *et al.* 1994; Roberts *et al.* 1994), as well as for tectonic studies (e.g. Wright & Walcott 1986; Roberts 1992; Wilson & McGuire 1995). Correlation among coeval successions, and with the international time scale, is a vital part of such studies.

Magnetostratigraphy has been long recognized as a powerful tool for global correlation, but early studies of sediments were hampered by the insensitivity of magnetometers and their generally weak magnetization of sediments (see Kennett 1980). Despite this, magnetostratigraphic studies were completed for several important sequences in New Zealand (e.g. Kennett *et al.* 1971; Lienert *et al.* 1972; Kennett & Watkins 1974; Seward *et al.* 1986). With the development of cryogenic magnetometers and the widespread use of more rigorous schemes for interpreting demagnetization data, further magnetostratigraphic studies have been completed (e.g. Wright & Vella 1988; Turner & Kamp 1990; Pillans et al. 1994; Roberts et al. 1994). However, the detrimental impact of pervasive reductive diagenesis on the magnetic records carried by these sediments has been recognized only relatively recently (e.g. Roberts & Pillans 1993; Roberts & Turner 1993) and must be taken into account in interpreting palaeomagnetic results from such sequences. Rapidly accumulating continental shelf sediments generally undergo sulphate reduction during shallow burial, as a result of degradation of organic matter (e.g. Berner 1984; Canfield & Berner 1987). Dissolution of detrital Fe-Ti oxide minerals has been widely documented in magnetic studies of marine sediments (e.g. Kobayashi & Nomura 1972; Karlin & Levi 1983, 1985; Canfield & Berner 1987; Channell & Hawthorne 1990; Karlin 1990a,b; Leslie et al. 1990a,b; Roberts & Pillans 1993; Roberts & Turner 1993; Tarduno 1994). Most of the marine sediments that now crop out above sea level in New Zealand accumulated rapidly under sulphate-reducing conditions, and widespread dissolution of detrital magnetic phases is the most likely explanation for the weak magnetizations of these sediments (Roberts & Pillans 1993; Roberts & Turner 1993). The problem is exacerbated by the fact that these sediments, as well as most of the sedimentary units from which the sediments were derived, have been recycled through successive episodes of erosion and deposition (MacKinnon 1983) in sulphate-reducing environments (Smale 1990). The result is that a large portion of the available iron from detrital iron-bearing grains has been dissolved to form authigenic pyrite (see Berner 1970, 1984), thereby obliterating a large part of the detrital magnetic signal. Few studies have been conducted to determine the origin of the palaeomagnetic signal in sediments subjected to dissolution during multiple cycles of erosion and redeposition. A rock magnetic and sedimentary petrographic study has therefore been undertaken to determine the origin of the palaeomagnetic signal in this Pliocene sequence. These results have important implications for future studies of such sequences in New Zealand, as well as relevance to other sequences where the magnetic mineral fraction is strongly influenced by reductive diagenesis.

Geological setting

The Wanganui basin lies in a back-arc position with respect to the Australia-Pacific plate boundary zone. The basin is submerged in the south and has been tectonically uplifted above sea level in its northern and eastern parts, in response to uplift of the adjacent axial ranges of the North Island (Fig. 1). The basin has an area of $c.\,40\,000\,\text{km}^2$ and contains a stratigraphic thickness of about 6 km of sediment that ranges from upper Miocene age in the north to present-day sediments that onlap to the south of the basin (Anderton 1981). Stern et al. (1992) described the basin as a lithospheric downwarp that resulted from frictional coupling at the interface of the underlying subduction thrust. The portion of the basin that lies above sea level is deeply dissected by rivers, which permits access to a nearly continuously exposed sequence of mid-outer shelf to shoreface sediments. The Jurassic-Cretaceous greywackes and argillites of the Torlesse Supergroup are the dominant source of siliciclastic sediments deposited in the marginal basins of New Zealand. Widespread early diagenetic pyritization and dissolution of detrital iron-bearing phases has been documented in Torlesse rocks (e.g. Smale 1990), which were subjected to multiple cycles of erosion and redeposition during Mesozoic time (MacKinnon 1983). Further cycles of erosion and redeposition occurred in the Wanganui basin during Neogene time (e.g. Stern et al. 1992), which led to dissolution of much of the available iron in detrital magnetic grains. Development of arc-related volcanism in Quaternary time in the Central Volcanic Region of the North Island (Fig. 1) provided a ready source of volcaniclastic magnetic minerals that give rise to a generally stronger palaeomagnetic signal in Pleistocene sediments of the Wanganui basin (Turner & Kamp 1990; Roberts & Pillans 1993; Pillans et al. 1994).

Methods

About 360 conventional palaeomagnetic cores (25 mm diameter) were collected from 114 sites that crop out in the Wanganui and Rangitikei Rivers (Fig. 1), to determine an early to middle Pliocene magnetostratigraphy (Wilson 1993). Where possible, cores were drilled from fresh, unweathered mudstones, sandy mudstones or muddy sandstones. Magnetic remanence measurements were made using a cryogenic magnetometer. Thermal demagnetization has repeatedly been shown to be more suitable than alternating field (AF) demagnetization for isolating characteristic remanence components in Neogene marine sequences in New Zealand (Kennett & Watkins 1974; McGuire 1989; Turner et al. 1989; Turner & Kamp 1990; Roberts & Turner 1993; Pillans et al. 1994; Roberts et al. 1994). At least three samples from each site were subjected to stepwise thermal demagnetization at 50°C intervals, usually to a maximum of 300-350°C. Above this temperature, the natural remanent magnetization (NRM)

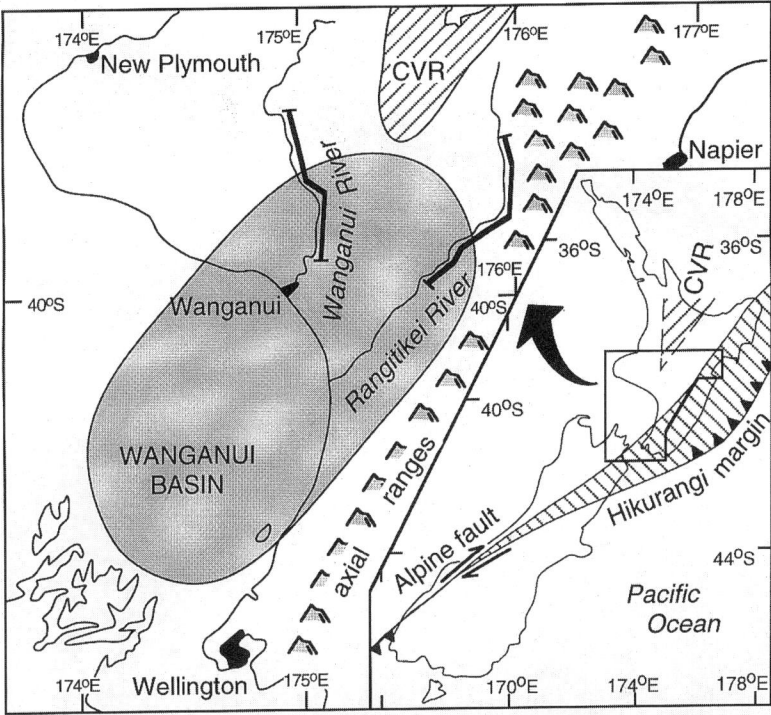

Fig. 1. Location map of Wanganui basin, New Zealand. The Wanganui River and Rangitikei River sections were sampled for this study. The tectonic setting of the New Zealand plate boundary zone is shown in the inset. The Wanganui basin lies to the west of the Hikurangi margin and axial ranges. CVR, Central Volcanic Region.

was greatly reduced and thermochemical alteration masked the remaining NRM.

Optical microscope and scanning electron microscope examinations of resin-impregnated polished sections of bulk sediment and magnetic separates, including electron probe microanalysis (EPMA) and back-scattered electron imaging (BEI), were performed to identify mineral types and proportions in sediments from the Wanganui basin and to relate the mineralogy to the observed magnetic behaviour. Magnetic separates were obtained by mechanically and ultrasonically disaggregating the sediment into a slurry with deionized water. The disaggregated sediment was then dried and sieved, and grain-size specific separates were prepared using a Frantz electromagnetic separator. A second set of separates was obtained by recirculating a slurry of disaggregated sediment past a permanent magnet with a peristaltic pump, using a device similar to that described by Petersen *et al.* (1986). Subsamples from the first set of separates were suspended within non-magnetic Plexiglas cylinders. Isothermal remanent magnetizations (IRMs) were imparted using a pulse magnetizer (up to peak fields of 800 mT) to assess the magnetic characteristics of the samples. After saturation was achieved, stepwise increasing backfields were applied to determine the coercivity of remanence (B_{cr}). Polished grain mounts were then made out of the cylinders to directly determine the mineralogy of the grains that gave rise to the observed remanence. Grains from the second set of separates were subjected to magnetic hysteresis analysis up to maximum fields of 1 T. Subsamples from these separates were subjected to temperature-dependent susceptibility measurements. Polished sections were also made from the same grains from which hysteresis measurements were made, to allow direct determination of the magnetic minerals responsible for the observed hysteresis behaviour.

Results

Palaeomagnetic behaviour

During thermal demagnetization, much of the signal was lost after heating to 150°C or 200°C. A stable linear remanence component was often observed between 150°C and 300°C, until thermal alteration prevented acquisition of useful data (Fig. 2). A weak remanence component commonly remained above 300–350°C. This component was inferred to be present in many samples because the demagnetization data were not directed to the origin of the vector component plot (Fig. 2c and d). Stable characteristic

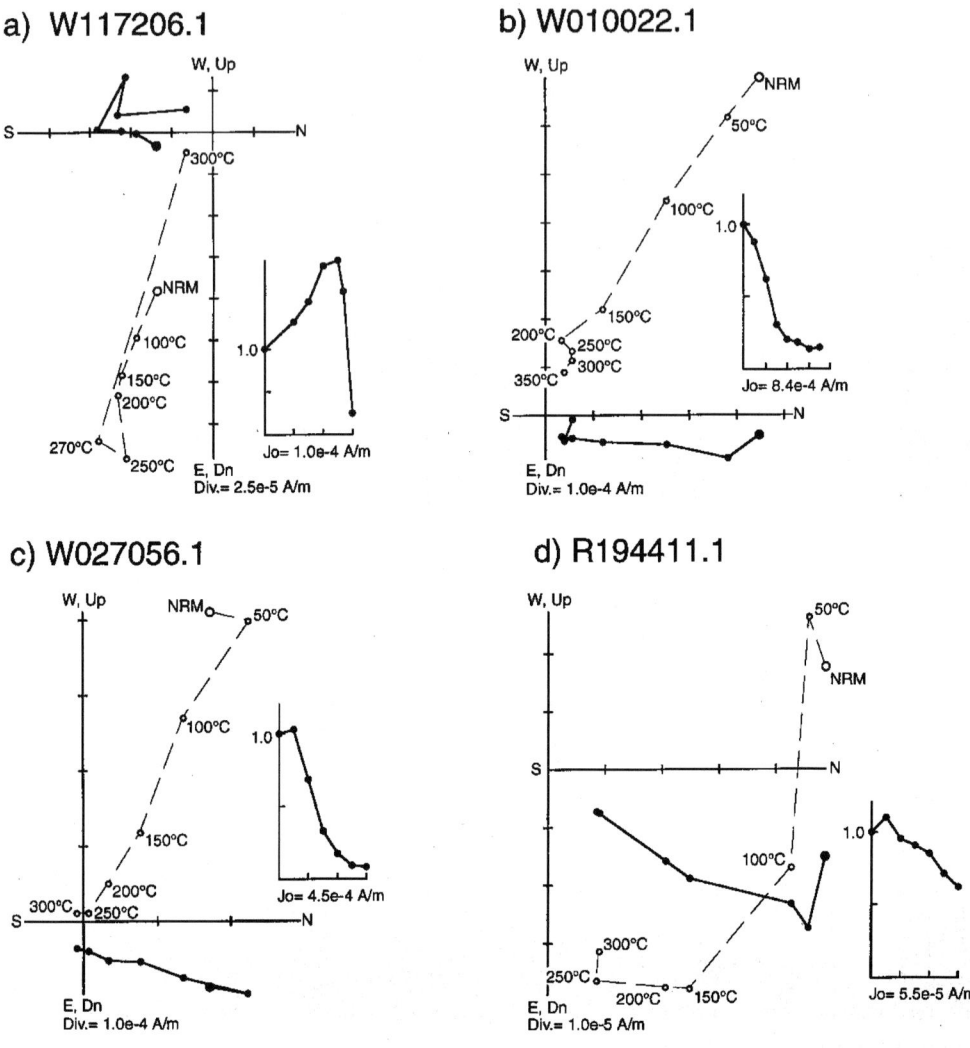

Fig. 2. Orthogonal vector component plots of thermal demagnetization results for representative samples from the Wanganui basin. Filled (open) symbols indicate projections onto the horizontal (vertical) plane. (**a**) Reversed polarity sample from the Wanganui River section; (**b** and **c**) normal polarity samples from the Wanganui River section; (**d**) reversed polarity sample from the Rangitikei River section.

Fig. 3. Photomicrographs of pyrite and ilmenite grains. (**a**) Reflected light photomicrograph (crossed nicols) of pyrite framboids (brassy yellow) in a quartzofeldspathic muddy sandstone from the Wanganui basin. Field of view is 80 μm. (**b**) Reflected light photomicrograph (crossed nicols) of a large ilmenite grain (dull metallic yellow) from the Wanganui basin, with solution weathering and pitting along weakness planes in the grain. The grain is 90 μm in height. (**c**) Back-scattered electron image of ilmenite grains in a magnetic extract from a sandy mudstone from the Wanganui basin with exsolution pitting (right-hand side) and weathering along cleavage and fracture planes (left-hand side). Field of view is 130 μm in height. (**d** and **e**) Reflected light photomicrographs (crossed nicols) of rims of microcrystalline hematite around pyrite framboids in muddy sandstones from the Wanganui basin. Field of view is 80 μm in height. (**f**) Back-scattered electron image of chromite grain (bright blue–white). The chromite is a late phase igneous mineral and is holding together several feldspar grains. The chromite grain is 90 μm in long dimension.

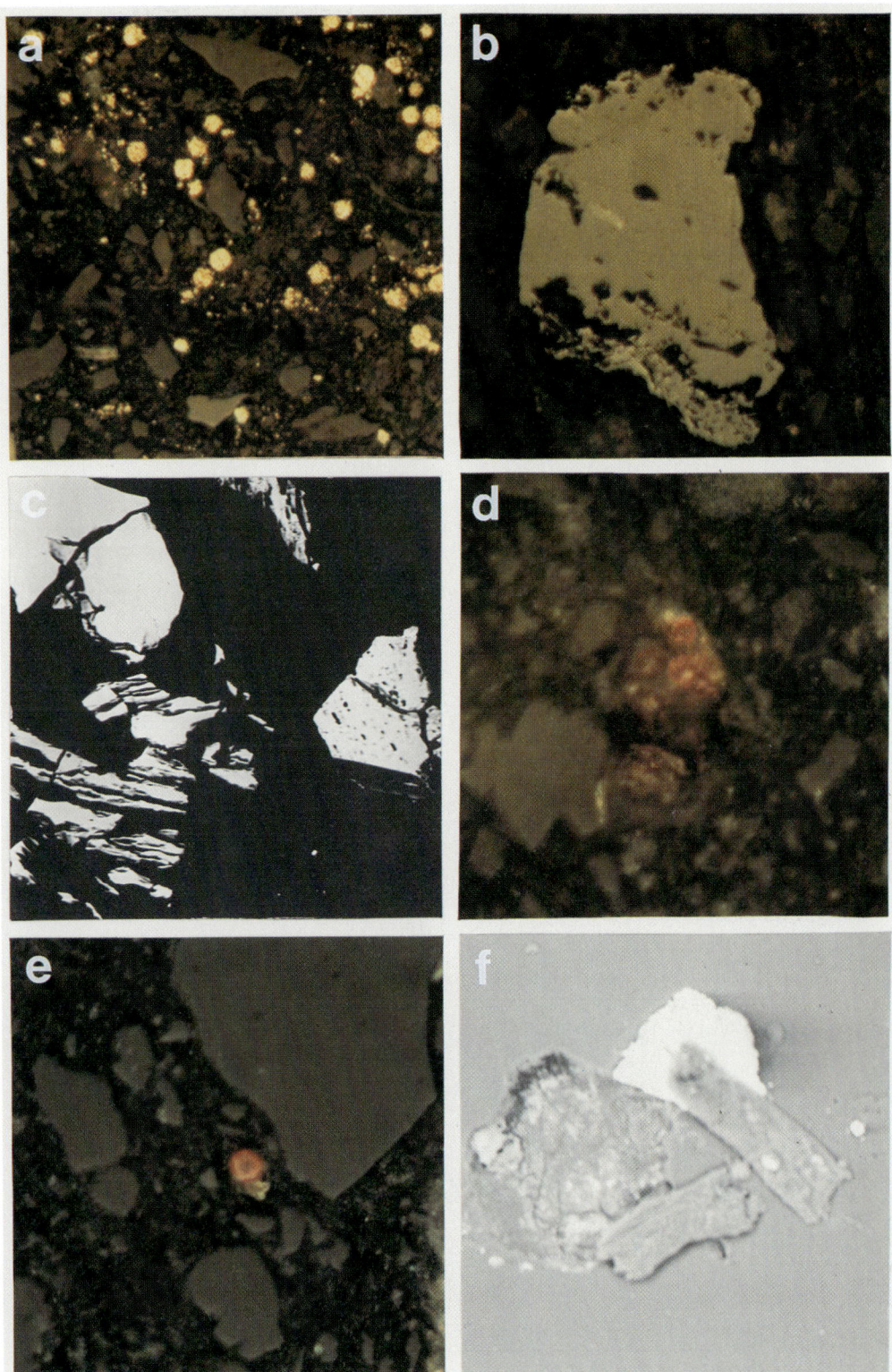

remanence components were isolated from 47% of the sample collection, and characteristic remanence directions were obtained from another 48% of the samples by means of remagnetization great circles analysis (Wilson 1993; Wilson & McGuire 1995).

Sedimentary petrography

Identification of stable remanence-bearing particles from thin sections of sediments is usually not feasible because the size and concentration of such grains is small. Sedimentary petrography can be extremely valuable, however, because it provides important information concerning grain types and concerning sediment diagenesis and authigenesis. Optical microscope observations indicate that 5–10% of the grains are opaque. Pyrite is common (Fig. 3), and is particularly abundant within foraminiferal tests. It occurs as framboids and as euhedral grains: individual grains are smaller than 5 μm and they have a brassy yellow colour with metallic lustre. No other iron sulphide phases were identified. The pyrite (FeS_2) appears to be of uniform composition under BEI. Iron to sulphur ratios, measured using EPMA, vary between 1:1.91 and 1:2.12 (Table 1). The abundance of pyrite indicates that these sediments were deposited under sulphate-reducing conditions (see Berner 1970, 1984). Dissolution of detrital iron-bearing magnetic minerals should be ubiquitous in sediments that support active sulphate reduction and H_2S formation (Canfield & Berner 1987). Evidence for dissolution of detrital iron-bearing minerals is clear, particularly in ilmenite grains

Table 1. *EPMA analyses of individual pyrite grains*

Fe	46.02	46.28	46.62	46.85	45.16	47.52	47.1	46.25	44.17	45.06	45.66
Mn	0.01	0	0	0.03	0.35	0.03	0.15	0.07	0.49	0.02	0
Cu	0.02	0.09	0.01	0.03	0.05	0.01	0	0.02	0.05	0.02	0.14
Sb	0.02	0	0.03	0.03	0.06	0	0.04	0	0.01	0.05	0.02
As	0.02	0.09	0.15	0.07	0.19	0.09	0.09	0.09	0.12	0.08	0.12
S	53.25	52.46	51.59	53.72	52.83	52.12	52.48	53.69	52.9	52.78	52.27
Total	99.54	98.91	98.38	100.7	98.64	99.77	99.87	100.1	97.73	98	98.2
Fe:S	1:2.02	1:1.97	1:1.93	1:2.00	1:2.04	1:1.91	1:1.94	1:2.02	1:2.09	1:2.04	1:1.99
Fe	44.56	45	45.7	46.14	45.91	45.49	46.67	45.36	46.41	46.49	46.42
Mn	0.48	0.39	0.11	0.04	0.05	0.35	0.02	0.04	0.01	0.03	0.18
Cu	0.04	0.07	0.05	0.12	0.07	0.09	0	0.05	0.02	0.02	0.05
Sb	0.01	0.03	0.03	0.01	0.01	0	0.01	0	0	0.01	0.02
As	0.1	0.13	0.12	0.19	0.07	0.14	0.08	0.02	0.06	0.12	0.16
S	54.29	53.05	52.58	53.51	53.69	53.39	53.08	53.06	52.35	52.32	52.21
Total	99.48	98.66	98.59	100	99.8	99.45	99.85	98.53	98.85	98.98	99.04
Fe:S	1:2.12	1:2.05	1:2.00	1:2.02	1:2.04	1:2.04	1:1.98	1:2.04	1:1.96	1:1.96	1:1.96

Table 2. *EPMA analyses of single detrital chromite grains*

SiO_2	0.1	0.13	0.11	0.05	0.18	2.51	0.01
TiO_2	0.19	0.35	0.13	0.3	0.13	0.11	0.01
Al_2O_3	24.86	8.46	10.76	24.08	25.49	11.77	29.19
FeO	21.82	26.98	27.29	23.92	19.32	24.32	16.22
Fe_2O_3*	2.27	13.06	15.74	0.46	7.1	20.35	0.89
MnO	0.42	0.58	0.55	0.36	0.18	2.51	0.21
MgO	8.95	2.35	3.8	8.81	11.04	7.31	13.02
CaO	0.11	0.07	0.02	0.01	0.1	0.07	0.03
Cr_2O_3	41.68	41.78	42.15	47.94	37.32	34.63	39.89
NiO	0.11	0.19	0.08	0.08	0.17	0.04	0.21
ZnO	0.44	0.56	0.41	0.48	0	0	0
Total	101	94.51	101.1	106.5	101	103.6	99.68
(total FeO	23.86	38.73	41.46	24.33	25.71	42.63	17.02)
% Ulvöspinel	77.84	19.29	5.9	99.34	17.86	2.37	47.43

* Recalculated after Stormer (1983).

(within magnetic extracts), which contain evidence of exsolution pitting and dissolution along cleavage and fracture planes (Fig. 3b and c). During later diagenesis, framboidal pyrite grains were oxidized, and abundant microcrystalline hematite formed on the surface of the pyrite grains (Fig. 3d and e). Rare chromite was also identified in sediments from the Wanganui basin (Table 2), and is generally associated with aluminosilicate grains (Fig. 3f).

No spinel phase iron oxide (ulvöspinel–magnetite solid solution series) grains were observed (except for rare chromite grains). However, rhombohedral iron oxide (ilmenite–hematite solid solution series) grains were abundant within magnetic extracts. All such rhombohedral grains contain 46.7–52.3% TiO_2 and up to 9.4% MnO (Table 3a and b). Recalculation of the Fe^{2+} and Fe^{3+} ratios was carried out using the method of Stormer (1983), which indicates that the compositions of these grains are close to that of ilmenite ($Ilm_{88-99\%}$, $Hem_{12-1\%}$; Fig. 4).

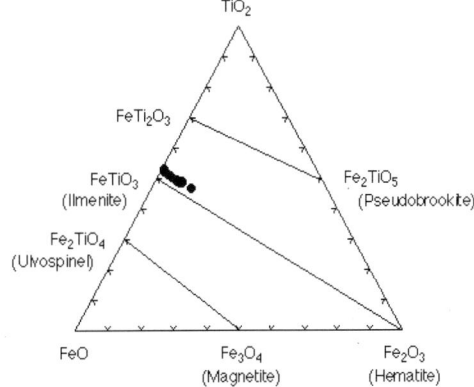

Fig. 4. The iron–titanium oxide system with compositions of ilmenite grains from the Wanganui basin measured by electron probe microanalysis. Oxide compositions were determined after Stormer (1983). Only iron and titanium components are plotted, although individual grains contained up to 9.4% MnO.

Table 3a. *EPMA analyses of homogeneous individual ilmenite grains*

SiO_2	0.09	0.09	0.04	0.01	0.12	0.14	0.06	0	0	0
TiO_2	48.83	52.5	49.99	49.84	52.5	50.41	49.42	50.25	52.43	49.51
Al_2O_3	0.05	0.05	0.05	0	0.16	0.1	0.14	0.05	0.06	0.07
FeO	41.4	39.53	41.94	38.37	36.78	38.7	42.43	43.56	45.65	42.14
Fe_2O_3*	7.09	0.49	6.83	6.48	0.92	4.43	7.96	2.59	0.83	0.88
MnO	2.09	6.58	2.72	5.83	9.35	6.44	2.04	1.38	1.37	1.96
MgO	0.25	0.07	0.11	0.22	0.04	0.06	0.08	0.11	0.01	0.2
CaO	0.04	0.71	0.06	0.06	0.7	0.06	0.05	0.03	0.01	0.03
NiO	0	0.1	0.04	0.09	0.14	0.1	0.01	0	0.08	0
Total	99.84	100.1	101.7	100.9	100.7	100.4	102	97.97	100.4	100.8
(total FeO	47.78	39.96	48.08	44.2	37.61	42.68	49.39	45.89	46.3	48.33)
% Ilmenite	93.04	99.5	93.39	93.43	99.02	95.46	92.37	97.33	99.2	93.33

Table 3b. *EPMA analyses of homogeneous individual ilmenite grains from magnetic extracts*

SiO_2	0.06	0.03	0.05	0.04	0.06	0.07	0.07	0	0.04	0.01
TiO_2	50.71	49.68	46.7	49.7	48.14	49.11	49.25	49.45	49.01	51.63
Al_2O_3	0.01	0.04	0.07	0.07	0.11	0.19	0.07	0.08	0.05	0.06
FeO	43.39	43.61	40.61	41.72	40.42	38.89	41.89	38.98	42.18	45.46
Fe_2O_3*	3.41	6.7	11.87	5.93	8.2	7.13	6.16	7.23	7.99	3.6
MnO	1.84	0.87	0.35	2.58	2.48	5.04	2.01	4.96	0.68	0.6
MgO	0.02	0.06	0.6	0.11	0.11	0.1	0.25	0.05	0.68	0.16
CaO	0.02	0.04	0	0.05	0	0.06	0	0.29	0.03	0.07
NiO	0.04	0.06	0.02	0.15	0.24	0	0	0	0	0
Total	99.68	101.1	100.3	100.4	99.76	100.6	99.7	101	100.7	101.6
(total FeO	46.46	49.64	51.29	47.06	47.8	45.3	47.43	45.49	49.37	48.70)
% Ilmenite	96.66	93.61	88.55	94.18	91.9	92.82	93.96	92.75	92.3	96.59

* Calculated after Stormer (1983).

Compositions more ilmenite rich than $Ilm_{73\%}$, $Hem_{27\%}$ within the hematite–ilmenite solid solution series fall within the range of values that are expected to be paramagnetic at room temperature (see Nagata 1961), therefore one would not expect the observed grains to be able to carry a remanence. It should be noted that ilmenite has been widely documented in magnetic extracts from similar New Zealand marine sequences (Roberts & Pillans 1993; Roberts & Turner 1993), although the common presence of ilmenite has been attributed to the relative insensitivity of magnetic separation techniques to discriminate adequately between ferrimagnetic and paramagnetic phases.

Rock and mineral magnetic studies

IRM acquisition and back-field experiments were conducted to compare the magnetic properties of bulk sediment samples (e.g. Fig. 5a) with those of magnetic extracts. The IRM was weak and saturates at low inductions (<200 mT), which suggests that the magnetic carriers occur in low concentrations and have low coercivities. Saturation at fields below 300 mT is consistent with the dominance of ferrimagnetic grains. The magnetic extracts contain higher concentrations of magnetic grains, which therefore give rise to higher IRM intensities, but the range of B_{cr} values is similar to those of the bulk sediment samples (Fig. 5), suggesting that the extracts are representative of the magnetic grains that are responsible for the magnetization. To directly compare mineralogy and magnetic properties, magnetic extracts, obtained using a Frantz electromagnetic separator, were suspended within Plexiglas cylinders using a non-magnetic glue as a fixative and were subjected to IRM acquisition and back-field demagnetization studies. A blank sample of Plexiglas was subjected to the same analysis to confirm that the blank did not hold a measurable remanence. Results from the extracts were consistent with the magnetic properties of bulk sediment samples. The samples were then made into polished grain mounts to allow direct mineralogical identification of the remanence-bearing grains. EPMA and BEI analyses indicated that all samples contained only grains with compositions close to ilmenite (Table 3a and b, Figs 3c and 4). BEI analysis indicated that the grains (typically >40 μm in size) are compositionally homogeneous on the scale of the observations (the electron beam is about 1 μm in diameter). Because of uncertainty as to whether small concentrations of undetected ferrimagnetic particles were giving rise to the observed magnetic

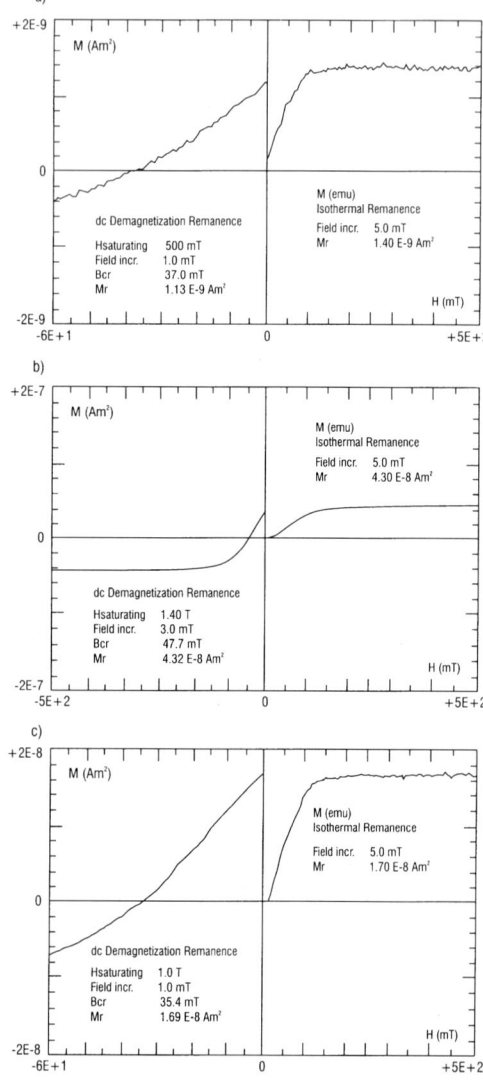

Fig. 5. IRM acquisition and back-field demagnetization curves. (**a**) Bulk sediment (W069) from the Wanganui River section; (**b**) magnetic extract (from W069); (**c**) magnetic extract (from W085) from the Wanganui River section.

behaviour, further analyses were conducted on magnetic separates that were obtained using the method of Petersen et al. (1986). The grains obtained by this separation technique were dominantly large (>40 μm), which meant that magnetic measurements could be made directly on individual grains using a Micromag alternating gradient magnetometer. The single grains display magnetic behaviour that is indistinguishable from that of the bulk magnetic separates

(Fig. 6) and EPMA analyses of carefully prepared polished grain mounts confirmed that the grains consist of ilmenite.

Hysteresis parameters (M_r/M_s and B_{cr}/B_c) are consistent (Fig. 6) with those repeatedly observed in marine sedimentary rocks (e.g. Roberts & Pillans 1993; Tarduno 1994). The hysteresis behaviour of the bulk sediment is not identical to that of the magnetic extract. This is probably due to contributions from (super)paramagnetic phases that are present in the bulk sediment but which are absent in the extracts. Using the grain-size dependent framework of Day et al. (1977), which is strictly applicable only to magnetite or titanomagnetite compositions, the hysteresis data fall in the field for pseudo-single-domain grains. Temperature-dependent susceptibility measurements were made on bulk sediment samples, as well as on magnetic extracts, to obtain more information concerning magnetic mineralogy. As expected, low-field magnetic susceptibilities for most bulk samples are weak. On heating, no Hopkinson peaks are evident, but susceptibility decreases to near-zero values at about 580°C (Fig. 7a and b), which is consistent with the presence of magnetite (see Hunt et al. 1995). On cooling, susceptibilities are much higher, which is consistent with thermochemical alteration of iron-bearing minerals (clays or pyrite). Analyses of magnetic extracts produced noisy results because of the weak susceptibility; however, a decay in susceptibility is evident between 580 and 610°C (Fig. 7c), which is nearly reproducible on cooling, probably because of the absence of thermally unstable phases such as clays and pyrite grains within the magnetic extracts.

There are several anomalies in the above-described data. First, the magnetic behaviour is consistent with that expected for ferrimagnetic particles that are known to be faithful recorders of the geomagnetic field on geological time scales. However, the mineralogy of the same grains is demonstrably close to an ilmenite composition, which would be expected to be paramagnetic at ambient temperatures, and, therefore, would be unable to retain a remanent magnetization on any time scale. Second, the grains that displayed this anomalous magnetic behaviour are usually large (>40 μm), and such

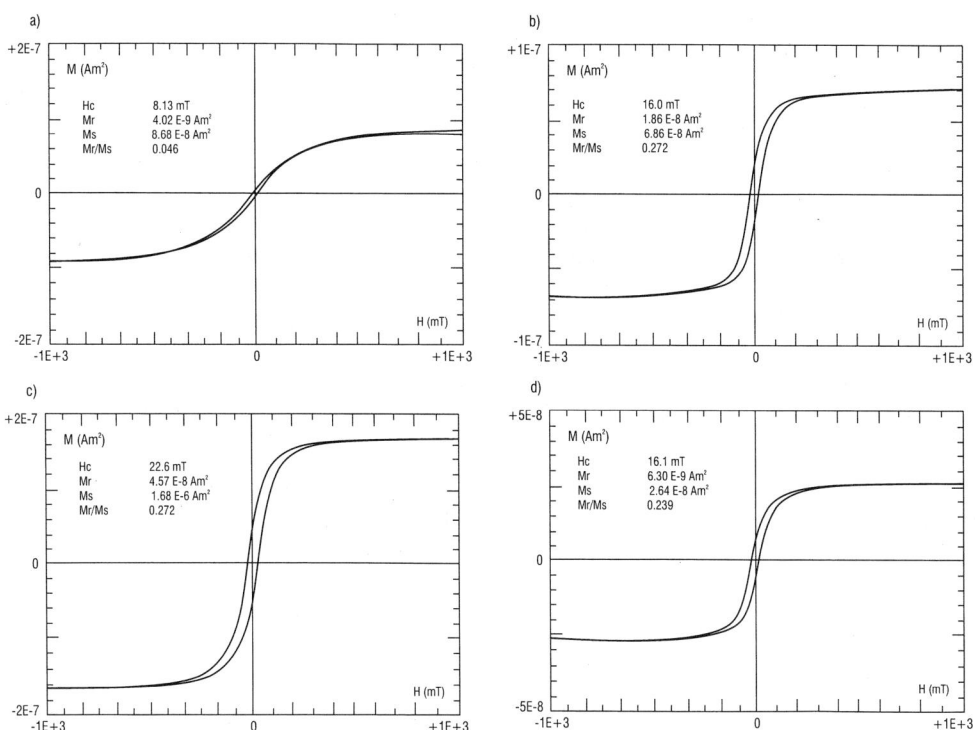

Fig. 6. Hysteresis loops for bulk samples and extracts. (**a**) Bulk sediment sample (W085); (**b**) magnetic extract (from W085); (**c**) magnetic extract (from W069); (**d**) single ilmenite grain (from W069), from the Wanganui River section.

large grains would normally be expected to display multi-domain behaviour, which is not what is indicated by the hysteresis data (Fig. 6).

Third, grains with apparently paramagnetic compositions display thermomagnetic behaviour that is consistent with the presence of magnetite.

Discussion

Reconciling the apparent conflict between magnetic behaviour and mineralogical identifications

Data from magnetic extracts and bulk samples point to the unexpected result that grains with apparently paramagnetic compositions are contributing the dominant ferrimagnetic signal in Pliocene sediments of the Wanganui basin. This observation has been validated by making careful magnetic and mineralogical observations from the same grains, as outlined above. It is therefore concluded that this result is a genuine reflection of the magnetic mineralogy of the Wanganui basin Pliocene sediments and that this conflict needs to be reconciled. Lawson & Nord (1984) synthesized well-crystalline, optically homogeneous $Ilm_{80\%}$, $Hem_{20\%}$ samples with an average grain size of $40\,\mu m$ that falls within the range of compositions that is classically considered to be paramagnetic at room temperature. They found that these particles carry a measurable and stable remanence, which they attributed to the presence of Fe-enriched transformation-induced microstructural domain boundaries that arise during the ordering transition when the materials originally cooled from high temperatures (Lawson & Nord 1984). Dark-field transmission electron microscopy indicated that these Fe-enriched zones are variable, but submicron, in size (Fig. 8), and that they may be as small as a few atoms across in some cases (Lawson et al. 1981; Lawson & Nord 1984; Nord & Lawson 1989). The possibility of a self-reversed thermal remanent magnetization (TRM) within ferrian ilmenites (Ishikawa & Syono 1963; Allen & Shive 1974; Lawson et al. 1981; Lawson & Nord 1984; Nord & Lawson 1989; Hoffman 1992) is not directly relevant in the present study because, as detrital particles, the dominant remanence component, which might have been originally self-reversed, would align with the ambient geomagnetic field in the depositional environment. Although the grains identified as the likely remanence carriers in this study have somewhat higher Ti (and Mn) contents than those in the above-cited studies, the magnetic properties of the ilmenite grains in the Wanganui basin sediments can be attributed to a similar mechanism to that discussed by Lawson & Nord (1984) and Nord & Lawson

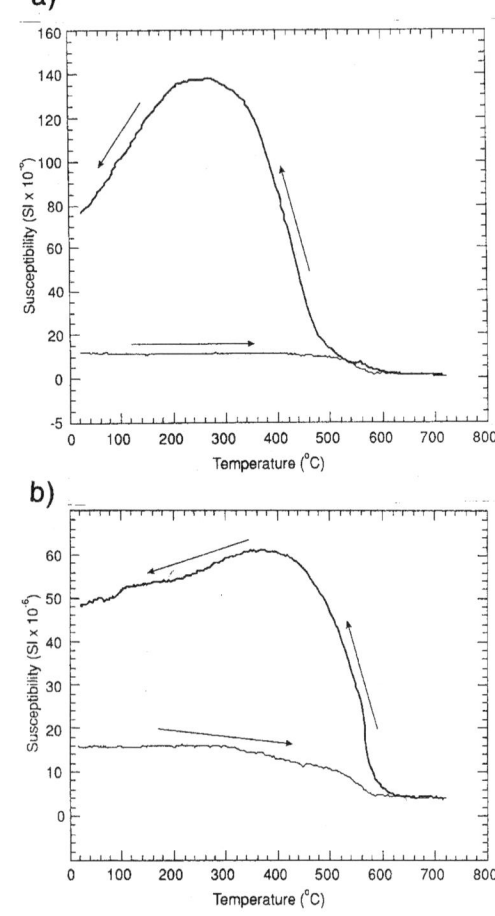

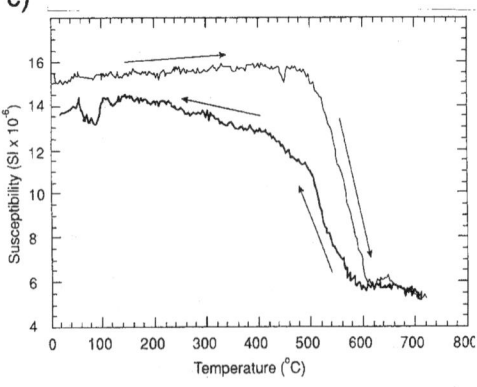

Fig. 7. High-temperature susceptibility results. (**a**) muddy sandstone; (**b**) sandy mudstone; (**c**) a magnetic extract from the Wanganui basin.

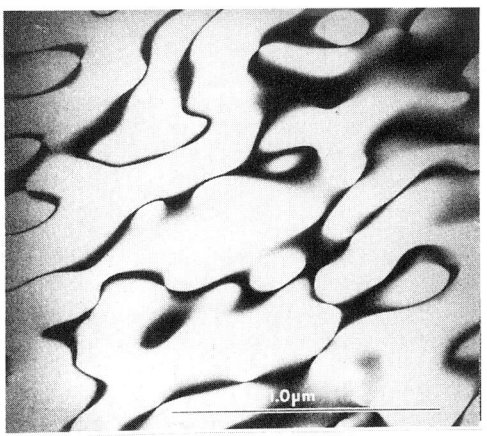

Fig. 8. Dark-field image of transformation-induced domains and domain boundaries. The transformation-induced domains are the light regions (Ti enriched) and the domain boundaries are dark (Fe enriched) in a sample of $Ilm_{80}Hem_{20}$ (from Lawson & Nord 1984).

(1989). This interpretation is preferred to the possibility that the magnetic properties are due to discrete inclusions of magnetic iron oxides, because such inclusions were not observed under BEI. EPMA results (Table 3a and b) indicate a near end-member ilmenite composition because this is the average composition detected across the width of the 1 μm electron beam, as was observed by Lawson & Nord (1984). Smaller, Fe-rich zones were detected by Lawson & Nord (1984) and Nord & Lawson (1989) only when dark-field transmission electron microscopy was used. The hysteresis properties measured in this study (Fig. 6) are also consistent with the observation of stable remanences in similarly large grains ($c. 40\,\mu$m) by Lawson & Nord (1984). These properties are best explained by the presence of submicron Fe-enriched zones within the large grains. Relatively low values of M_r/M_s of $c. 0.26$ (Fig. 6) probably result from a mixture of single-domain (SD) and superparamagnetic (SP) grains within these submicron zones. If the photomicrograph of Lawson & Nord (1984) is taken as a guide (Fig. 8), one would expect grain sizes within these zones to vary within the normal range of SD and SP sizes (i.e. a few atoms thickness to hundreds of nanometres). There is probably a relatively broad range of sizes within the SP and SD range in these Fe-rich zones, although the range is not sufficient to give rise to irregularly shaped hysteresis loops (Fig. 6) (see Roberts et al. 1995; Tauxe et al. 1996). A distribution of grain sizes in the SD range can also account for the distributed unblocking temperatures observed in stepwise thermal demagnetization data (Fig. 2), as has also been suggested by Roberts & Pillans (1993). A distribution of such grains would produce an overlap between the coercivity spectra of the primary and secondary remanence components, which may explain the frequent failure of AF demagnetization to isolate primary remanence directions in similar sediments in New Zealand. Also, a narrow grain-size distribution would be expected to give a well-defined Hopkinson peak near the Curie temperature for the phase under investigation: the absence of Hopkinson peaks in the high-temperature susceptibility results (Fig. 7) may therefore be attributable to the distributed grain-size spectrum.

Lawson & Nord (1984) suggested that the Fe-enriched zones within such ilmenite grains are enriched in a phase that lies toward the hematite end of the solid solution series. An additional puzzle concerning the magnetic properties of the Wanganui ilmenite grains is that high-temperature results suggest that the Fe-enriched zones have thermomagnetic properties that are close to those expected for magnetite (Fig. 7). As ilmenite is a product of the solid solution series between hematite and ilmenite, it would be surprising if a product of another solid solution series (i.e. the ulvöspinel series) were present. Temperature-dependent susceptibility results display somewhat variable results, with susceptibility decreasing to zero in a range of temperatures between 580 and 610°C (Fig. 7). Susceptibility values that persist to 610°C in a magnetic extract (Fig. 7c) are important because they suggest that the Fe-enriched material within the ilmenite grains may lie within the ferrimagnetic range of the hemo-ilmenite series rather than in the ulvöspinel series. Temperature-dependent susceptibility results also preclude maghemite as the mineral in the Fe-enriched zones. The lower magnetic moment of hemo-ilmenites with respect to magnetite (see Nagata 1961) is also consistent with the generally weak magnetizations documented in the Wanganui basin. Given the above evidence, it is concluded that the magnetization of the Wanganui basin Pliocene sediments is dominated by ilmenite grains that contain submicron domains of a SD ferrimagnetic hemo-ilmenite phase.

Impact of diagenesis on magnetic mineral assemblages in redeposited sediments

Diagenesis has clearly had a major impact on the magnetization of Pliocene sediments in the

Wanganui basin. This is attested to by the abundance of authigenic pyrite (Fig. 3a) which formed at the expense of detrital iron-bearing minerals. Detailed magnetic and petrographic studies of bulk sediments, and of magnetic extracts, failed to identify any detrital magnetite or titanomagnetite grains. The only positively identified spinel phases were rare chromite grains (Fig. 3f). Ilmenite is the only abundant detrital Fe–Ti phase recovered in magnetic extracts and these grains have been extensively etched and pitted, presumably by sulphidic pore waters during early burial (Fig. 3b and c). Ilmenite is one of the least reactive iron-bearing minerals under diagenetic sulphide-reducing conditions; it is three orders of magnitude less reactive than magnetite and six orders of magnitude less reactive than hematite (Canfield et al. 1992). From these observations, it appears that all of the commonly occurring detrital magnetic minerals (magnetite, titanomagnetite and hematite) have been completely dissolved during sulphidization in repeated cycles of erosion and redeposition in the Wanganui basin (as well as in earlier cycles of erosion and redeposition in the parent Torlesse rocks). A significant amount of ilmenite has survived pyritization, as has also been documented in other studies of sediments that have undergone extensive diagenetic dissolution of magnetic iron-bearing minerals (e.g. Reynolds 1977; Roberts & Pillans 1993; Roberts & Turner 1993; Hounslow et al. 1995; Reynolds et al. in press). Chromite also appears to be resistant to dissolution, as reported elsewhere (e.g. Leslie et al. 1990a; Hounslow et al. 1995; Hounslow 1996). Chromite, with similar compositions to those from the Wanganui basin (Table 2), has been reported to carry a strong and stable remanence (Kumar & Bhalla 1984a, b; Refai et al. 1989). Given the ability of chromite to resist diagenetic dissolution, it may be responsible for the remanence in some Wanganui basin sediments. However, a dominant magnetic signal due to chromite is considered unlikely, because chromite grains are rare in magnetic extracts and there is no widespread source of chromite in rocks from which the Wanganui sediments were derived.

The dominant magnetic mineral in Pliocene sediments of the Wanganui basin therefore appears to be a ferrimagnetic hemo-ilmenite phase that occurs as microcrystalline domains within ilmenite grains. This possibility has not been widely considered because such particles are not normally considered to be a source of remanent magnetization in sediments. However, in environments such as the Pliocene sediments of the Wanganui basin, where all other magnetic detrital Fe–Ti oxide minerals have been dissolved, hemo-ilmenite domains within ilmenite grains might make a significant contribution to the sediment magnetism. The widespread occurrence of ilmenite grains within magnetic extracts in similar sediments from New Zealand (Roberts & Pillans 1993; Roberts & Turner 1993) may therefore be due to the presence of hemo-ilmenite domains. In addition to the dominant contribution of hemo-ilmenite domains within ilmenite grains, microcrystalline hematite has been identified on the surfaces of pyrite framboids in the Wanganui basin sediments (Fig. 3d and e). The hematite occurs on pyrite that did not form until early burial, therefore, the most likely time of hematite formation is during late-stage oxidation, probably after uplift when pore fluids may have been oxic. The hematite is extremely fine grained (Fig. 3d and e) and may be superparamagnetic, but an SD hematite fraction may be responsible for the weak high-temperature component that persists in many samples after thermal demagnetization up to 350°C (e.g. Fig. 2c).

Conclusions

Repeated magnetic and mineralogical measurements on magnetic grains from Pliocene marine sediments in the Wanganui basin indicate that the dominant carrier of remanent magnetization is hemo-ilmenite that occurs as submicron domains within large ilmenite grains. Ilmenite is much more resistant to dissolution during early diagenetic sulphidization than most ferrimagnetic Fe–Ti oxide minerals, which appear to have been completely dissolved. Ilmenite grains with compositions of those documented in this study are normally considered to be paramagnetic. The ferrimagnetic properties of these grains are attributed to microstructural inhomogeneities that arose during cooling of the original igneous source rock, following the mechanism described by Lawson & Nord (1984). Although this remanence mechanism is unexpected, and would probably be unrecognized in sediments that contain significant magnetite, it may be widespread in other sediments that have undergone extensive diagenetic modification as a result of prolonged sulphidization.

We thank C. Lawson and the American Geophysical Union for permission to reproduce Fig. 8. We are grateful to K. Verosub for use of facilities at the Paleomagnetism Laboratory, University of California, Davis, and to G. Turner and E. Broughton for assistance and for use of facilities at Victoria University, Wellington. R. Grapes gave invaluable assistance with microprobe analyses and interpretation. Funding for fieldwork was provided by the Victoria University Internal Grants Committee.

References

ABBOTT, S. T. & CARTER, R. M. 1994. The sequence architecture of mid-Pleistocene ($c. 1.1–0.4$ Ma) cyclothems from New Zealand: facies development during a period of orbital control on sea level cyclicity. *In*: DE BOER, P. L. & SMITH, D. G. (eds) *Orbital Forcing and Cyclic Sequences*. International Association of Sedimentologists, Special Publication, **19**, 367–394.

ALLEN, J. L. & SHIVE, P. N. 1974. Mössbauer effect observations of the 'x-phase' in the ilmenite-hematite series. *Journal of Geomagnetism and Geoelectricity*, **26**, 329–333.

ANDERTON, P. W. 1981. Structure and evolution of the South Wanganui Basin, New Zealand. *New Zealand Journal of Geology and Geophysics*, **24**, 39–63.

BERNER, R. A. 1970. Sedimentary pyrite formation. *American Journal of Science*, **268**, 1–23.

—— 1984. Sedimentary pyrite formation: an update. *Geochimica et Cosmochimica Acta*, **48**, 605–615.

CANFIELD, D. E. & BERNER, R. A. 1987. Dissolution and pyritization of magnetite in anoxic marine sediments. *Geochimica et Cosmochimica Acta*, **51**, 645–659.

——, RAISWELL, R. & BOTTRELL, S. 1992. The reactivity of sedimentary iron minerals toward sulfide. *American Journal of Science*, **292**, 659–683.

CARTER, R. M., ABBOTT, S. T., FULTHORPE, C. S., HAYWICK, D. W. & HENDERSON, R. A. 1991. Application of global sea level and sequence stratigraphic models in Southern Hemisphere Neogene strata from New Zealand. *In*: MACDONALD, D. I. M. (ed.) *Sedimentation, Tectonics and Eustasy: Sea Level Changes at Active Margins*. International Association of Sedimentologists, Special Publication, **12**, 41–65.

CHANNELL, J. E. T. & HAWTHORNE, T. 1990. Progressive dissolution of titanomagnetites at ODP Site 653 (Tyrrhenian Sea). *Earth and Planetary Science Letters*, **96**, 469–480.

DAY, R., FULLER, M. D. & SCHMIDT, V. A. 1977. Hysteresis properties of titanomagnetites: grain-size and compositional dependence. *Physics of the Earth and Planetary Interiors*, **13**, 260–267.

HOFFMAN, K. A. 1992. Self-reversal of thermoremanent magnetization in the ilmenite–hematite system: order–disorder, symmetry, and spin alignment. *Journal of Geophysical Research*, **97**, 10883–10895.

HORNIBROOK, N. DE B., BRAZIER, R. C. & STRONG, C. P. 1989. *Manual of New Zealand Permian to Pleistocene Foraminiferal Biostratigraphy*. New Zealand Geological Survey Paleontological Bulletin, **56**.

HOUNSLOW, M. W. 1996. Ferrimagnetic Cr and Mn spinels in sediments: residual magnetic minerals after diagenetic dissolution. *Geophysical Research Letters*, **23**, 2823–2826.

——, MAHER, B. A. & THISTLEWOOD, L. 1995. Magnetic mineralogy of sandstones from the Lunde Formation (late Triassic), northern North Sea: origin of the palaeomagnetic signal. *In*: TURNER, P. & TURNER, A. (eds) *Palaeomagnetic Applications in Hydrocarbon Exploration and Production*. Geological Society, London, Special Publication, **98**, 119–147.

HUNT, C. P., MOSKOWITZ, B. M. & BANERJEE, S. K. 1995. Magnetic properties of rocks and minerals, rock physics and phase relations. *American Geophysical Union Reference Shelf*, **3**, 189–204.

ISHIKAWA, Y. & SYONO, Y. 1963. Order–disorder transformation and reverse thermo-remanent magnetism in the $FeTiO_3–Fe_2O_3$ system. *Journal of the Physics and Chemistry of Solids*, **24**, 517–528.

KAMP, P. J. J. & TURNER, G. M. 1990. Pleistocene unconformity-bounded shelf sequences (Wanganui basin, New Zealand), correlated with global isotope record. *Sedimentary Geology*, **68**, 155–161.

KARLIN, R. 1990a. Magnetite diagenesis in marine sediments from the Oregon continental margin. *Journal of Geophysical Research*, **95**, 4405–4419.

—— 1990b. Magnetic mineral diagenesis in suboxic sediments at Bettis site W-N, NE Pacific Ocean. *Journal of Geophysical Research*, **91**, 4421–4436.

—— & LEVI, S. 1983. Diagenesis of magnetic minerals in Recent haemipelagic sediments. *Nature*, **303**, 327–330.

—— & —— 1985. Geochemical and sedimentological control of the magnetic properties of hemipelagic sediments. *Journal of Geophysical Research*, **90**, 10373–10392.

KENNETT, J. P. (ed.) 1980. *Magnetic Stratigraphy of Sediments*. Benchmark Papers in Geology Series, Dowden, Hutchison & Ross, **54**.

—— & WATKINS, N. D. 1974. Late Miocene–Early Pliocene paleomagnetic stratigraphy, paleoclimatology and biostratigraphy in New Zealand. *Geological Society of America Bulletin*, **85**, 1385–1398.

——, —— & VELLA, P. P. 1971. Paleomagnetic chronology of Pliocene–Early Pleistocene climates and the Plio-Pleistocene boundary in New Zealand. *Science*, **171**, 272–274.

KOBAYASHI, K. & NOMURA, M. 1972. Iron sulphides in the sediment cores from the Sea of Japan and their geophysical implications. *Earth and Planetary Science Letters*, **16**, 200–208.

KUMAR, A. & BHALLA, M. S. 1984a. Source of stable remanence in chromite ores. *Geophysical Research Letters*, **11**, 177–180.

—— & —— 1984b. Palaeomagnetism of Sukinda chromites and their geological implications. *Geophysical Journal of the Royal Astronomical Society*, **77**, 863–874.

LAWSON, C. A. & NORD, G. L., JR. 1984. Remanent magnetization of a 'paramagnetic' composition in the ilmenite–hematite solid solution series. *Geophysical Research Letters*, **11** 197–200.

——, ——, DOWTY, E. & HARGRAVES, R. B. 1981. Antiphase domains and reverse thermoremanent magnetism in ilmenite–hematite minerals. *Science*, **213**, 1372–1374.

LESLIE, B. W., HAMMOND, D. E., BURLESON, W. M. & LUND, S. P. 1990a. Diagenesis in anoxic sediments from the California continental borderland and its influence on iron, sulfur, and magnetite behavior. *Journal of Geophysical Research*, **95**, 4453–4470.

——, LUND, S. P. & HAMMOND, D. E. 1990b. Rock magnetic evidence for dissolution and authigenic growth of magnetic minerals within anoxic marine sediments of the California continental borderland. *Journal of Geophysical Research*, **95**, 4437–4452.

LIENERT, B. R., CHRISTOFFEL, D. A. & VELLA, P. P. 1972. Geomagnetic dates on a New Zealand Upper Miocene–Pliocene section. *Earth and Planetary Science Letters*, **16** 195–199.

MACKINNON, T. C. 1983. Origin of the Torlesse terrane and coeval rocks, South Island, New Zealand. *Geological Society of America Bulletin*, **94**, 967–985.

MCGUIRE, D. M. 1989. *Paleomagnetic stratigraphy and magnetic properties of Pliocene strata, Turakina River, North Island, New Zealand*. PhD thesis, Victoria University of Wellington.

NAGATA, T. 1961. *Rock Magnetism*. Maruzin, Tokyo.

NORD, G. L., JR. & LAWSON, C. A. 1989. Order–disorder transition-induced twin domains and magnetic properties in ilmenite–hematite. *American Mineralogist*, **74**, 160–176.

PETERSEN, N., VON DOBENECK, T. & VALI, H. 1986. Fossil bacterial magnetite in deep-sea sediments from the S. Atlantic Ocean. *Nature*, **320**, 611–615.

PILLANS, B. J., ROBERTS, A. P., WILSON, G. S., ABBOTT, S. T. & ALLOWAY, B. V. 1994. Magnetostratigraphic, lithostratigraphic and tephrostratigraphic constraints on Middle/Lower Pleistocene sea level changes, Wanganui basin, New Zealand. *Earth and Planetary Science Letters*, **121**, 81–98.

REFAI, E., WASSIF, N. A. & SHOAB, A. 1989. Stability of remanence and palaeomagnetic studies of some chromite ores from Barramiya and Allawi occurrences, Eastern Desert, Egypt. *Earth and Planetary Science Letters*, **94**, 151–159.

REYNOLDS, R. L. 1977. *Magnetic titanohematite minerals in uranium-bearing sandstones*. US Geological Survey Open File Report, **77-355**.

——, ROSENBAUM, J. G., SWEETKIND, D. S., LANPHERE, M. A., ROBERTS, A. P. & VEROSUB, K. L. Recognition of primary and diagenetic magnetizations to determine the magnetic polarity record and timing of deposition of the moat-fill rocks of the Oligocene Creede Caldera, Colorado. *Geological Society of America Special Paper*, in press.

ROBERTS, A. P. 1992. Paleomagnetic constraints on the tectonic rotation of the southern Hikurangi margin. *New Zealand Journal of Geology and Geophysics*, **35**, 311–323.

—— & PILLANS, B. J. 1993. Rock magnetism of Middle/Lower Pleistocene marine sediments, Wanganui basin, New Zealand. *Geophysical Research Letters*, **20**, 839–842.

—— & TURNER, G. M. 1993. Diagenetic formation of ferrimagnetic iron sulphide minerals in rapidly deposited marine sediments, New Zealand. *Earth and Planetary Science Letters*, **115**, 257–273.

——, CUI, Y. L. & VEROSUB, K. L. 1995. Wasp-waisted hysteresis loops: mineral magnetic characteristics and discrimination of components in mixed magnetic systems. *Journal of Geophysical Research*, **100**, 17 909–17 924.

——, TURNER, G. M. & VELLA, P. P. 1994. Magnetostratigraphic chronology of Late Miocene to Early Pliocene biostratigraphic and oceanographic events in New Zealand. *Geological Society of America Bulletin*, **106**, 665–683.

SEWARD, D., CHRISTOFFEL, D. A. & LIENERT, B. R. 1986. Magnetic polarity stratigraphy of a Plio-Pleistocene marine sequence of North Island, New Zealand. *Earth and Planetary Science Letters*, **80**, 353–360.

SMALE, D. 1990. Distribution and provenance of heavy minerals in the South Island: a review. *New Zealand Journal of Geology and Geophysics*, **33**, 557–571.

STERN, T. A., QUINLAN, G. M. & HOLT, W. E. 1992. Crustal shortening and basin formation behind an active subduction zone: Wanganui basin, New Zealand. *Basin Research*, **4**, 197–214.

STORMER, J. C. 1983. The effects of recalculation on estimates of temperature and oxygen fugacity from analyses of multi-component iron–titanium oxides. *American Mineralogist*, **68**, 586–594.

TARDUNO, J. A. 1994. Temporal trends of magnetic dissolution in the pelagic realm: gauging paleoproductivity? *Earth and Planetary Science Letters*, **123**, 39–48.

TAUXE, L., MULLENDER, T. A. T. & PICK, T. 1996. Potbellies, wasp-waists, and superparamagnetism in magnetic hysteresis. *Journal of Geophysical Research*, **101**, 571–583.

TURNER, G. M. & KAMP, P. J. J. 1990. Paleomagnetic location of the Jaramillo and the Matuyama–Brunhes transition in the Castlecliffian stratotype section, Wanganui basin, New Zealand. *Earth and Planetary Science Letters*, **100**, 42–50.

——, ROBERTS, A. P., LAJ, C., KISSEL, C., MAZAUD, A., GUITTON, S. & CHRISTOFFEL, D. A. 1989. New paleomagnetic results from Blind River: revised magnetostratigraphy and tectonic rotation of the Marlborough region, South Island, New Zealand. *New Zealand Journal of Geology and Geophysics*, **32**, 191–196.

WILSON, G. S. 1993. *Ice induced sea level change in the late Neogene*. PhD thesis, Victoria University of Wellington.

—— & MCGUIRE, D. M. 1995. Distributed deformation due to coupling across a subduction thrust: Mechanism of young tectonic rotation within the south Wanganui basin, New Zealand. *Geology*, **23**, 645–648.

WRIGHT, I. C. & VELLA, P. P. 1988. A New Zealand Late Miocene magnetostratigraphy: glacioeustatic and biostratigraphic correlations. *Earth and Planetary Science Letters*, **87**, 193–204.

—— & WALCOTT, R. I. 1986. Large tectonic rotation of part of New Zealand in the last 5 Ma. *Earth and Planetary Science Letters*, **80**, 348–352.

Remanence acquisition and magnetostratigraphy of the Leman Sandstone Formation: Jupiter Fields, southern North Sea

P. TURNER,[1] P. CHANDLER,[2] D. ELLIS,[3] G. P. LEVEILLE[4] & M. L. HEYWOOD[1,*]

[1] *School of Earth Sciences, The University of Birmingham, Edgbaston, Birmingham B15 2TT, UK*
[2] *Pole Position, 45 Court Farm Avenue, Ewell, Surrey KT19 OHD, UK*
[3] *Mobil North Sea Ltd, Grampian House, Union Row, Aberdeen AB1 1SA, UK*
[4] *Conoco (UK) Ltd, Rubislaw House, North Anderson Drive, Aberdeen AB2 4AZ, UK*

Abstract: The Leman Sandstone Formation, which forms part of the unfossiliferous Rotliegend sediments in the Jupiter Fields, has been cored at various location in the southern North Sea. Samples from the finer-grained sediments have measurable remanences that were classified into four main types. Some of these characteristics can be related to the petrological features that demonstrate that they are likely to have originated during deposition or very early diagenesis. This allows the age of the sediments to be determined using magnetostratigraphy, indicating that the sediments are of post-Kiaman age as they contain both normal and reversed polarities. This magnetostratigraphy can also be used to determine different rates of deposition in different areas.

The Jupiter Fields in the UK sector of the southern North Sea comprise five gas fields: Callisto, Europa, Ganymede, Sinope and Thebe. They are centred on Block 49/22 and extend into Blocks 49/16 and 49/17. The field geometries are controlled by predominantly NW–SE trending faults of the Swarte Bank Hinge Zone (Glennie 1984) (Fig. 1). These faults are considered to have been active during Permian to Early Cretaceous subsidence and during Late Cretaceous to Tertiary inversion through oblique compression (George & Berry 1994; Leveille *et al.* 1997). The reservoir sandstones in the Leman Sandstone Formation are of Upper Rotliegend (Permian) age, and, in this area, comprise mainly aeolian dune and dry sandsheet depositional facies together with minor fluvial sandstones, playa lake mudstones and sabkha deposits. The current stratigraphy and reservoir zonation of these unfossiliferous Rotliegend sediments is based on the recognition of facies stacking patterns, interpreted from core and wireline logs. These are interpreted by many workers as climatically controlled drying-upward cycles (George & Berry 1993, 1994; Howell 1992). On the basis of these cycles the Leman Sandstone Formation is divided into five main reservoir units (Fig. 2), which have been interpreted as regionally correlatable time stratigraphic units (Leveille *et al.* 1997). The current stratigraphic subdivision is considered more than adequate for the field development and infill exploration programme continuing in the area. However, doubts exist with regard to the chronostratigraphic equivalence of the reservoir zones and the key bounding surfaces from field to field. If the stratigraphic and sedimentological model is to be of maximum use in evaluating the controls on sedimentation and predicting the occurrence, thickness and quality of reservoir facies within other parts of the southern North Sea then an accurate method of dating and correlating these sediments is needed. To attempt to address this correlation problem in the Leman Sandstone Formation, and to test the existing facies-based stratigraphy, detailed studies have been made of the magnetization of the Leman Sandstone in the Jupiter Fields area. The aim has been to determine if magnetostratigraphy is a viable correlation technique in these Permian continental red beds.

Magnetostratigraphy is a well-documented method of correlating barren and fossiliferous

---

* Matthew Heywood died suddenly in September 1996. This paper is dedicated to his memory.

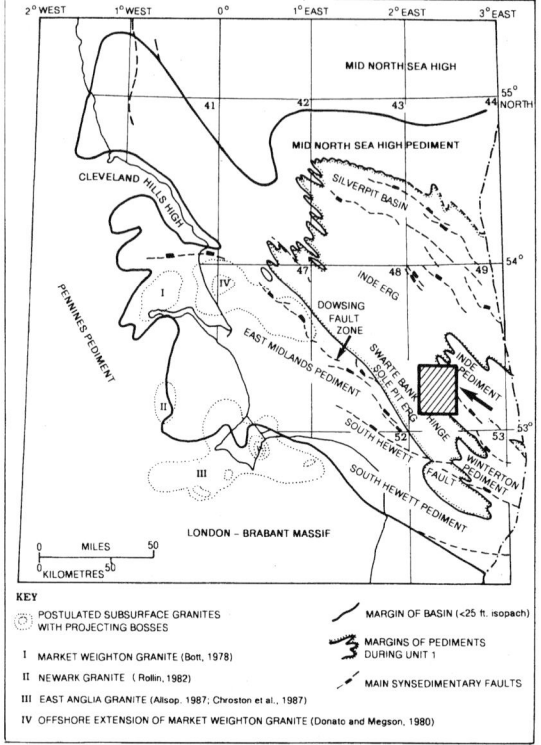 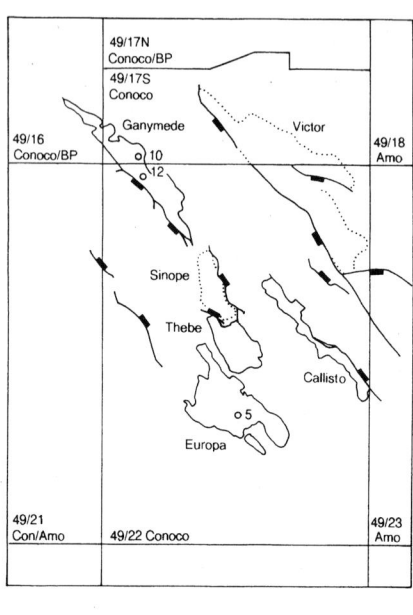

Fig. 1. Regional geological setting of the Jupiter Fields, UK southern North Sea (after George & Berry 1993). The inset shows the location of the studied wells in detail.

sedimentary sequences both in the cored subsurface and at outcrop (Turner 1987; Hailwood 1989; Turner et al. 1989; Rey et al. 1993). However, there is no previously published record of a systematic analysis of the cored Rotliegend of the southern North Sea. Previous workers have regarded Permian continental red beds to be exclusively of reversed polarity and thus to lie within the Kiaman Superchron, which may have lasted for about 50 Ma, from 310 to 265 Ma (Menning et al. 1988; Menning 1995; Nawrocki 1997). However, land-based studies of supposedly Lower Permian red sandstones in the UK have shown that their magnetic properties are complex, with normal polarity components generally present (Maslanyi & Collinson 1988; Turner et al. 1995). Preliminary studies of Rotliegend cores indicated the presence of magnetizations with apparent normal and reversed directions, and correlation studies by Menning et al. (1988) and Gebhardt et al. (1991) had indicated that the Leman Sandstone Formation was most likely to have been deposited after the Kiaman Superchron. On this basis, an initial study of the palaeomagnetism of the Leman Sandstone Formation was undertaken using 124 samples from c. 500 m of core from three wells, 49/17-10, 49/22-5 and 49/22-12 (Fig. 1). The samples were also studied petrologically to assess the likely origin of the magnetic remanence and its validity in magnetostratigraphic studies.

Sampling and measurement

Cored intervals from the three wells, 49/17-10 (215 m), 49/22-5 (108 m) and 49/22-12 (144 m), totalling 466 m were sampled at c. 3 m intervals, with sampling being focused in the finer-grained units, such as wet interdune and lake margin sabkha facies, which are more likely to preserve detrital remanences and are less likely to be altered by subsurface diagenetic processes or formation waters. For each of the samples the cut surface of the slabbed half-round core was marked with an orientation line parallel to the long axis of the core and an uphole arrow was marked on this line. This orientation line is referred to as reference (0) throughout this

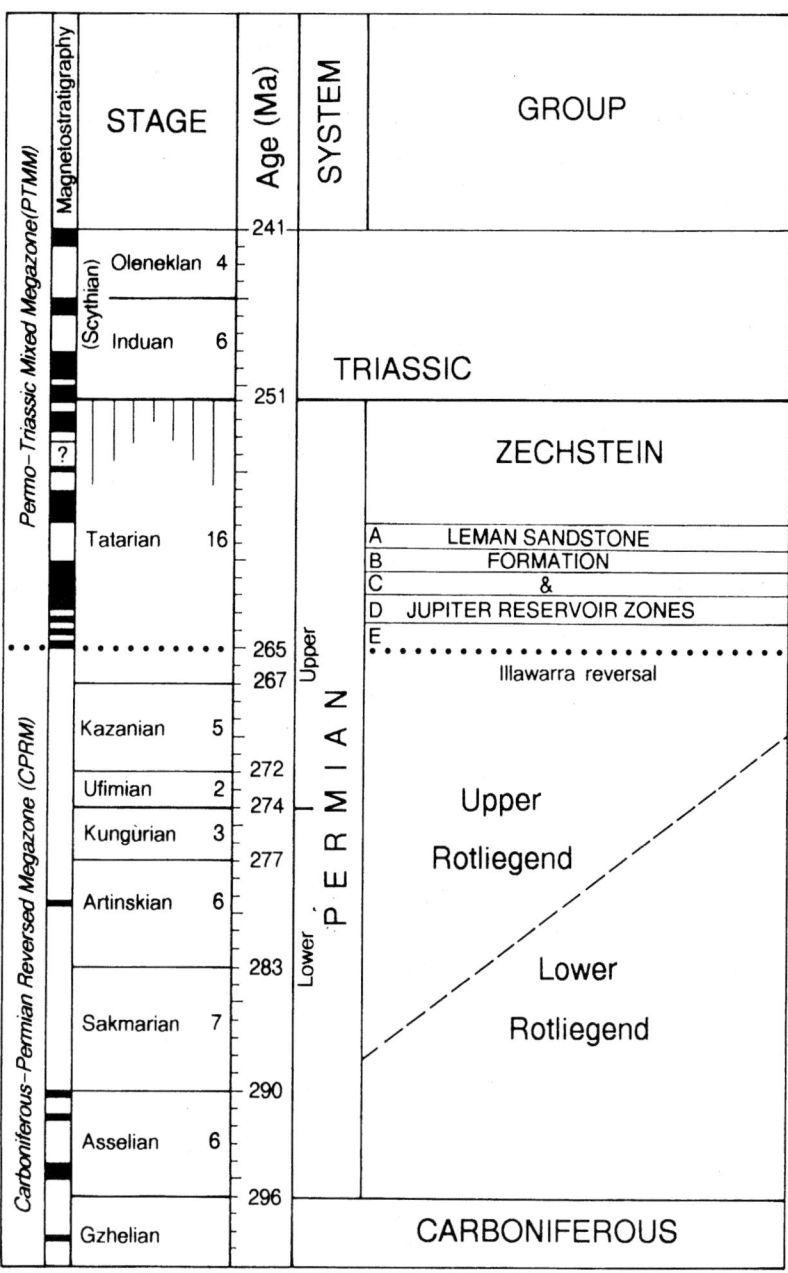

Fig. 2. Generalized Permian stratigraphy and magnetostratigraphy showing the position of the Leman Sandstone Formation and the reservoir zonation in the Jupiter area.

paper, and all palaeomagnetic directions are quoted relative to it. However, as the orientation of the bedding relative to the core is known, the absolute magnetic inclination could be determined. A total of 124 samples were taken in this way. (Well 49/17-10 was sampled at two different times to test the reproducibility of the results.) A 2.54 cm diameter core plug was then drilled at each sampling point (Fig. 3). The plugging was undertaken using a drill bit

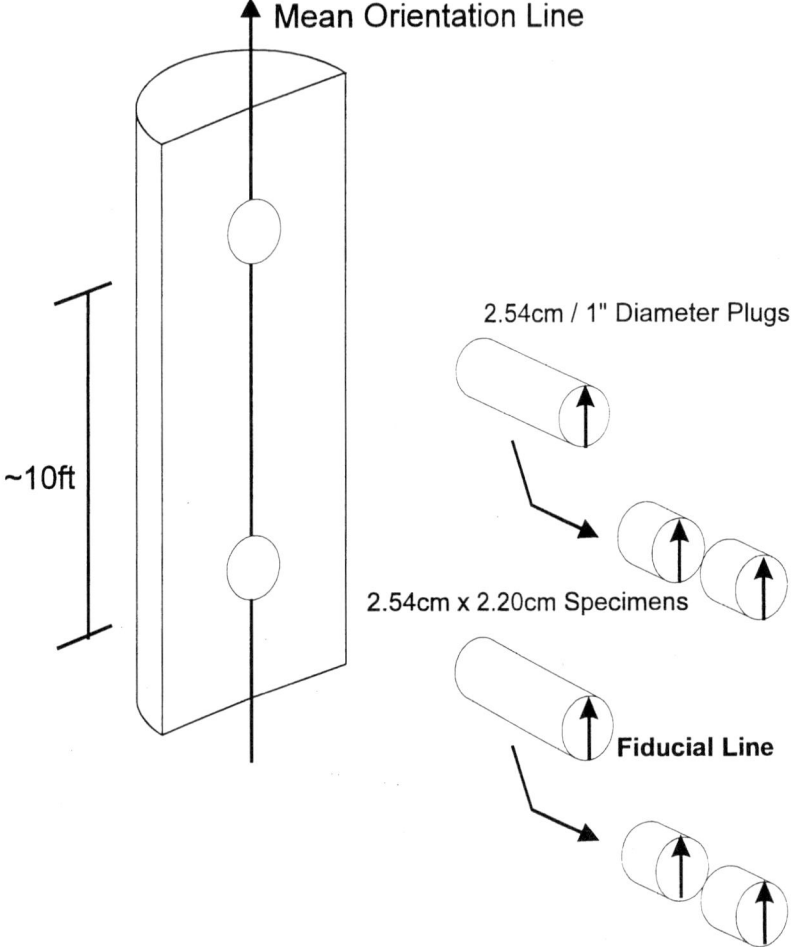

Fig. 3. Palaeomagnetic sampling technique for stabbed half-round core. The directions are later corrected so that inclination is reported relative to the palaeohorizontal.

with a non-magnetic chrome–alumel barrel and a diamond-impregnated phosphor bronze cutting edge. The plugs were then trimmed to a 2.2 cm length suitable for palaeomagnetic measurement.

Measurements of the natural remanent magnetization (NRM) and magnetic susceptibility were made on one specimen from each of the 124 sampled horizons. The initial susceptibilities were measured with a susceptibility bridge and the NRM with a modified Molspin balanced fluxgate magnetometer. All specimens were then subjected to stepwise thermal demagnetization at 50°C intervals up to 650°C. This was with a Magnetic Measurements thermal demagnetizer which comprises a non-inductive furnace and continuous degaussed mu-metal shielding.

The remanent magnetization of each sample was measured after each heating and cooling step, paying close attention to changes in susceptibility. Increases in low-field susceptibility during heating indicate magnetic mineral transformations and the possible acquisition of laboratory imparted remanence. After correction of the directional data for tectonic dip and well deviation, the data were then interpreted using a variety of techniques. Initially the magnetizations were plotted on a stereographic projection to identify temperature-dependent changes and to define the characteristic magnetic inclinations of the specimens. However, stereographic projections represent only directional changes and to isolate discrete components of magnetization, intensity information must be incorporated so that individual vectors can be recognized. Different components of magnetization can be isolated using orthogonal plots that combine directional and intensity changes (see Dunlop 1979; Collinson 1983; Butler 1992). On the orthogonal plots, individual components of magnetization are identified from straight-line segments (Fig. 4) determined by principal component analysis (Kirschvink 1980).

Palaeogeographic reconstructions for NW Europe indicate that the characteristic magnetic inclinations for the Permian deposits of the southern North Sea are ±20°, corresponding to a palaeolatitude of about 10° (Johnson et al. 1995). Secondary components of magnetization of Tertiary or Recent age, acquired during diagenesis, oil residence and burial, may overprint the primary Permian magnetization. The most common secondary components are viscous magnetizations with an inclination very close to the present-day geomagnetic field (69°), but other secondary components of magnetization can also be induced during coring or later plugging in the laboratory (Hailwood & Ding 1995). This technique allowed an analysis of the different types of magnetization present in core samples from their characteristic inclinations and blocking temperature spectra. The orientations of the characteristic magnetization were plotted alongside the core logs to reveal which parts of the section were magnetized during periods of normal magnetic polarity and which during periods of reversed polarity. The patterns of normal and reversed polarity, together with the facies stacking patterns, were then used to correlate key surfaces and sediment packages from well to well.

Palaeomagnetic results

The NRM intensities of over 99% of the samples ranged between 0.33 and 4.65 mA/m (Fig. 5a–c). Samples with high intensities were generally from wet interdune facies and lake margin sabkhas, whereas the lower intensities are observed in the drab-coloured, grey sandy intervals found in the uppermost sections of the cored intervals. The magnetic susceptibility ranged between 357×10^6 and 141×10^6 SI. These values were also directly related to colour, grain size and facies. For well 49/17-10 (Fig. 5a), there is a wide range of NRM directions (relative declination and true inclination) and many specimens showed relatively shallow inclinations. In contrast, the initial NRM results from wells 49/22-5 (Fig. 5b) and 49/22-12 (Fig. 5c) showed directions with steeper inclinations and less dispersion. During incremental thermal demagnetization the specimens showed a variety of directional and intensity changes. These can be grouped into three types of magnetization (Figs 6–10) as follows.

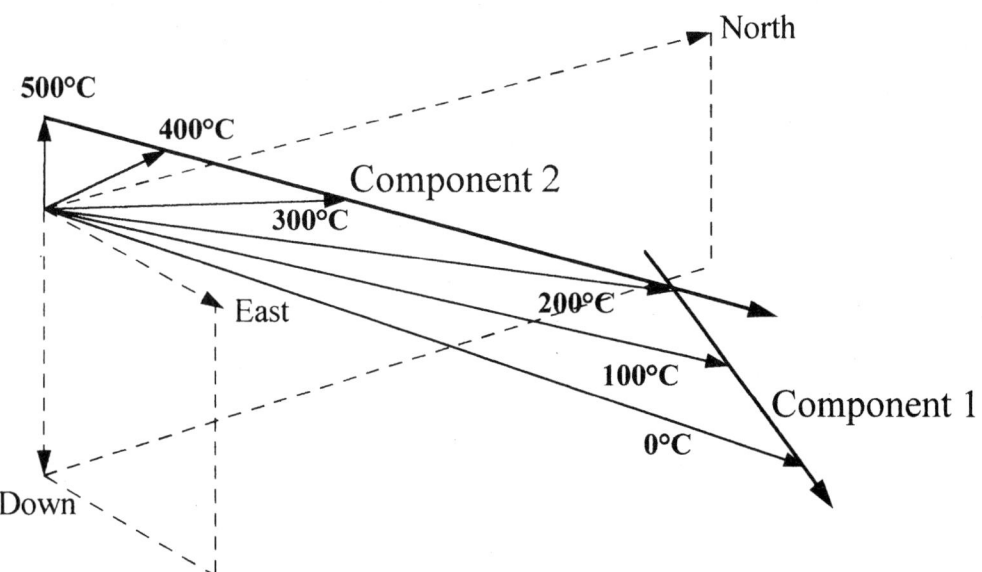

Fig. 4. The separation of discrete components of magnetization using orthogonal vector plots. Component 1 has the lower unblocking temperature range (20–200°C) and Component 2 has the higher unblocking temperature range (>2000°C).

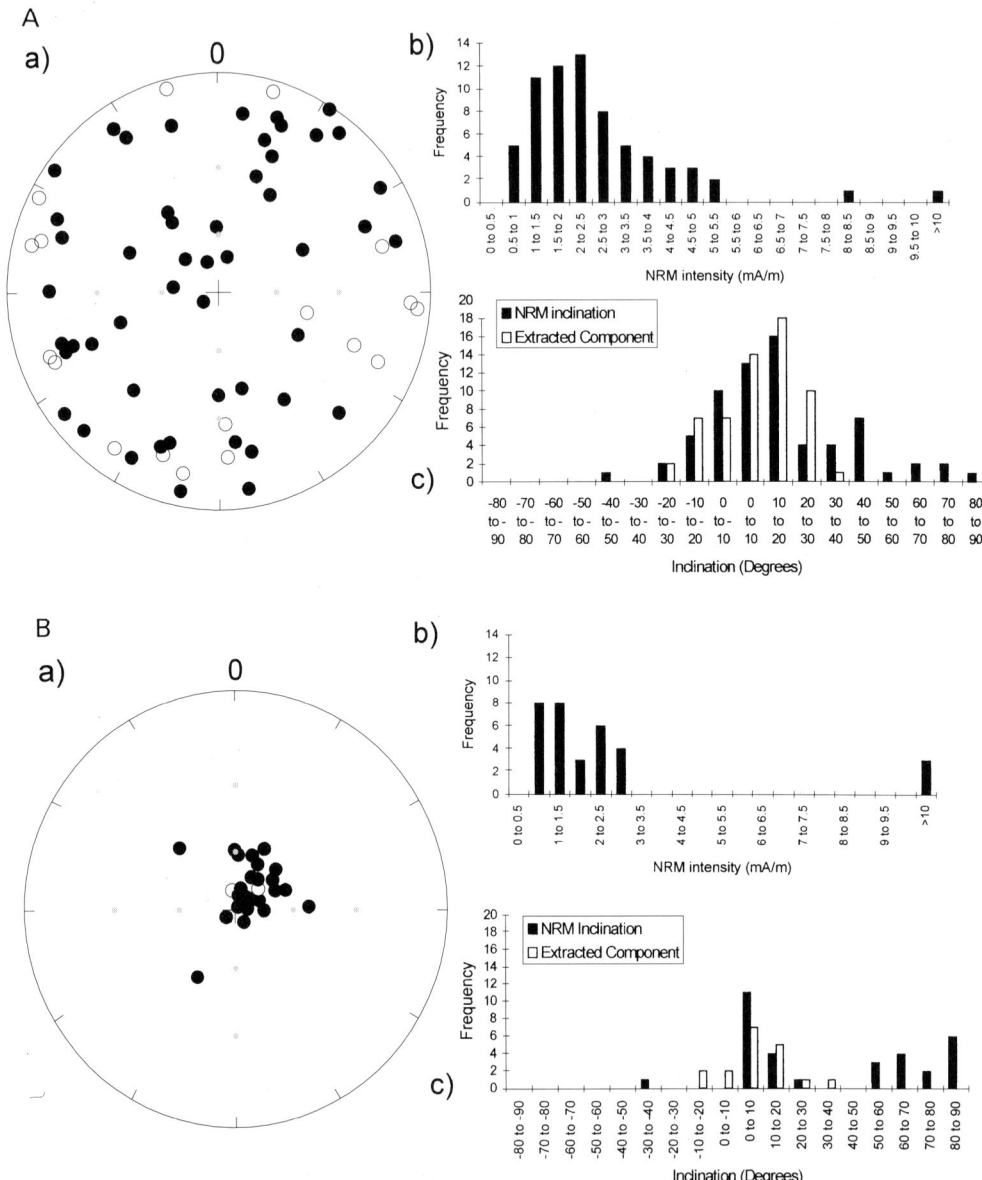

Fig. 5. Stereographic projections showing the initial NRM inclinations of the three wells. (**a**) Well 49/17-10; (**b**) well 49/22-5; (**c**) well 49/22-12. Open symbols indicate negative inclination; filled symbols, positive inclination. The bar charts show the distribution of the NRM intensifies in mA/m and the inclinations of the ChRM.

Type I: These magnetizations were characterized by shallow inclinations, averaging about 20°, and have a wide range of unblocking temperatures. Discrete components may be isolated over a narrow temperature range but more typically the unblocking temperatures were distributed up to a maximum of 600°C. The characteristic inclinations may be positive (Type 1A) or negative (Type IB). In some cases (Fig. 7) there appeared to be two closely overlapping components, whereas in others only a single component is present (Fig. 7).

Type II: These were more complex magnetizations which comprised two distinct components; a low unblocking temperature component (20–150°C) of variable direction, which was

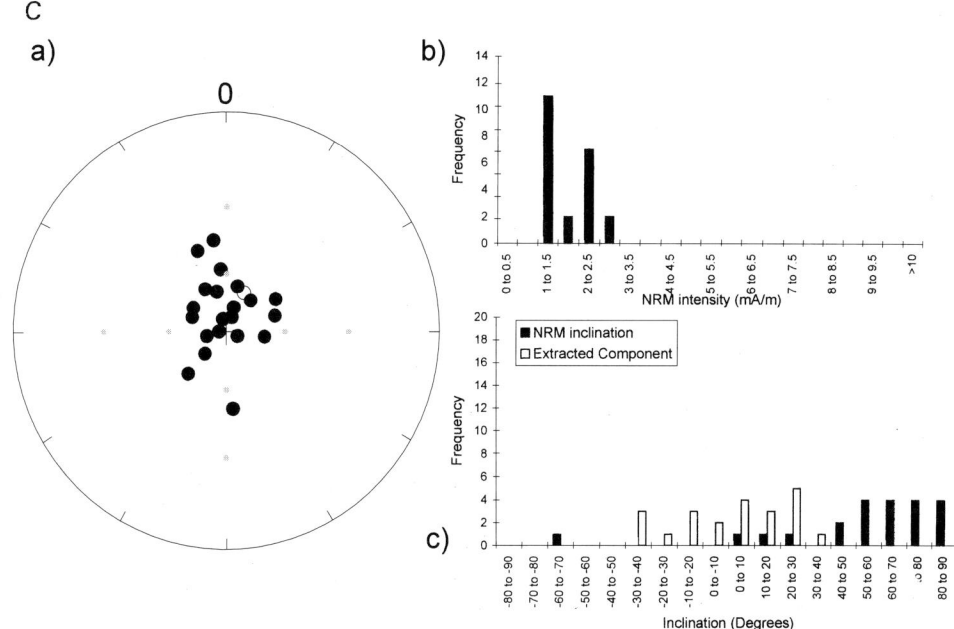

Fig. 5. (*continued*).

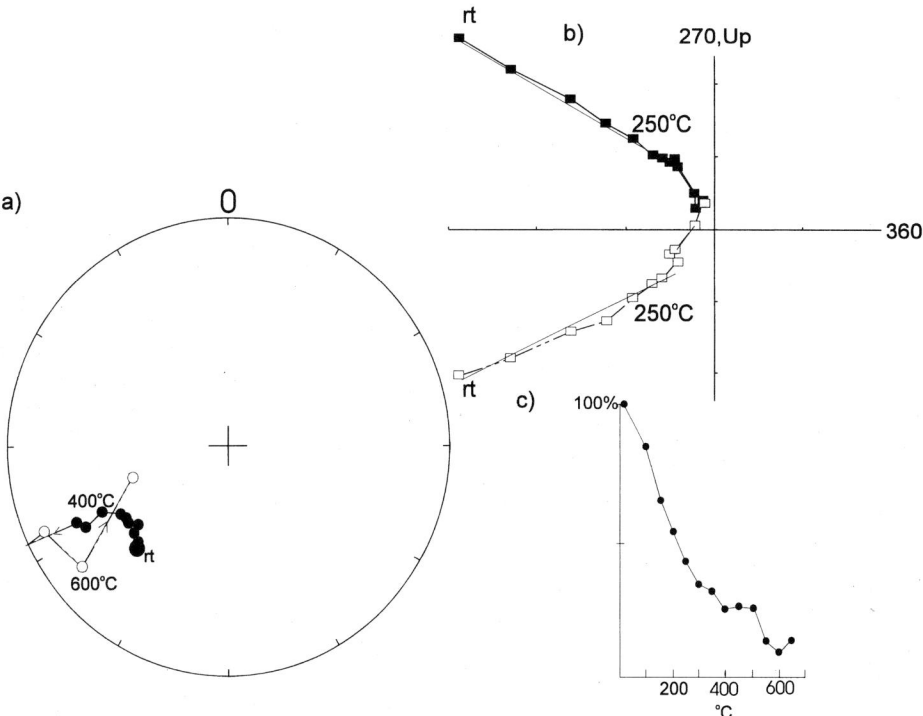

Fig. 6. Typical example of Type Ia magnetization. (**a**) Stereographic projection. (**b**) Orthogonal vector plot. Filled symbols indicate the horizontal plane (declination) and open symbols the vertical plane (apparent inclination). (**c**) Normalized intensity decay curve. Sample H8, interpreted to have Permian normal polarity magnetization.

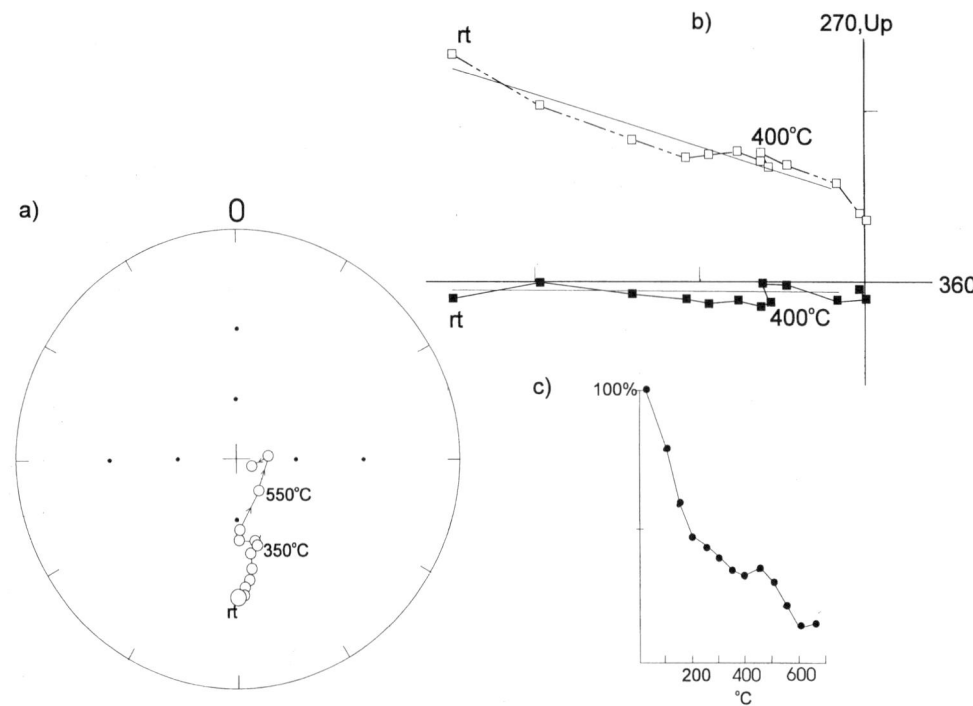

Fig. 7. Typical example of Type Ib magnetization. Other details as in Fig. 7. Sample H7, interpreted to have Permian reversed polarity magnetization.

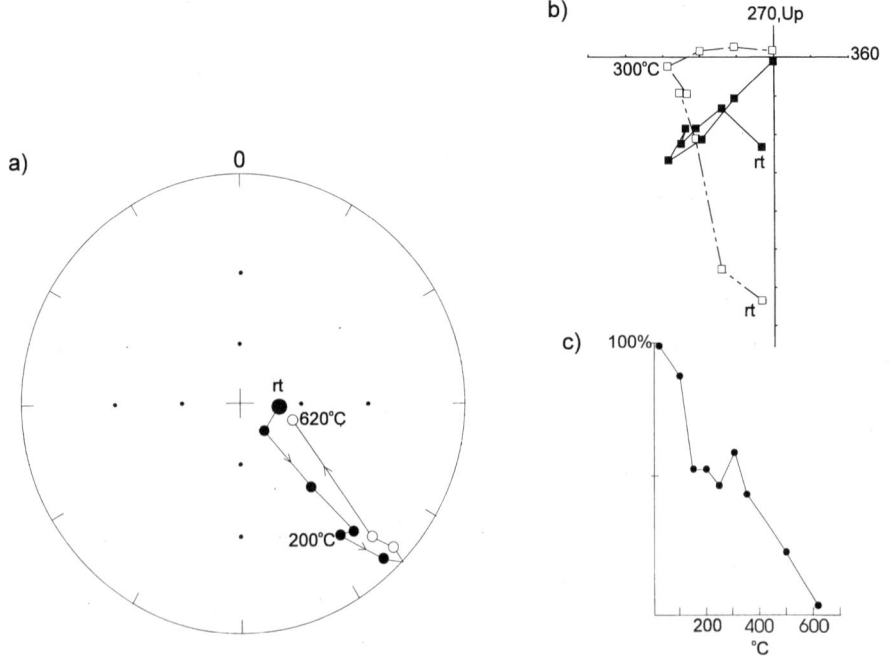

Fig. 8. Typical example of Type II magnetization. Other details as Fig. 7. Sample F6, which comprises a steep normal component of secondary origin (Dec 324°; Inc +76°) superimposed on a higher unblocking temperature component (Dec 139°, Inc +18°) isolated between 250 and 550°C, which is interpreted as a Permian normal polarity magnetization.

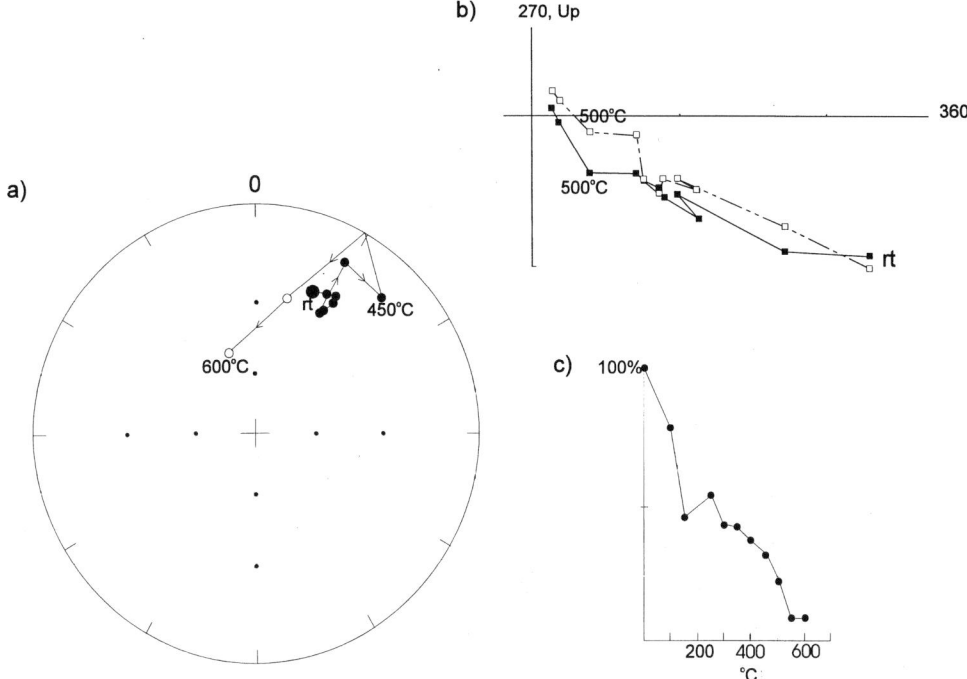

Fig. 9. Typical example of Type III magnetization. Other details as in Fig. 7. Sample R33, interpreted as Permian normal polarity magnetization between 250 and 500°C.

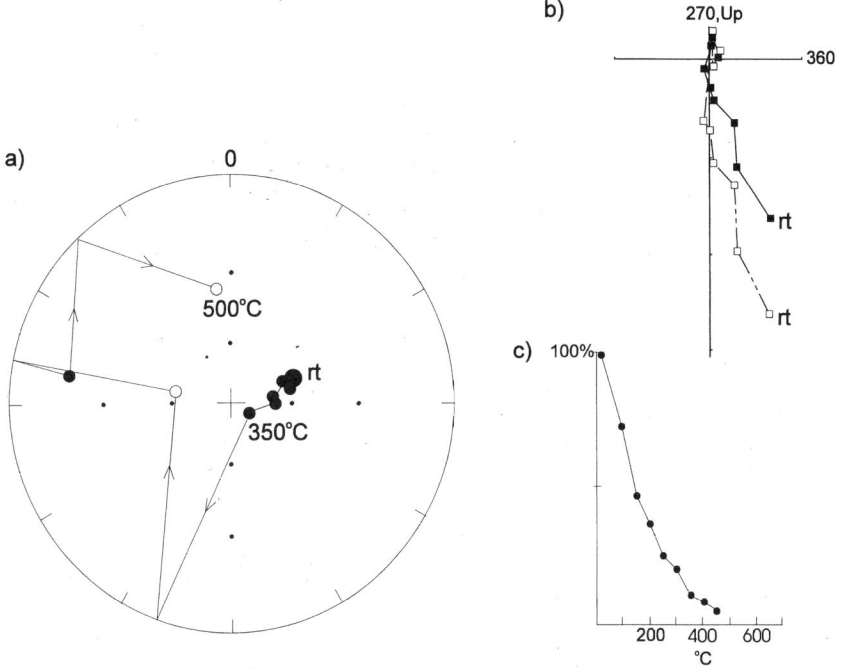

Fig. 10. Typical example of Type IV magnetization. Other details as in Fig. 7: Sample G6. No Permian ChRM is preserved.

overprinted on a Type I magnetization. In Type II magnetizations the break in the unblocking temperature spectra was very sharp and the two components could be easily distinguished on orthogonal vector plots (Fig. 8).

Type III: These magnetizations also comprised two discrete components of magnetization: a relatively steeply inclined low-temperature component (20–150°C), which was overprinted on a weak, higher-temperature component of shallower inclination. The inclination of the shallower component was difficult to determine precisely because of magnetomineralogical changes at higher temperature (Fig. 9). However, in most cases there was a narrow window between the unblocking of the lower-temperature component and the onset of mineralogical changes, and this could be used to assess whether the higher-temperature component was of normal or reversed polarity.

Type IV: These were magnetizations consisting of one or more low unblocking temperature components (20–150°C). At higher temperatures the behaviour became erratic and there was no evidence of any stable components of magnetization. Generally, Type IV magnetizations had very steep positive inclinations near the ambient geomagnetic inclination (Fig. 10).

Palaeomagnetic interpretation

The origin of the remanence

The palaeomagnetic results show that the magnetization of the Leman Sandstone is relatively complex with a low unblocking temperature spectrum. Discrete individual components are frequently difficult to separate and the high-temperature chemical changes associated with laboratory heating further complicate the matter. However, a number of factors indicate that the characteristic remanence is not associated with widespread remagnetization and was probably acquired during early diagenesis. These factors include the presence of normal and reverse polarity components, the thermally discrete difference between viscous remanent magnetization (VRM) and the characteristic remanent magnetization (ChRM) and the consistently shallower inclination of the ChRM with respect to the VRM. Reflected light photomicrographs show that the most abundant typical iron oxides in the Leman Sandstone are hematite grains (specularite) usually in the size range 5–50 μm. The detrital nature of these grains is clearly indicated by their roundness and aerodynamic equivalence when compared with the lighter siliciclastic grains, which may be up 500–600 μm (Fig. 11). Some grains show martitization textures typical of the oxidation of magnetite (Fig. 11c and d), and may be derived from Lower Permian volcanics rocks. The secondary magnetizations (VRMs) are probably carried by pigmentary hematite particles with sizes less than 2 μm usually present within the clay matrix or as grain coatings (Turner 1980). Some samples show slightly coarser authigenic hematites (Fig. 11e and f).

Magnetostratigraphic background

The duration of normal (north-seeking) and reversed (south-seeking) time intervals (chrons) ranges between $c.$ 0.1 and 30 Ma. In the last two decades a large number of land-based studies have extended geomagnetic polarity time scales back into early Mesozoic and Palaeozoic time but it is only recently that significant contributions have been made to the Permian time scale (Menning 1995; Nawrocki 1997). In applying magnetostratigraphy to the subsurface Rotliegend there are a number of potential problems. The first of these is the absence of a well-constrained Permian geomagnetic polarity time scale (GPTS) to allow the independent correlation of any results because of the absence of available intact Permian oceanic crust. Land-based studies indicate the presence of the long

Fig. 11. Reflected light photomicrographs (oil immersion) illustrating the characteristic magnetic minerals in the Leman Sandstone Formation of the Jupiter Fields area. (**a**) Partially rounded hematite grain with a diameter of 45 mm. The hematite grains are typically smaller than the associated clastic grains because of their greater density (aerodynamic equivalence). Plane-polarized light (PPL). (**b**) The same area seen in partially crossed nicols. The internal structure of the grain is homogeneous. (**c**) In right centre is a hematite grain with euhedral crystal faces. The grain diameter is 54 μm. PPL. (**d**) The same area in partially crossed nicols. (Note the characteristic triangular alteration texture of martite and the pinkish red internal reflections, which indicate the formation of iron and titanium oxides as a result of secondary alteration of titanomagnetite or ilmenite.) The grain left of centre with bright cream-coloured internal reflections is titanium oxide. (**e**) Large hematite grain with authigenic pore-lining crystals. The diameter of the detrital grain is 240 μm. The crystals are probably tabular in shape so that the short axis is the *c*-axis. Apart from pigmentary material, authigenic hematite is rare in the Jupiter Fields area. (**f**) The same area in partially crossed nicols. The internal structure of the detrital grain is essentially homogeneous.

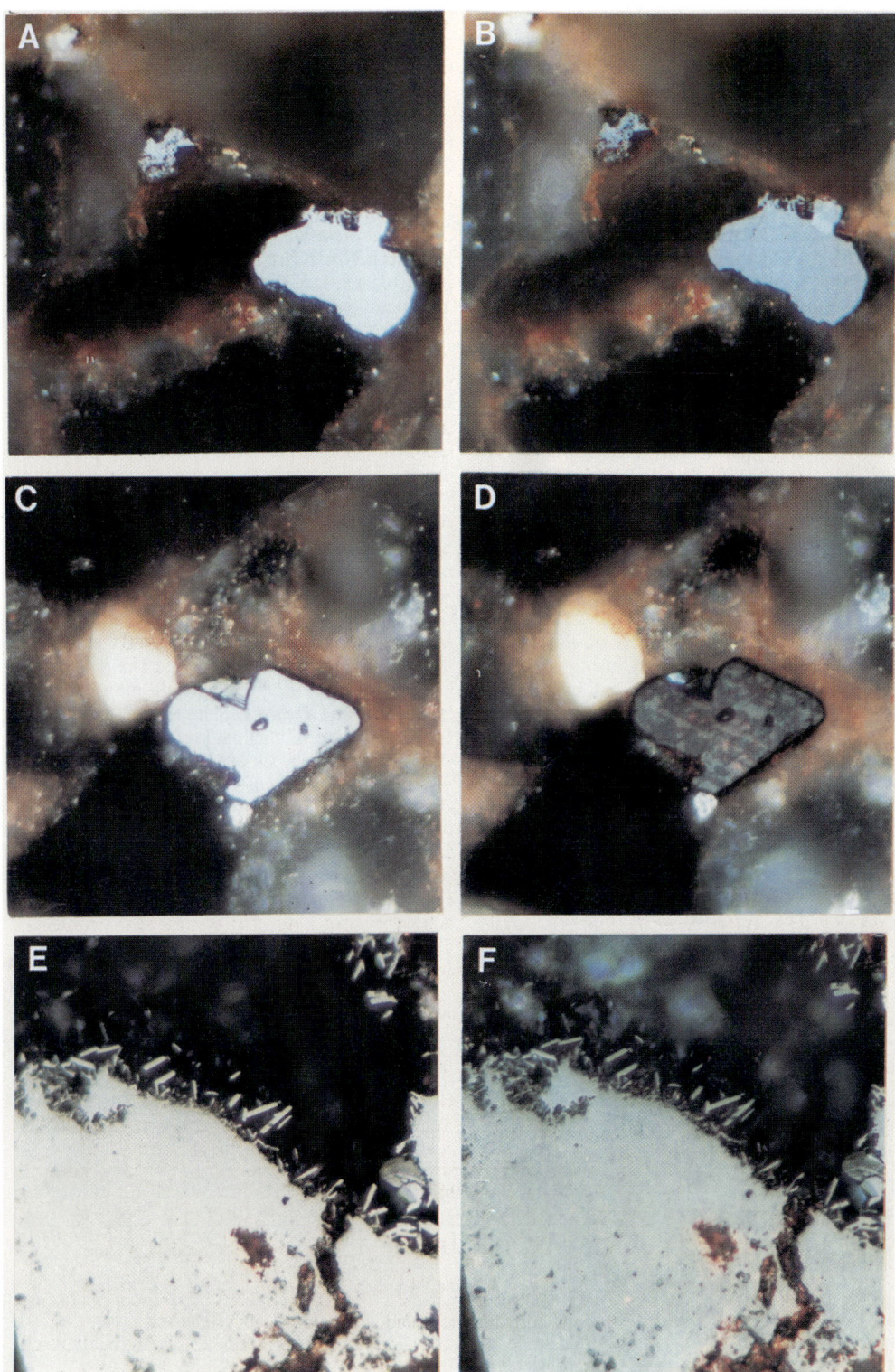

Kiaman Superchron during late Carboniferous and early Permian time. Hailwood (1989) suggested that all rocks of Permian age were deposited during the Kiaman Superchron but studies by Gebhardt *et al.* (1991) and Menning *et al.* (1988) have shown data from eastern Europe suggesting that the base of the Kiaman Superchron lies near the Westphalian A–B boundary and that this reverse period extends well into late Permian time (up to 265 Ma). Perroud *et al.* (1995) have presented data indicating that the Lower Rotliegend of the offshore Netherlands is exclusively of reversed polarity and considered it to be well within the Kiaman Superchron (Fig. 2). Nawrocki (1997) presented data for Poland which are consistent with the global scale (Opdyke 1995), suggesting that the end of the Kiaman Superchron is probably close to the Lower–Upper Rotliegend boundary. The new palaeomagnetic results, presented here, strongly suggest that the Leman Sandstone lies above the Kiaman Superchron.

The Jupiter Fields magnetostratigraphies

Magnetostratigraphies for each of the studied wells have been constructed using the characteristic magnetization extracted by principal components analysis of Type I and Type II magnetizations. In addition, quantitative determination of the likely characteristic polarity from some Type III magnetizations has been used to help constrain the magnetostratigraphies. The magnetic polarities are conventionally shown with normal polarity zones in black and reversed polarity zones in white. Samples interpreted as exhibiting Type IV magnetizations (no Permian remanence preserved) are marked by a hatched ornament in Fig. 12. There are clearly potential problems associated with this technique. One of the main problems arises from the relatively low palaeolatitude of the southern North Sea in Permian times. As it lay only 10° north of the equator, the corresponding magnetic inclination is expected to be around +20° for normal polarity and −20° for reversed polarity. When secular variations and other factors are taken into consideration this leaves a relatively narrow arc within which to define the magnetic polarity. This problem is particularly serious in subsurface data of the type considered here, as the measurements necessarily have only relative declination values. However, in a number of cases it is possible to demonstrate that the characteristic remanence can be reorientated using the VRM, allowing the polarity determinations to be verified. Other potential problems include the effects of the passage through the sediments of acidic, reducing fluids associated with gas migration and charge (Leveille *et al.* 1997; Turner *et al.* 1995). These may cause the reduction and removal of magnetic minerals or the acquisition of chemical remanent magnetizations (CRMs) associated with the precipitation of magnetic sulphides such as pyrrhotite or

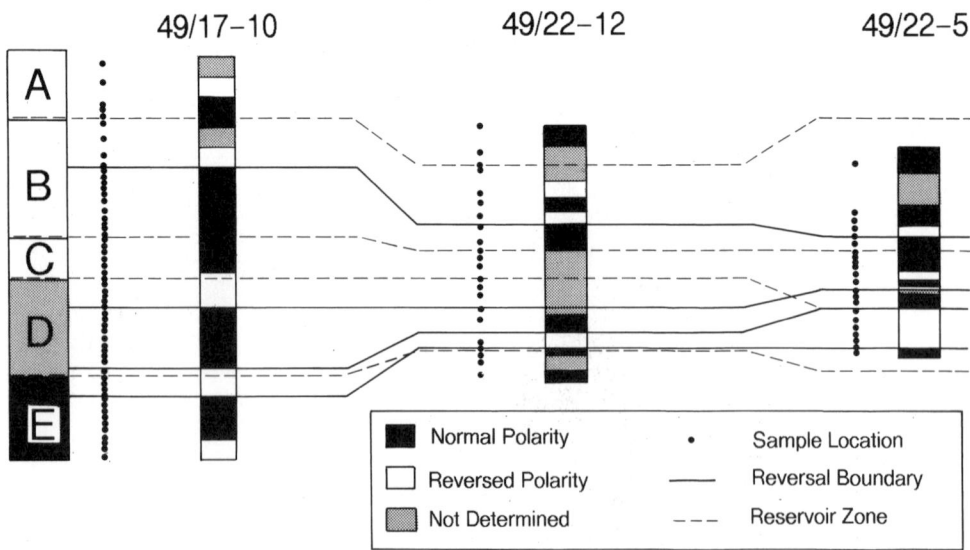

Fig. 12. Magnetostratigraphy of the three studied wells in relation to reservoir Zones A–E.

greigite. However, using petrographic techniques, it is possible to demonstrate that, at least in the reddened sections, this process has not been a problem. The sampling intervals in wells 49/17-10 and 49/22-5 are considered sufficiently close for the construction of well-constrained magnetostratigraphies for the cored intervals within these two wells, but the magnetostratigraphy for well 49/22-12 can only be considered preliminary due to the wider sampling interval.

Well 49/17-10. The cored interval is characterized by four normal polarity magnetozones and five reversed zones. The lowermost section of the core is interpreted as carrying a reversed polarity ChRM, which is isolated from three samples. This is followed by a period of normal polarity through reservoir Zone E. The top of this reservoir zone carries a reverse polarity ChRM, which extends upwards into reservoir Zone D. The latter zone is dominated by normal polarity but is capped by a reverse polarity interval, which extends up into reservoir Zone C. A normal polarity interval extends from near the base of reservoir Zone C through to the upper half of Zone B. In the uppermost section of the core, characterized by the grey sediments of the Weissliegend, a reverse polarity interval is interpreted, followed by a short interval for which no stable remanences could be extracted. A thin normal, and then reversed, interval within the uppermost reservoir Zone B and A is then followed by another section for which the ChRM polarity could not be defined.

Well 49/22-5. Six intervals of normal polarity and four of reversed polarity have been interpreted. A ChRM could not be isolated from one long interval in the uppermost part of the core. The base of the cored section is of normal polarity, as interpreted from one sample level. This is followed by a long period of reversed polarity defined by six sample locations. Both of these magnetozones lie within reservoir Zone D. Above this is an interval of predominantly normal polarity with two thin reversed intervals, which correspond to reservoir Zone C and the lower part of Zone B. Within Zone B is a thin reversed interval overlain by rocks with a normal polarity ChRM. The uppermost part of the core was not sampled because of the nature of the lithologies, which were drab coloured and correspond to the Weissliegend.

Well 49/22-12. ChRMs could be extracted from only ten sample horizons, so the resulting magnetostratigraphy is not as complete as for the other two wells. Two of the three samples in reservoir Zone A have normal polarity. At the base of reservoir Zone D is a thin reversed magnetozone followed by an interval with normal polarity ChRM. The uppermost section of reservoir Zone D and the whole of reservoir Zone C has not yielded any stable magnetizations. However, reservoir Zone B is characterized by two normal and two reversed intervals. The uppermost section of Zone B and the lowermost section of Zone A have no stable ChRM.

Magnetostratigraphic correlation

Of the wells studied (Fig. 12), the magnetostratigraphy interpreted for Well 49/17-10 is considered to be the best constrained in terms of the quality of the remanence and the numbers of samples. As this well also provides the most complete section through all five reservoir zones, it is used as the basis for the correlation with the other wells. On this basis, three main observations are made:

(1) The reversed magnetozone observed in the uppermost part of reservoir Zone E in well 49/17-10 can be correlated with the reversed interval at the base of reservoir Zone D in well 49/22-12. This may suggest that there is some diachroneity along this D–E reservoir zone boundary inasmuch as the bounding surface in well 49/17-10 is younger than that in 49/22-12. The magnetozone can also be correlated through to well 49/22-5, where it is significantly thicker. This interpretation suggests that the sedimentation rate during this reversed period was faster in well 49/22-5, in the southern part of the area.

(2) The reversed interval in the upper part of reservoir Zone D in well 49/17-10 has been correlated with the two reversed intervals observed in well 49/22-5 (Fig. 13). At the same stratigraphic level in well 49/22-12 no magnetostratigraphy could be defined.

(3) The reversed interval observed in reservoir zone B of well 49/17-10 can be identified in the other two wells.

If the palaeomagnetic data are integrated with lithological and wireline log data (Fig. 13), then there are some key consistencies between the magnetostratigraphy and lithological correlations. The main feature concerns the regional supersurface at the top of reservoir zone C (units 5 + 7); above this surface is normal polarity and below reversed polarity. Another key feature of the interpretation concerns the erg lithofacies, which corresponds to reservoir Zone D (unit 4). The magnetostratigraphy strongly suggests that this may represent two or more smaller dunefields of significantly different ages in the three

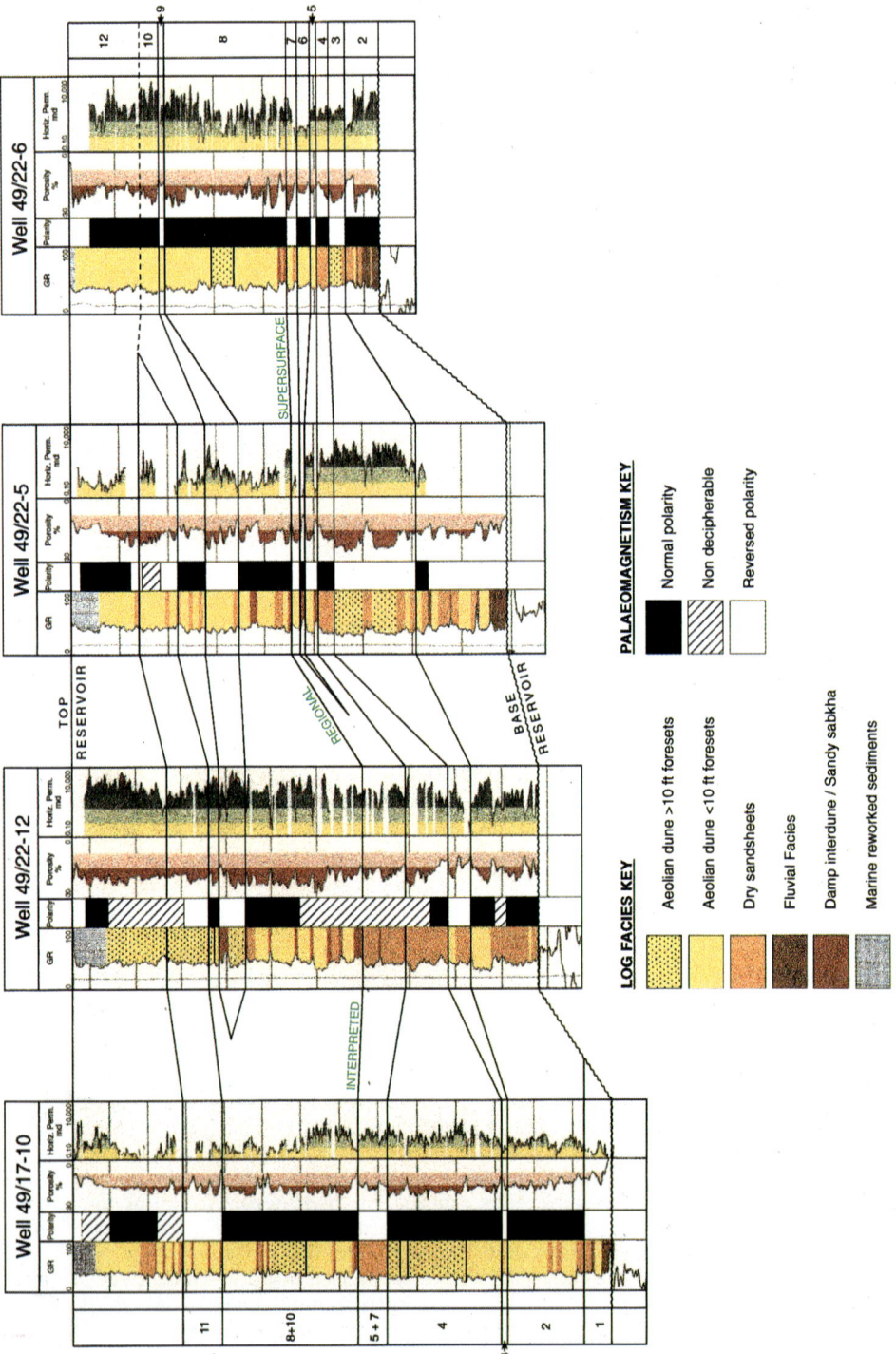

Fig. 13. Correlation of the Lower Leman Sandstone Formation in the Jupiter Fields using combined lithostratigraphy and magnetostratigraphy.

wells rather than a single continuous dune-field. The interpretation suggests also that the reversed magnetozones in the lowermost part of the wells studied (reservoir Zone E) are in a 'wetter' facies in the NW of the area (well 49/17-10) whereas in well 49/22-5, in the SE of the area, the same reversed magnetozone lies within a dry aeolian dune facies. The magnetostratigraphy therefore provides independent evidence that the depositional facies within the Rotliegend sequence are strongly diachronous (Glennie 1984). Thus, effective time lines can be delineated within the reservoir sequence, to help constrain the lithofacies architecture in relation to a chronostratigraphic framework.

Conclusions

The Leman Sandstone Formation has a ChRM thought to have been acquired during initial burial and diagenesis. On this basis, the main conclusions from this study are as follows:

(1) It is possible to recognize palaeomagnetic reversals within the cored intervals, which indicate that the Leman Sandstone Formation lies above the Kiaman Superchron.

(2) This suggests that the Leman Sandstone Formation in this area is of lower Upper Permian age.

(3) Interpretation of the thermal demagnetization analysis has defined a record of nine magnetozones in the cored section of the Jupiter Fields area. The recognition of magnetic reversals allows the development of a new chronostratigraphic correlation framework, within which the existing stratigraphic correlation schemes can be tested.

(4) It is possible to refine predictive facies models in the light of these results. In particular, the data indicate that aeolian erg developments in the area may be of smaller areal extent and more variable in time and space than previously thought.

We thank our colleagues at Mobil, Conoco and BP for stimulating discussions regarding the significance of this work. Mobil North Sea Ltd kindly gave us permission to publish the work.

References

BUTLER, R. F. 1992. *Paleomagnetism: Magnetic Domains to Geological Terranes*. Blackwell Scientific, Oxford.

COLLINSON, D. W. 1983. *Methods in Rock Magnetism and Palaeomagnetism*. Chapman & Hall, London.

DUNLOP, D. J. 1979. On the use of Zijderveld vector diagrams in multicomponent paleomagnetic studies. *Physics of the Earth and Planetary Interiors*, **20**, 12–24.

GEBHARDT, U., SCHNEIDER, J. & HOFFMAN, N. 1991. Modelle zur Stratigraphie und Beckenentwicklung im Rotliegenden der Norddeutschen Senke. *Geologisches Jahrbuch*, **127**, 405–427.

GEORGE, G. T. & BERRY, J. K. 1993. A new lithostratigraphy and depositional model for the Rotliegend of the UK sector of the southern North Sea. *In:*, NORTH, C. P. & PROSSER, D. J. (eds) *Characterisation of Fluvial and Aeolian Reservoirs*. Geological Society, London, Special Publication, **73**, 291–319.

—— & ——1994. A new palaeogeography and depositional model for the Upper Rotliegend, offshore The Netherlands. *First Break*, **1**, 147–158.

GLENNIE, K. 1984. *Introduction to the Petroleum Geology of the North Sea*. Blackwell Scientific, Oxford.

HAILWOOD, E. A. 1989. *Magnetostratigraphy*. Geological Society, London, Special Report, **19**.

—— & DING, F. 1995. Paleomagnetic reorientation of cores and the magnetic fabric of hydrocarbon reservoir sands. *In:* TURNER, P. & TURNER, A. (eds) *Palaeomagnetic Applications in Hydrocarbon Exploration and Production*. Geological Society, London, Special Publication, **98**, 245–258.

HOWELL, J. A. 1992. *Sedimentology of the Rotliegend Group of the UK Southern North Sea*. PhD thesis, University of Birmingham.

JOHNSON, S. A., TURNER, P., HARTLEY, A. & REY, D. 1995. Palaeomagnetic implications for the timing of hematite precipitation and remagnetization in the Carboniferous Barren Red Measures, UK southern North Sea. *In:* TURNER, P. & TURNER, A. (eds) *Palaeomagnetic Applications in Hydrocarbon Exploration and Production*. Geological Society, London, Special Publication, **98**, 97–117.

KIRSCHVINK, J. L. 1980. The least-squares line and plane and the analysis of palaeomagnetic data. *Geophysical Journal of the Royal Astronomical Society*, **62**, 699–718.

LEVEILLE, G. P., KNIPE, R., MORE, C., *et al.* 1997. Compartmentalisation of Rotliegendes gas reservoirs by sealing faults, Jupiter Fields area, southern North Sea. *In:* ZIEGLER, K., TURNER, P. & DAINES, S. (eds) *Petroleum Geology of the Southern North Sea: Future Potential*. Geological Society, London, Special Publication, **123**, 87–104.

MASLANYI, M. P. & COLLINSON, D. W. 1988. A magnetic and petrological study of some Permian aeolian red sandstones. *Geophysical Journal of the Royal Astronomical Society*, **92**, 421–430.

MENNING, M. 1995. A numerical time scale for the Permian and Triassic Periods: an integrated time analysis. *In:* SCHOLLE, P. A., PERYT, T. M. & ULMER-SCHOLLE, D. S. (eds) *The Permian of Northern Pangea. Vol. 1*. Springer, Berlin, 77–97.

——, KATZUNG, G. & LUTZNER, H. 1988. Magnetostratigraphic investigations in the rotliegendes

(300–252 Ma) of Central Europe. *Zeitschrift für Geologische Wissenschaften*, **16**, 1045–1063.

NAWROCKI, J. 1997. Permian to Early Triassic magnetostratigraphy from the central European Basin in Poland: implications on regional and worldwide correlations. *Earth and Planetary Science Letters*, **152**, 37–58.

OPDYKE, N. D. 1995. Magnetostratigraphy of Permo-Carboniferous time. *In*: BERGGREN, W. A., KENT, D. V., AUBRY, M.-P. & HARDENBOL, J. (eds) *Geochronology Time Scales and Global Stratigraphic Correlation*. Society of Economic Paleontologists and Mineralogists Special Publication, **54**, 41–47.

PERROUD, H., CHAUVIN, A. & REBELLE, M. 1995. Hydrocarbon seepage dating through chemical remagnetisation. *In*: TURNER, P. & TURNER, A. (eds) *Palaeomagnetic Applications in Hydrocarbon Exploration and Production*. Geological Society, London, Special Publication, **98**, 33–41.

REY, D., TURNER, P. & YALIZ, A. 1993. Palaeomagnetism and magnetostratigraphy of the Skagerrak Formation, Crawford Field, North Sea. *In*: NORTH, C. P. & PROSSER, D. J. (eds) *Characterization of Fluvial and Aeolian Reservoirs*. Geological Society, London, Special Publication, **73**, 339–420.

TURNER, P. 1980. *Continental Red Beds*. Elsevier Developments in Sedimentology, 29. Elsevier, Amsterdam.

——1987. Magnetostratigraphy. *Geology Today*, August, 127–132.

——, BURLEY, S. D., REY, D. & PROSSER, J. 1995. Burial history of the Penrith Sandstone (Lower Permian) deduced from the combined study of fluid inclusion and paleomagnetic data. *In*: TURNER, P. & TURNER, A. (eds) *Palaeomagnetic Applications in Hydrocarbon Exploration and Production*. Geological Society, London, Special Publication, **98**, 43–78.

——, TURNER, A., RAMOS, A. & SOPENA, A. 1989. Palaeomagnetism of PermoTriassic rocks in the Iberian Cordillera, Spain: acquisition of secondary and characteristic remanence. *Journal of the Geological Society, London*, **146**, 61–76.

Characterizing pore fabrics in sediments by anisotropy of magnetic susceptibility analyses

E. A. HAILWOOD,[1] D. BOWEN,[2] F. DING,[1] P. W. M. CORBETT[2] & P. WHATTLER[3]

[1] *Core Magnetics, The Green, Sedbergh, Cumbria LA10 5JS, UK*
[2] *Department of Petroleum Engineering, Heriot Watt University, Edinburgh, UK*
[3] *Enterprise Oil plc, London, UK*

Knowledge of the directional characteristics of fluid flow in hydrocarbon reservoirs is important for developing refined reservoir models. These characteristics are influenced strongly by the shapes of individual pores (i.e. the pore geometry) and by the statistical alignment of pore long and short axes within the sediment (i.e. pore fabric). In reservoir characterization, one of the most important pore fabric parameters is the direction of preferred orientation of pore long axes, which is likely to exert a strong control on the direction of maximum permeability.

Quantitative information on reservoir pore fabric characteristics is notoriously difficult to acquire. In the past, information has been obtained from 2D sections, limited 3D views of small numbers of pores using scanning electron microscopy, or by examining resin casts of pore

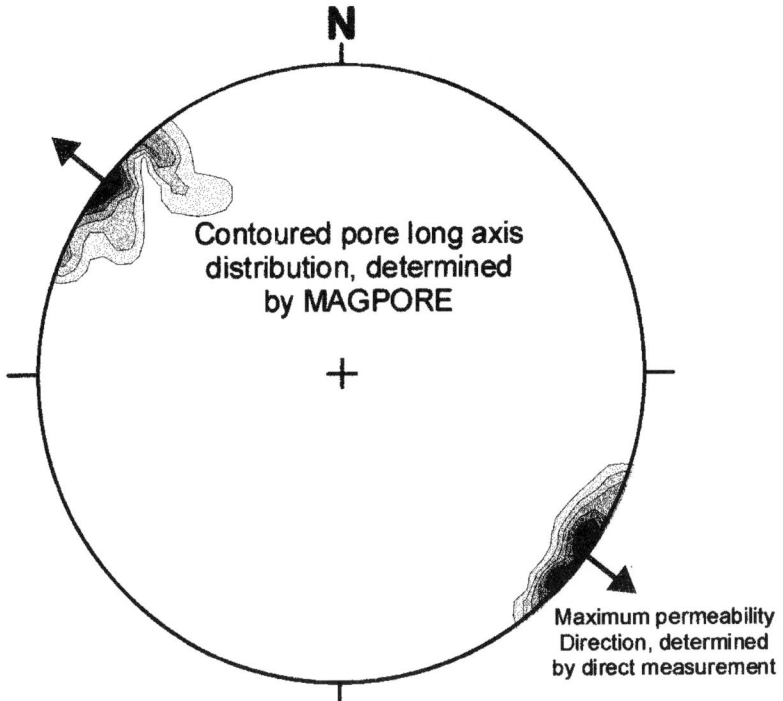

Fig. 1. Comparison between maximum permeability axis determined from Hassler cell measurements (arrows) and mean pore long-axis alignment determined from MAGPORE measurements (contours). In this figure, the arrows at the perimeter of the plot represent the average declination of the maximum permeability axes (which is near horizontal). The contours represent the directions of pore long axes, determined from MAGPORE measurements on sets of oriented plug samples. These results are from a deep-water sand deposit.

HAILWOOD, E. A., BOWEN, D., DING, F., CORBETT, P. W. M. & WHATTLER, P. 1999. Characterizing pore fabrics in sediments by anisotropy of magnetic susceptibility analyses. *In*: TARLING, D. H. & TURNER, P. (eds) *Palaeomagnetism and Diagenesis in Sediments*, Geological Society, London, Special Publications, **151**, 125–126

structure, after dissolving the rock. However, these methods are time-consuming and expensive, and do not readily lend themselves to providing usable quantitative information on the 3D pore fabric. Furthermore, it is rarely possible to analyse sufficient numbers of samples or volume of rock to provide a meaningful statistical representation of key pore fabric characteristics.

An alternative method for specifying pore fabrics is based on anisotropy of magnetic susceptibility (AMS) analyses. This method involves filling the pore space of small core samples with ferrofluid (a suspension of ultrafine magnetite particles in a suitable carrier fluid), using a specially developed saturation cell. The directional properties of the pore network are then quantified from AMS measurements on the ferrofluid-saturated samples. This method, termed 'MAGPORE', provides a rapid specification of the pore fabric in three dimensions, in a fraction of the time taken for standard petrofabric analyses.

In the first phase of the project, the method has been successfully applied to samples from a Permian aeolian dune sand reservoir and to an Early Tertiary deep-water fan reservoir in the northern North Sea. An important target of this phase was to quantify the relationship between pore fabric and permeability anisotropy in these reservoirs. This required constructing a specialized Hassler permeability cell, to allow fluid-permeability measurements to be made across the faces of cubic samples. By cutting samples with appropriate orientations, measurements could be made along orthogonal principal axes and diagonal axes, to specify fully the 3D permeability tensor.

Results from the deep-water fan reservoir (Fig. 1) show that the mean pore long-axis alignment in this reservoir conforms closely to the maximum permeability direction. This method is now being applied to several other types of reservoir, including turbidite sands and fluvial channel deposits. Clearly, determination of fabrics associated with such migration routes has important consequences for the studies of fluid movement through sedimentary rocks. It is also probable that where mineralization is associated with past fluid migration the combination of MAGPORE analyses and palaeomagnetic dating may allow dating of when such minerals were deposited, probably towards the end of the fluid migration.

Magnetic anisotropy indications of deformations associated with diagenesis

F. HROUDA[1,2] & J. JEŽEK[3]

[1] AGICO Inc., Ječná 29a, Box 90, CZ-621 00 Brno, Czech Republic
[2] Institute of Petrology and Structural Geology, Faculty of Science, Charles University, Albertov 6 CZ-128 43 Praha 2, Czech Republic
[3] Institute of Applied Mathematics and Computer Science, Faculty of Science, Charles University, Albertov 6 CZ-128 43 Praha 2 Czech Republic

Abstract: Anisotropy of magnetic susceptibility (AMS), as revealed by mathematical modelling and empirical studies, sensitively reflects initial ductile deformation associated with diagenesis and early tectonism. In rocks shortened vertically because of the gravitational loading by the weight of overlying strata, the degree of AMS increases and the magnetic fabric becomes more oblate. The magnetic foliation remains parallel to the bedding and the magnetic lineation remains near the water current directions. In rocks shortened laterally, usually by tectonic forces, the degree of AMS and the magnetic fabric oblateness initially decrease and only later increase. The magnetic lineation deflects from the water current direction and the magnetic foliation from the bedding, with its poles creating a girdle pattern. The AMS changes are due to the physical rotation of magnetic grains, which has serious implications for palaeomagnetism, but are major indicators of the change from a diagenetic to a tectonically controlled regime.

Diagenetic processes are traditionally considered to be those occurring during the burial of sediments which result in initial mineral changes and pore space reduction. The deformation associated with these processes is considered to be predominantly represented by vertical shortening because of the loading of the rocks by the weight of overlying strata. However, during the last 30 years, studies of the anisotropy of magnetic susceptibility (AMS) of the flysch-like sediments of the Rheno-Hercynian Zone of the Variscan orogeny of the NE Bohemian massif and rocks of the Flysch Belt of the West Carpathians (as summarized by Hrouda (1979, 1990), Hrouda & Stráník (1985) and Hrouda & Potfaj (1993)) have shown that the situation is not always so simple. Many cases are dominated by vertical shortening, but many examples have been found of lateral shortening and even more complex deformations that have affected the rocks immediately after their deposition. To better understand the origin of these observed magnetic fabrics, mathematical modelling of the magnetic fabrics developed under various regimes of strain associated with diagenesis has been examined (Hrouda 1991, 1994; Hrouda & Hrušková 1990). This paper summarizes the results of such mathematical modelling and provides examples of their assumed natural equivalents found in the magnetic fabrics of rocks in the Rheno-Hercynian Zone and the Flysch Belt where the modelled strain regimes can be compared with natural strain histories.

Magnetic variables and fabrics in sediments

The conventional variables defining the magnetic fabric, P, T, q and f have been used here. These are defined as follows:

$$P = k_1/k_3$$

$$T = 2\ln(k_2/k_3)/\ln(k_1/k_3) - 1$$

$$q = (k_1 - k_2)/[(k_1 + k_2)/2 - k_3]$$

and f is the angle between the magnetic foliation and the bedding plane; where k_1, k_2 and k_3 are the orthogonal maximum, intermediate and minimum principal susceptibilities of the susceptibility ellipsoid of each individual sample (see Tarling & Hrouda (1993) for further details). P, also called the degree of AMS, indicates the intensity of the preferred net orientation of magnetic minerals, and T (Jelínek 1981) and q

(Granar 1958) indicate their average shape. T is generally used in structural–tectonic studies of lithified rocks and varies between -1 (perfectly linear) and $+1$ (perfectly planar), with $-0.05 < T < 0.05$ corresponding to the transitional values. q (Granar 1958) has conventionally been used in sedimentary studies and varies from zero (perfectly planar) to two (perfectly linear), the transition value between planar and linear magnetic fabric being $c.\,0.7$.

The nature of the magnetic fabrics generated during the deposition of a sedimentary rock has

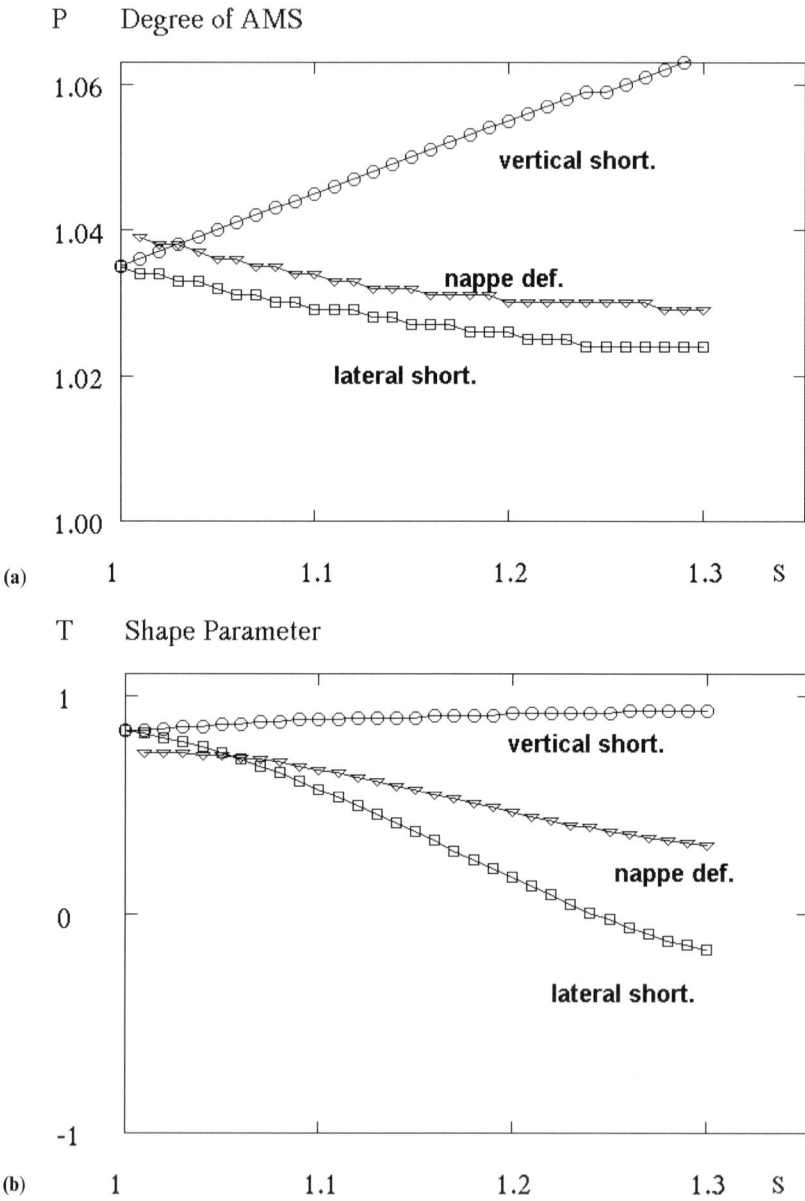

Fig. 1. The effects of increasing post-depositional deformations on the depositional magnetic fabric. These values are determined mathematically based on a line–plane model following Owens (1974). (**a**) The degree of AMS (P) as a function of increasing strain (S) measured as the maximum to minimum stretch. (**b**) The shape of the magnetic fabric (T) as a function of increasing strain (S). ○, Vertical shortening; □, lateral shortening; △, nappe deformation processes.

been investigated in many laboratory experiments, using a variety of depositional regimes that, as far as possible, simulate natural depositional conditions (for a summary, see Rees & Woodall (1976) and Rees (1983)). These have shown three main characteristics.

(1) When deposition is from a dispersed suspension, or from grain-by-grain deposition from still or running water, onto a flat or sloping bottom, then q is less than <0.5 and the magnetic foliation is gently imbricated (i.e. it dips by $<15°$ from the bedding plane in a direction opposite to the direction of flow) and the magnetic lineations are parallel to the direction of flow and similarly plunging slightly towards the source of flow.

(2) When deposition is from a medium concentrated suspension ($c.8\%$ in Rees' (1983) experiments), usually at flow velocities above 1 cm/s, $q < 0.3$ and the magnetic foliation still dips at $<15°$ towards the origin of flow. However, the magnetic lineation becomes perpendicular to both the direction of flow and the dip of magnetic foliation. (Such transverse orientation of the magnetic lineation can also originate through synsedimentary pure shear deformation (Rees 1983) but, in these conditions, the q values lie in the range 0.5–1.)

(3) During deposition from a very concentrated grain dispersion, usually requiring a current velocity significantly greater than 1 cm/s, onto a sloping bottom, the q values are higher, reaching 0.7, and the magnetic foliation dips 25–30° towards the origin of flow and the magnetic lineation is parallel both to the flow and to the dip of magnetic foliation, and plunges, within the foliation plane, towards the origin of flow.

These initial depositional magnetic fabrics may persist but can be rapidly modified by superimposed ductile deformations, as modelled by Owens (1974), Hrouda & Hrušková (1990) and Hrouda (1991). If these deformations involve vertical shortening, as can occur during any stage of burial, then the degree of AMS increases and the magnetic fabric becomes more oblate and flattened with increasing strain (Fig. 1a and b). Hence, with increasing strain, the magnetic foliation becomes closer to the bedding and the angle between the magnetic foliation and bedding decreases, but the magnetic lineation does not change its orientation (Fig. 2a). Lateral (i.e. bedding parallel) shortening can also occur early after deposition. For weak lateral shortening, the degree of AMS initially decreases and becomes more planar, but when stronger shortening occurs, such as during late diagenesis or early metamorphism, the fabric becomes stronger and more triaxial, and even linear (figs 1–8 of Hrouda (1991)). The lineation increasingly deviates from the flow direction towards that of the maximum strain, often forming a bimodal pattern; the foliation similarly moves from the bedding plane to a plane perpendicular to the principal strain direction (Fig. 2b and c). There is thus a gradation in fabrics from those of the initial depositional conditions to those associated with the later increasing tectonic strains, but until these tectonic strains become dominant, the intermediate magnetic fabrics are still a partial reflection of the original depositional conditions. Thus the magnetic fabric variables are relatively easily determined characteristics that indicate the extent to which depositional properties have been altered by lithification and increasing tectonism. However, all such physically controlled factors may themselves be drastically modified by chemical changes, particularly when authigenic grain growths occur under conditions of strain. It is therefore necessary to compare the theoretical situation with those found in well-studied sediments and rocks exhibiting different degrees of weak tectonism, such as the Flysch Belts of the West Carpathians and the Bohemian Massif.

The Flysch belt of the West Carpathians

The mostly Paleogene rocks of the Flysch Belt (Fig. 3) in this area form an apparent continuation of the Flysch Zone of the Alps and have been involved in a series of complex Miocene thrust sheets (Andrusov 1968; Hrouda & Stráník 1985). They have undergone only very weak regional metamorphism, as indicated by the disseminated organic matter being transitional between brown and sub-bituminous coal grades (Müller et al. 1991), and often contain almost undeformed, or very weakly deformed, sedimentary structures (flute casts, groove casts, cross-bedding). In many localities of the Flysch Belt, the magnetic fabric is apparently depositional in origin (Fig. 4a and b); the magnetic foliation is near the bedding and the magnetic lineations are consistent with the water current directions derived from the flute casts (after simple tectonic correction for the tilt of the beds), and their AMS magnitudes are low and oblate. The dominance of depositional controls on the fabric is also indicated, for example, in the Bouma flysch sequence where the lowest member, A, has a magnetic lineation perpendicular to the flow direction (Fig. 4a), whereas higher levels show lineations parallel to the flow direction (Fig. 4b and c), consistent with the changes in flow regime indicated by the sediments themselves. However,

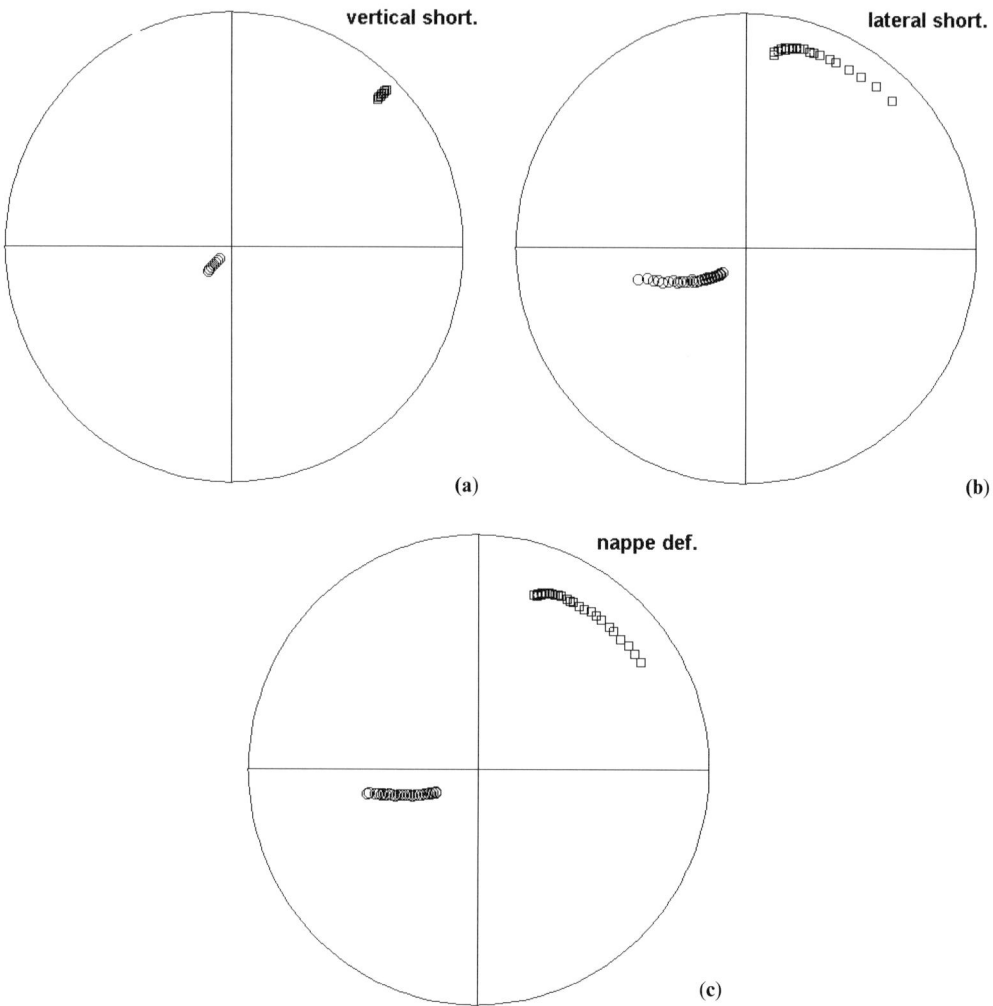

Fig. 2. The effect of increasing post-depositional deformations on the orientation of the magnetic fabric. These directions are determined using a line–plane model following Owens (1974) for cases of (**a**) vertical shortening, (**b**) lateral shortening, and (**c**) nappe deformation. □, Magnetic lineations; ○, foliation poles. Equal-area projection on lower hemisphere.

in some areas, particularly in the central areas, the magnetic fabric pattern is so different that it cannot be considered depositional in origin (Fig. 4d and e). Here the magnetic lineation may be still approximately parallel to the current direction, but the magnetic foliation is only rarely near the bedding and clearly deviates well away from it in most specimens, being almost perpendicular to it in some localities, where the poles to the magnetic foliation plane form conspicuous girdles (Fig. 4e). Such magnetic fabric patterns correspond to those modelled for the effect of lateral shortening on the depositional magnetic fabric which eventually combines with bedding parallel simple shear. These combined patterns are characteristic of the nappe strain expected from lateral shortening perpendicular to the long axis of the basin. In other areas within the Flysch Belt, the magnetic fabrics are similarly low and vary from clearly oblate to clearly prolate. Their magnetic foliation poles also have a girdle distribution, but the magnetic lineations are no longer parallel to the current direction, but are scattered widely, and the magnetic foliations are not parallel to the bedding (Fig. 4d). The coincidence of the current flow direction and the magnetic lineation in the former area (as shown in Fig. 4e) can therefore be explained

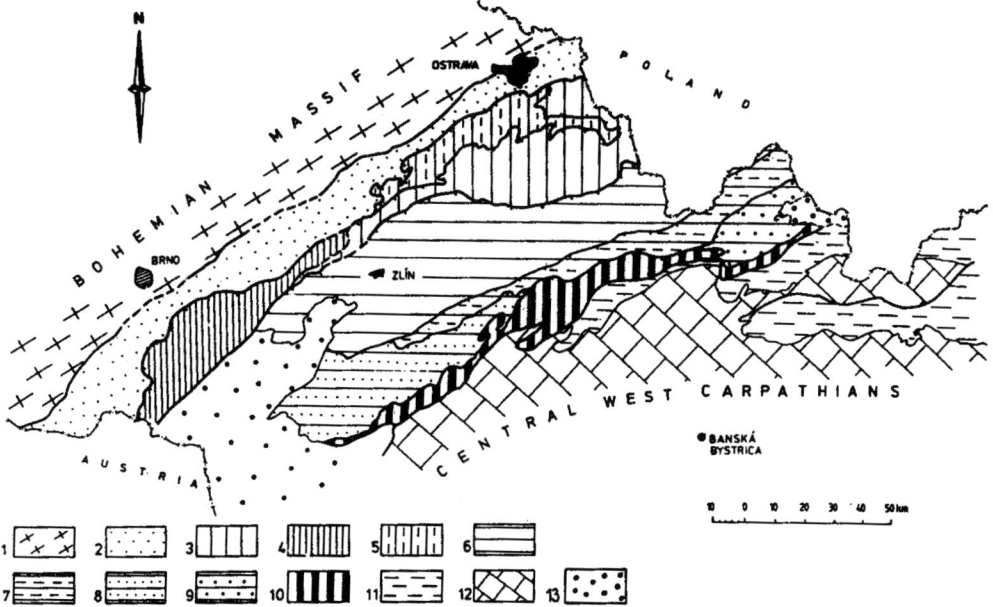

Fig. 3. An outline of the geology of the western sector of the Flysch Belt of the West Carpathians. 1, Bohemian Massif; 2, Carpathian Foredeep; 3–9, Flysch Belt thrust sheets; 10, Klippen Belt; 11, 12, Central Carpathians; 13, Vienna and Orava basins. After Hrouda & Potfaj (1993).

by the tectonic axis of the basin being the same as that for the orientation of the current flows at the time of deposition; the magnetic lineations have not been reorientated. In these virtually unmetamorphosed rocks, there is therefore clear magnetic evidence for a relatively strong effect of ductile deformation, which is hard to detect using conventional techniques.

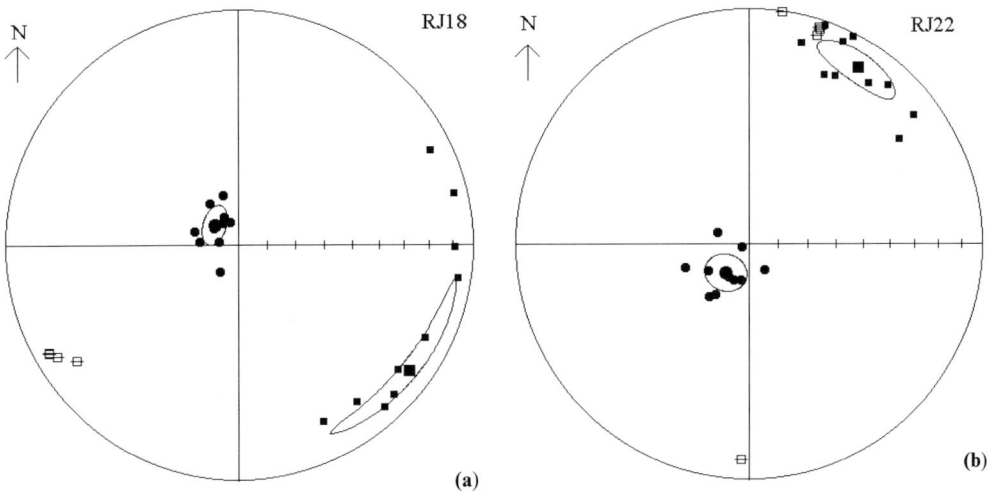

Fig. 4. Examples of magnetic foliations, lineations and palaeocurrents in the West Carpathian Flysch Belts. The magnetic foliation poles (●), magnetic lineations (■) and palaeocurrent directions (□) are shown on the lower hemisphere of an equal-area projection. The large symbols represent the mean values, with associated circles of 95% confidence. The orientations have been corrected for the tilt of the bedding.

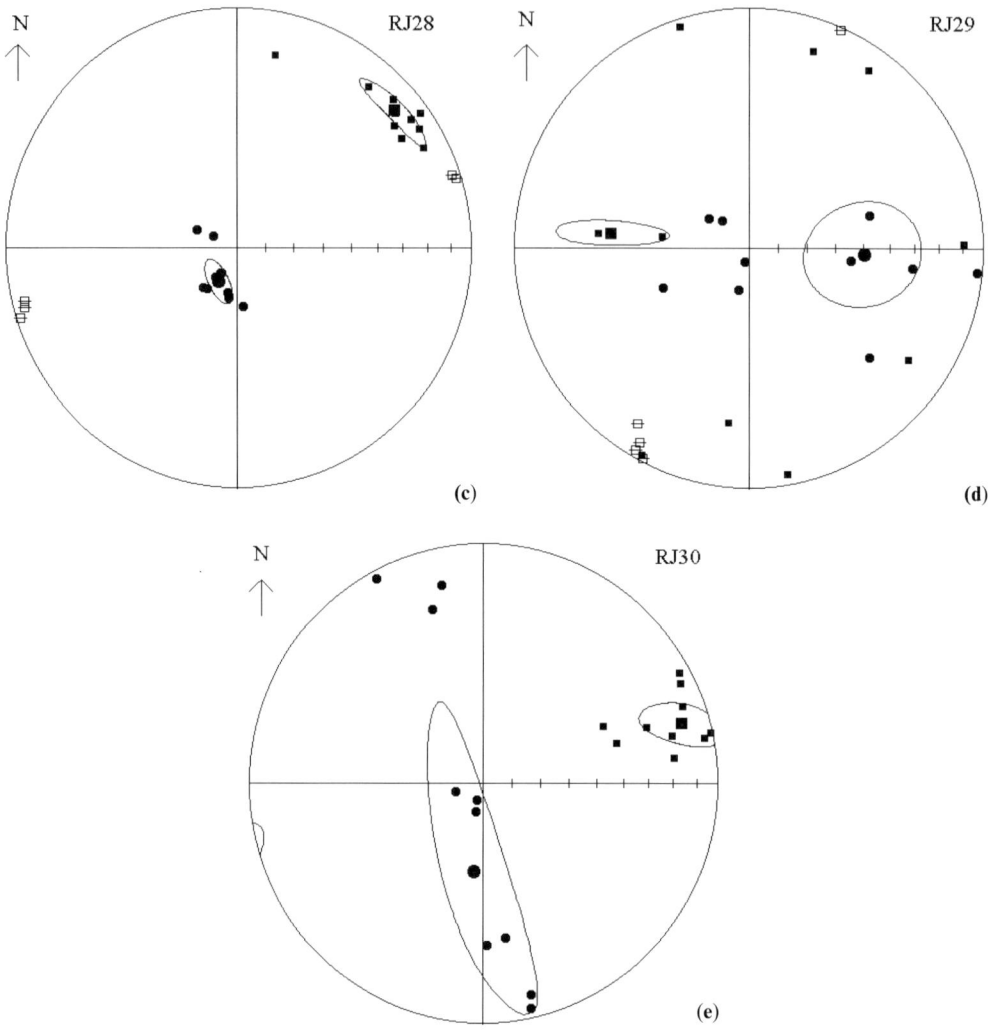

Fig. 4. (*continued*).

Flysch deposits in the Rheno-Hercynian Zone

Several different flysch formations occur in these sediments overlying the Bohemian Massif (Fig. 5), and these exhibit different degrees of metamorphism. In the Drahanská vrchovina Upland the eastern Myslejovice Formation is only very weakly metamorphosed, with the disseminated organic matter just transforming into brown and black coal, and virtually no ductile deformation, and the only observable fabric element is that of the bedding. The rocks of the western Protivanov Formation, in the same region, are slightly more metamorphosed and deformed to the extent that cleavage can be observed in some places. In the Nízký Jeseník Mts area, the easternmost formation is the Hradec–Kyjovice Formation, which underwent only weak metamorphism, although to a higher grade than the Myslejovice Formation, as the dispersed organic matter has been transformed into black coal (Müller 1987; Müller & Stráník 1991), and there are some SE vergent buckle folds of long, gentle wavelength. Further west, the metamorphic grade and ductile deformation gradually increase, so that the Moravice Formation shows some spaced cleavage and NW vergent buckle folds, and slaty cleavage and cleavage folds occur in the more westerly Horní Benešov Formation. The westernmost Andělská

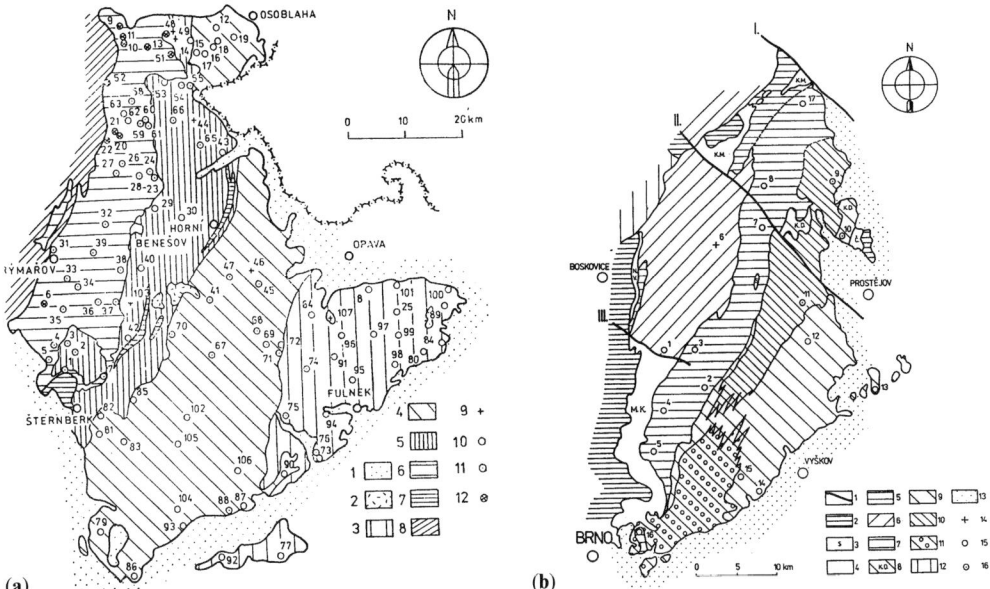

Fig. 5. Geological outline of the easternmost Rheno-Hercynian Zone. (**a**) The Nízký Jeseník Mts. 1, Tertiary and Quaternary; 2, young volcanic rocks; 3, Hradec–Kyjovice Formation; 4, Moravice Formation; 5, Horní Benešov Formation; 6, Andělská Hora Formation; 7, Šternberk–Horní Benešov Belt; 8, Vrbno Formation; 9–12, sites for AMS study. After Hrouda (1979). (**b**) The Drahanská vrchovina upland. 1, Transverse faults; 2, crystalline basement; 3, 4, Devonian rocks of the Moravian Karst; 5, 6, Protivanov Formation; 9–11, Myslejovice Formation; 13, Tertiary and Quaternary rocks; 14–16, sites for AMS study.

Hora Formation is characterized by NW vergent cleavage folds and a well-developed slaty cleavage, which, on its western border, becomes transformed into metamorphic schistosity. In this formation, the organic matter has been transformed to high coalification grade, anthracite or even graphite (Müller 1987; Müller & Stráník 1991). For the purpose of this paper, which is primarily concerned with late-stage diagenesis rather than tectonic structures, the examples have been selected only from the weakly deformed Myslejovice, Protivanov, Hradec–Kyjovice and Moravice Formations, which do not show slaty cleavage.

The degree of AMS (P) is lowest in the Myslejovice and the Protivanov Formations (Fig. 6a). In the Myslejovice Formation, the magnetic fabric is clearly planar (Fig. 6b) and the magnetic foliation is very close to the bedding (Fig. 7a), as is the magnetic lineation (Fig. 7b), with both foliations and lineations showing patterns compatible with those of the palaeocurrent directions (Kumpera 1984). Comparison of these features with those of the depositional experiments clearly shows that they can be considered to be fundamentally of depositional origin. In the Protivanov Formation, the degree of AMS is also low, but the magnetic fabric ranges from moderately planar to moderately linear (Fig. 6a and b). The magnetic foliations are parallel to the bedding only in some specimens. In most specimens, the magnetic foliations are considerably deflected from the bedding and their foliation poles tend to form a relatively narrow irregular girdle (Fig. 8a). The magnetic lineations create a conspicuous cluster perpendicular to the girdle in magnetic foliation poles (Fig. 8b). This magnetic fabric is no longer depositional in origin; the predominantly linear magnetic fabric in many specimens, the existence of a girdle in magnetic foliation poles and the cluster in magnetic lineations are not compatible with the magnetic fabric originating through deposition. On the other hand, this magnetic fabric is compatible with that modelled as originating through weak post-depositional deformation, mainly by lateral shortening and possibly combined with bedding parallel simple shear. The Hradec-Kyjovice Formation shows a higher magnitude AMS (Fig. 6a and b), but more variable fabric than in the previous zones. Most specimens show magnetic lineations that are scattered within the bedding plane and have

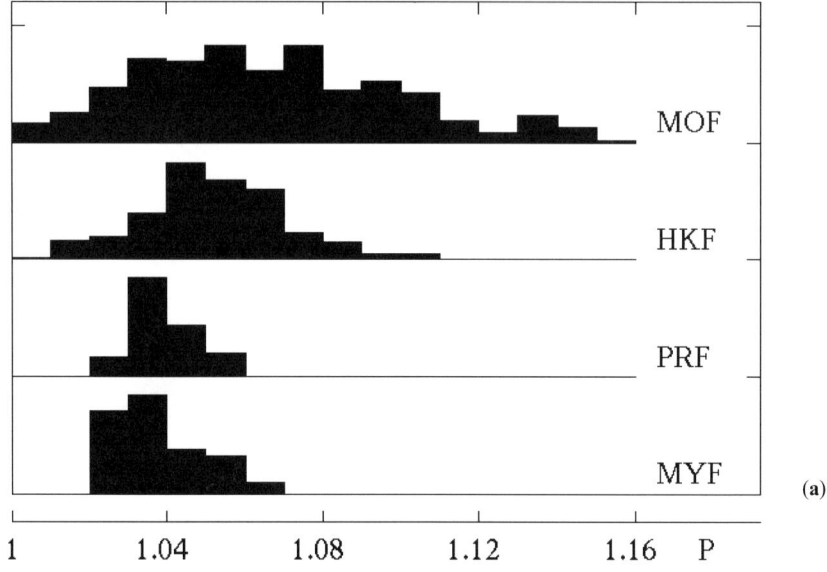

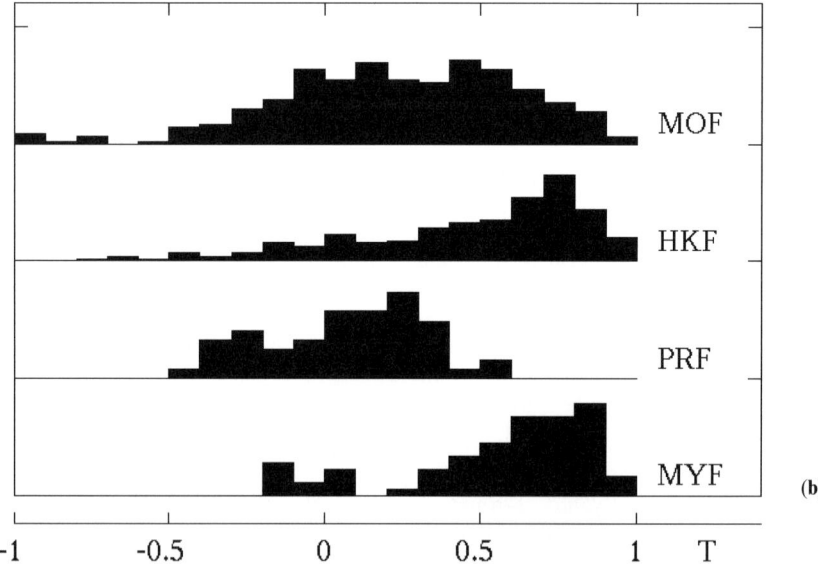

Fig. 6. The frequencies of (**a**) degree of AMS (P) and (**b**) shape parameters (T) in the Rheno-Hercynian Zone. MOF, Moravice Formation; HKF, Hradec–Kyjovice Formation; PRF, Protivanov Formation; MYF, Myslejovice Formation.

orientations that are compatible with the principal orientations of the current directions (Fig. 9b). The magnetic foliations are mostly close to the bedding, although there are some exceptions (Fig. 9a). These magnetic fabrics could, in principle, be regarded as depositional in origin, but the relatively high degree of AMS testifies against this. It seems more probable that

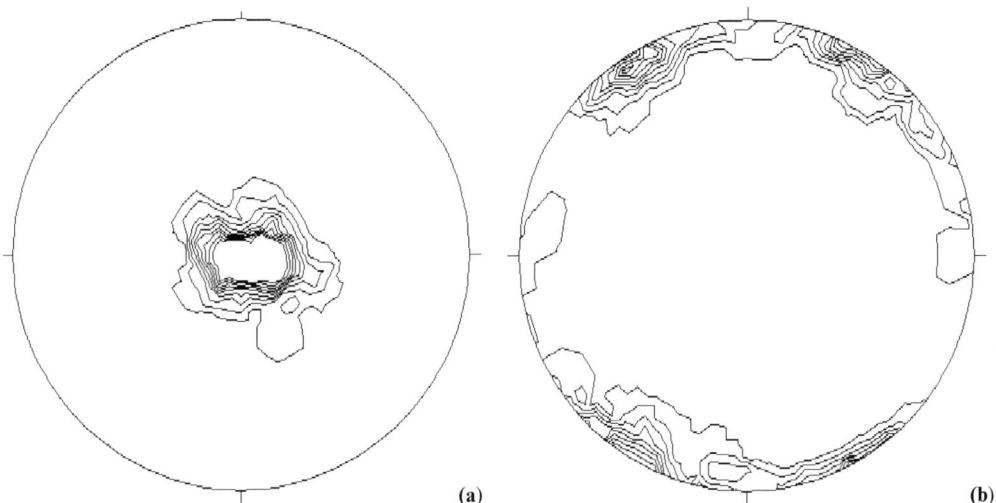

Fig. 7. Magnetic foliations and lineations in the Myslejovice Formation. (**a**) Magnetic foliation poles, and (**b**) the magnetic lineations plotted on equal-area lower hemisphere projections. The directions have been corrected for bedding tilt. The outermost contour corresponds to the 1% density; the other contours increase by 2%.

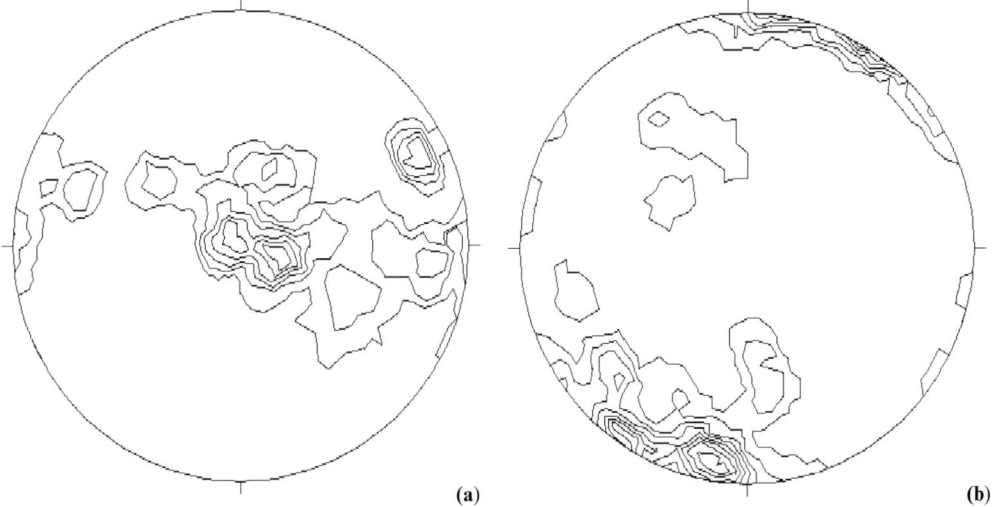

Fig. 8. Magnetic foliations and lineations in the Protivanov Formation. (**a**) Magnetic foliation poles, and (**b**) the magnetic lineations plotted on equal-area lower hemisphere projections. The directions have been corrected for bedding tilt. The outermost contour corresponds to the 1% density; the other contours increase by 2%.

the magnetic fabric, immediately after deposition, was shortened vertically, probably because of the loading by the weight of overlying strata. Such a conclusion is in agreement with the results of the mathematical modelling, and is also supported by the higher coalification grades in this formation than in the Myslejovice and Protivanov Formations in the Drahanská vrchovina upland. In the Moravice Formation, which shows spaced cleavage in some areas, the degree of AMS is on average high, but extremely variable; the magnetic fabric is also extremely variable, ranging from very planar to very linear (Fig. 6a and b). The magnetic foliations are close to the bedding in the great majority of specimens, but deflected in many specimens, creating an initial girdle in their poles (Fig. 10a). The magnetic lineations are near the bedding and create a cluster perpendicular to the initial girdle in magnetic foliation poles (Fig. 10b). The very

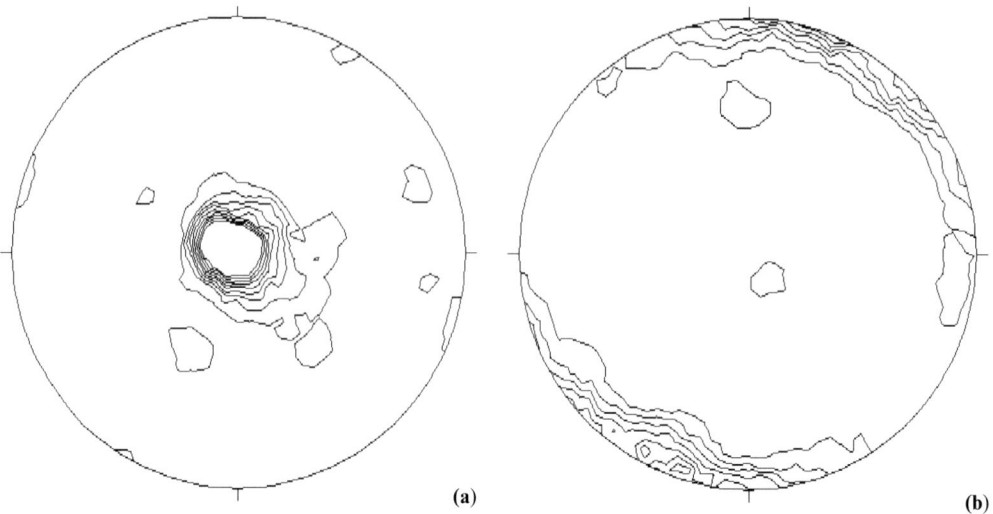

Fig. 9. Magnetic foliations and lineations in the Hradec-Kyjovice Formation. (**a**) Magnetic foliation poles, and (**b**) the magnetic lineations plotted on equal-area lower hemisphere projections. The directions have been corrected for bedding tilt. The outermost contour corresponds to the 1% density; the other contours increase by 2%.

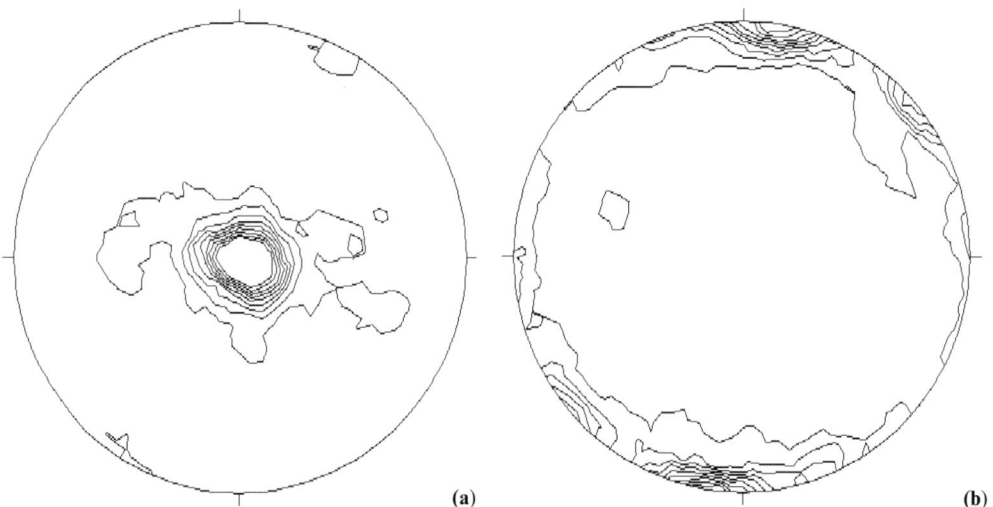

Fig. 10. Magnetic foliations and lineations in the Moravice Formation. (**a**) Magnetic foliation poles, and (**b**) the magnetic lineations plotted on equal-area lower hemisphere projections. The directions have been corrected for bedding tilt. The outermost contour corresponds to the 1% density; the other contours increase by 2%.

strong variation in the degree of AMS and in the magnetic fabric shape, as well as the existence of the initial girdle in the magnetic foliation poles, all demonstrate that the magnetic fabrics of this formation can be regarded as partially deformational even though some features of the original depositional fabrics persist. The deformation is most likely to be by lateral shortening associated with the creation of the spaced cleavage, which is likely to be younger than some of the diagenetic features.

Concluding remarks

Diagenesis is traditionally considered in terms of burial of sediments resulting in pore space reduction. During the burial, therefore, the sediments are implicitly assumed to be primarily affected by vertical shortening because of the loading by the weight of overlying strata. However, in these examples from the Flysch Belts in the West Carpathians and the RhenoHercynian Zone of the NE Bohemian Massif, the magnetic fabrics are far more complex than indicated by such a simple model. Mathematical modelling of the effects of the weak deformations, particularly bedding parallel shortening and bedding parallel simple shear, shows that when these are superimposed on a depositional magnetic fabric, there is relatively good agreement between the modelled and observed magnetic fabrics. This shows that, although vertical shortening remains important, lateral shortening and bedding parallel simple shear are both major contributors to the processes operating on sediments during their diagenesis.

P. Müller is thanked for providing the unpublished data on the coalification of the organic matter in the rocks investigated by his team.

References

ANDRUSOV, D. 1968. *Grundriss der Tektonik der Nördlichen Karpaten.* SAV, Bratislava.

GRANAR, L. 1958. Magnetic measurements on Swedish varved sediments. *Arkiv för Geofysik*, **3**, 1–40.

HROUDA, F. 1979. The strain interpretation of magnetic anisotropy in rocks of the Nízký Jeseník Mountains (Czechoslovakia). *Sbornik Geologickych Ved, Uzita Geofzika*, **16**, 27–62.

—— 1990. Variscan tectonic overprinting the magnetic fabric in the sedimentary and crystalline nappes in the NE Bohemian Massif. *Tectonics*, **12**, 507–518.

—— 1991. Models of magnetic anisotropy variations in sedimentary thrust sheets. *Tectonophysics*, **185**, 203–210.

—— 1994. Mathematical modelling of the behaviour of passive fabric elements (and corresponding AMS) in the transpression zone. *In*: BUNGE, H. J., SIEGESMUND, S., SKROTZKI, W. & WEBER, K. (eds) *Textures of Geological Materials.* DGM Verlag, Oberursel, 381–392.

—— & HRUŠKOVÁ, L. 1990. On the detection of weak strain parallel to the bedding by magnetic anisotropy: a mathematical model study. *Studia Geophysica et Geodaetica*, **34**, 327–341.

—— & POTFAJ, M. 1993. Magnetic anisotropy as an indicator of the weak ductile deformation of the Intra-Carpathian Palaeogene and the Magura Flysch (in Czech). *Západné Karpaty, geol.*, **17**, 121–134.

—— & STRÁNÍK, Z. 1985. The magnetic fabric of the ždánice thrust sheet of the Flysch Belt of the West Carpathians: sedimentological and tectonic implications. *Sedimentary Geology*, **45**, 125–145.

JELÍNEK, V. 1981. Characterization of magnetic fabric of rocks. *Tectonophysics*, **79**, T63–T67.

KUMPERA, O. 1984. *Geology of the Carboniferous of the Jeseníky block of the Bohemian Massif* (in Czech). Czech Geological Survey, ser. Knihovna ÚÚG, Prague.

MÜLLER, P. 1987. *Source rocks for the crude oil and gas in the SE slopes of the Bohemian Massif and in the Vienna basin* (in Czech). PhD thesis, Czech Geological Survey.

—— & STRÁNÍK, Z. 1991. *Crude oil and gas. Perspectives of the survey for the crude oil and gas in the Czech Republic.* Unpublished report, Czech Geological Survey.

——, KREJČI, D., ČIČEK, P., FRANCŮ, J. & ŠEVČIKOVÁ, E. 1991. *Origin, migration, and accumulation of the hydrocarbons in the area of the SE slopes of the Bohemian Massif.* Unpublished report, Czech Geological Survey.

OWENS, W. H. 1974. Mathematical studies on factors affecting the magnetic anisotropy of deformed rocks. *Tectonophysics*, **24**, 115–131.

REES, A. I. 1983. Experiments on the production of transverse grain alignment in a sheared dispersion. *Sedimentology*, **30**, 437–448.

—— & WOODALL, W. A. 1976. The magnetic fabric of sand and sandstones. *Earth and Planetary Science Letters*, **25**, 121–130.

TARLING, D. H. & HROUDA, F. 1993. *The Magnetic Anisotropy of Rocks.* Chapman and Hall, London.

Acquisition of anhysteretic remanence and tensor subtraction from AMS isolates true palaeocurrent grain alignments

GRAHAM J. BORRADAILE, PHILIP W. FRALICK & FRANCE LAGROIX

Geology Department, Lakehead University, Thunder Bay, Ont., Canada, P7B 5E1
(e-mail: borradaile@lakeheadu.ca)

Abstract: Anisotropy of magnetic susceptibility (AMS) is commonly used to detect subtle palaeocurrent directions. A Proterozoic subarkose shows extreme anomalous inverse AMS fabrics that cannot be attributed to the well-known effect of single-domain magnetite. This was confirmed by anisotropy of anhysteretic remanent magnetization (AARM). Further, by subtracting increasing proportions of the AARM tensor from the AMS tensor the paramagnetic/diamagnetic response of the silicate fabric was isolated. This reveals an inverse quartz fabric but analysis of the results shows that two grain alignments are present. A prominent inverse magnetic fabric is due to quartz long axes aligning with current flow revealed by the difference [AMS] – [AARM]. This is due to the avalanche flow of concentrated sand dispersions down the slip-face of cross-stratified units. Dispersive pressure caused by grain interactions and shear within the fluid produces a subhorizontal resultant force that aligns grains by frictional freezing during deposition. In some samples AARM isolates a secondary feeble alignment of magnetite nearly perpendicular to bedding. This is due to suspension rainout and sweeping by the reverse circulation separation eddy that supplies additional material to the top of the sand laminae. It particularly affects fine grains such as magnetite that infiltrate the framework's pores to produce a secondary population of vertically oriented grains.

Anisotropy of magnetic susceptibility (AMS) has been used to identify palaeocurrent directions from sedimentary rocks since about 1961. Tarling & Hrouda (1993) summarized the principal discoveries. Minimum principal susceptibility is normally almost perpendicular to bedding. Maximum susceptibility may define the current grain alignment at slow velocities (<1 cm/s) whereas faster currents may roll grains so that their long axes and maximum susceptibilities are perpendicular to flow. The magnetic foliation may also imbricate, tilting upstream depending on the hydrodynamic conditions. The effects of inverse magnetic fabrics have not been considered previously for sedimentary fabrics. From the viewpoint of fabrics and magnetic anisotropy these studies make one or two assumptions that have been resolved in studies of metamorphic rocks (Borradaile & Henry 1997). The first is that the rock's AMS is controlled by the orientation distribution of grains whose shape is coaxial and congruent with the grain's AMS ellipsoid. Considerable errors arise where inverse fabrics occur because of the presence of single-domain (SD) magnetite (Jackson & Tauxe 1991) or matrix minerals such as calcite (Rochette 1988). Secondly, it is assumed that AMS is due to one mineralogical or magnetic response. In the early literature, it was commonly assumed that traces of magnetite accounted for the entire AMS signal. Later, the importance of the paramagnetic contribution was realized (Borradaile 1987; Borradaile *et al.* 1990) and subsequently the importance of even a diamagnetic matrix has been realized (Hrouda 1986); Hrouda's case will be extended here. Commonly, two or even more equally significant magnetic sub-fabrics may coexist. These may be due to different paramagnetic, diamagnetic or ferromagnetic accessory minerals, and it is even possible that they may have different orientation distributions. For example, one mineral may be aligned parallel to current flow and another perpendicular to it, depending on their hydrodynamic behaviour. One can isolate the orientation distribution of the ferromagnetic component by applying anhysteretic remanent magnetizations (ARMs) in different directions to detect the anisotropy of ARM (AARM). The complexities that arise in a coarse-grained subarkose with palaeocurrent directions well known from macroscopic sedimentation structures will be illustrated.

BORRADAILE, G. J., FRALICK, P. W. & LAGROIX, F. 1999. Acquisition of anhysteretic remanence and tensor subtraction from AMS isolates true palaeocurrent grain alignments. *In*: TARLING, D. H. & TURNER, P. (eds) *Palaeomagnetism and Diagenesis in Sediments*, Geological Society, London, Special Publications, **151**, 139–145.

Sedimentary processes

Samples were obtained from a drill core through the Matinenda Formation, north of Elliot Lake, Ontario (Fralick & Miall 1989, fig. 1). Matinenda Formation fluvial sandstones and conglomerates form the basal portion of the Palaeo-Proterozoic Huronian Supergroup. Sediments were deposited by channel-dominated braided streams in a broad NNW-SSE orientated palaeo-valley (Fralick & Miall 1989). The subarkose comprises 90% quartz, largely strained, about 10% heavily altered feldspar and traces of opaque minerals. The coarse-grained sand, which dominates this portion of the formation, was transported via small- and medium-scale dunes on channel floors. All samples were of trough cross-stratified, coarse-grained sandstones produced by this dune migration (e.g. Fig. 1). Sampling was confined to five depositional events within a single channel, with two samples from each small-scale, trough cross-stratified set, $c.\,20\,cm$ thick. Major bedding planes were at 90° to the drill-core axis and foreset slopes vary from 26° to 32° and are shown dipping arbitrarily to 'north', on the stereogram (broken line, Fig. 2a). Magnetic core samples were drilled along the axis of the larger diameter field core.

The foreset laminae were deposited by avalanches originating near the crest lines of the dunes. Concentrated suspensions were maintained by dispersive pressure caused by collisions (Lowe 1982) or close approaches (Rees 1968). In these circumstances, grains will align to minimize angular momentum transfer caused by collision. This results in a preferred orientation with long axes parallel to the movement direction and the short axis near the direction of maximum velocity gradient (Rees 1968). An imbrication will also develop as the resistance to shear of the dispersion creates pressure (Bagnold

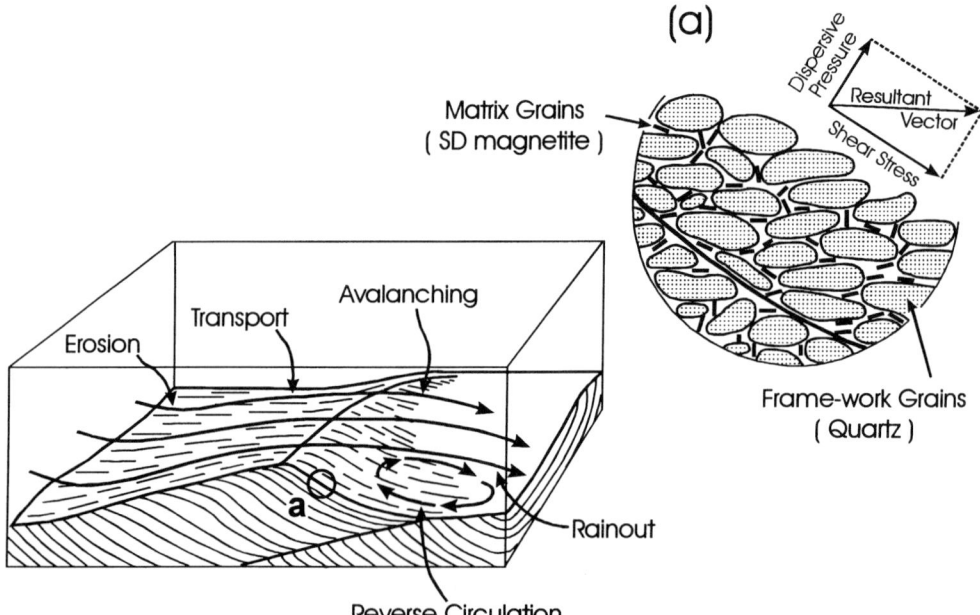

Fig. 1. Processes controlling grain orientation in trough cross-stratification formed by downstream migration of a discontinuously crested dune. Water flowing over the crest of a dune upstream from the dune shown loses contact with the bed. Where contact is regained the force results in a zone of erosion. Sand entrained from this scour and other grains reaching the bed from intermittent suspension travel as a traction carpet up the surface of the dune and avalanche down the lee slope depositing downstream-dipping laminae. Between sporadic avalanche events grains are delivered to the lee slope by rainout into the slower-moving water immediately behind the dune. The reverse circulation separation eddy that develops here serves to sweep material onto the surface of the laminae. During an avalanche grains align with long axes parallel to the resultant vector produced by the two main forces operating in the flow: shear stress and dispersive pressure. The enlargement in (**a**) shows that matrix grains, such as fine magnetite, added to the top of the laminae through rainout and traction by the separation eddy will infiltrate pore spaces and acquire steep orientations. Other matrix grains may align parallel to the framework grains during sedimentary compaction.

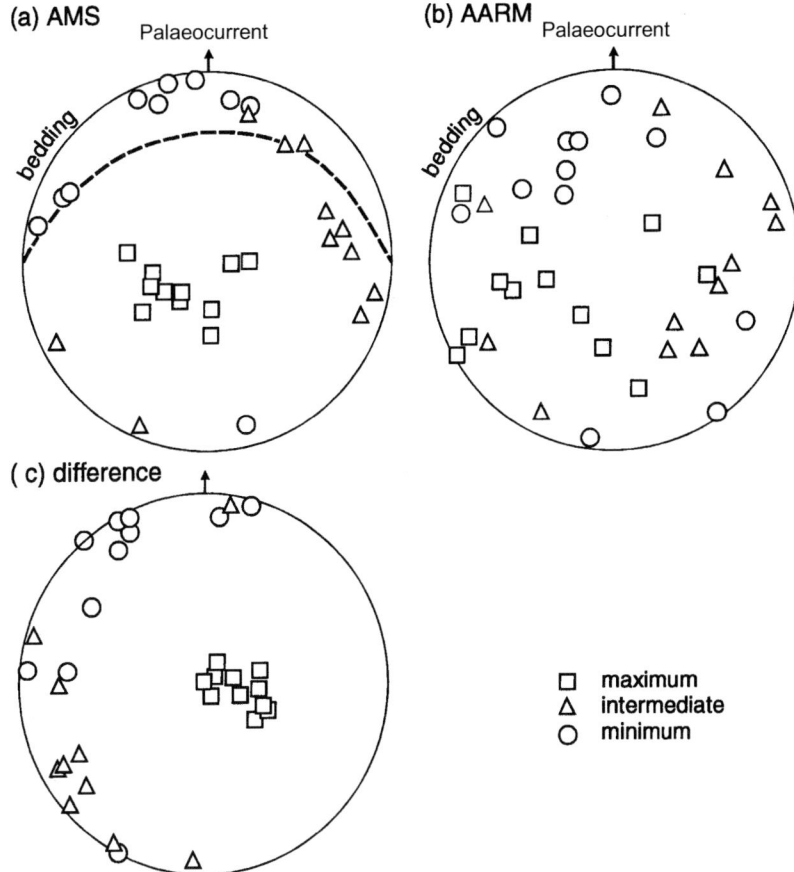

Fig. 2. The AMS fabric orientations shown on equal-area lower-hemisphere stereonets. Bedding is horizontal in all cases. (**a**) The AMS summing the induced magnetization anisotropy for all grains (quartz and magnetite). (Note the inverse fabric with maximum susceptibility perpendicular to bedding and minimum susceptibility parallel to movement direction.) The average avalanche slope is shown by a broken line. (**b**) The ARM anisotropy that records only the fabric contributions of very fine SD magnetite. This shows also an inverse fabric, but a weaker one. The grains' alignments are less well developed because some grains may have settled into steep pore spaces and adhered to the near-vertical pore walls (see text). (**c**) The AMS contribution of the framework (quartz) grains alone, revealed by subtracting (**b**) from (**a**). As quartz has an inverse fabric the maximum susceptibility is perpendicular to bedding and the minimum susceptibility is parallel to flow.

1954). This results in a shear stress (T) in the plane of shear, and a dispersive pressure (P) normal to that plane (Fig. 1). The resultant force on the grain acts at an average angle of $\tan^{-1}(T/P)$ to form an imbrication angle of 18°, $T/P = 0.32$, for purely inertial particle collisions. Where viscous drag of the intergranular fluid dominates, the imbrication angle is 37°, $T/P = 0.75$ (Hamilton et al. 1968; Rees 1968). Experimental results (Hamilton et al. 1968; Ellwood & Howard 1981), and data from non-lithified foreset laminae (Rees 1968; Taira & Lienert 1979) and lithified foreset laminae (Potter & Mast 1963) corroborate this theory. One minor, unexplained discrepancy is the tendency, within a foreset unit, for the long axis orientations to deviate consistently in one direction on the plane of layering, from the movement direction defined by the maximum slope of the foresets (Potter & Mast 1963).

The average preferred orientation discussed above should apply to all grains transported by the concentrated dispersions because the same forces are expected to affect all moving particles.

Thus, both the larger grains, resting on one another to form the framework, and the smaller grains infilling pores between framework grains will have, on average, similar orientations if the grains travelled in the same dispersion. However, matrix grains and some framework grains will probably be added to the top of the laminae after deposition (Fig. 1). This is accomplished through rainout into the separation eddy and reverse flow, traction transport by the separation eddy. During deposition of the avalanche, smaller grains concentrate near the bottom of the flow because of the effect of dispersive pressure (Lowe 1976, 1982). This may leave the upper portion of the laminae deficient in matrix so that small grains supplied by the separation eddy infiltrate the pore space. This may apply particularly to micron-sized magnetite. Since the average pore wall will be perpendicular to the layer, it may align the small particle in this direction, producing a sieve texture. This grain orientation will be perpendicular to the avalanche slope but will be represented by only a small percentage of suitably small grains (Fig. 1a).

One other sedimentary process has implications for the magnetic fabrics of the Matinenda sandstones. Early Palaeo-Proterozoic weathering, which supplied the detritus, occurred in an oxygen-deficient atmosphere (Cloud 1973; Holland 1973). This resulted in a very limited oxidation of iron in iron-bearing mineral phases during weathering. An indication of this is the large amount of detrital pyrite in the Matinenda Formation. Its contribution to the magnetic fabric is negligible because it is both a trace mineral and diamagnetic. However, the same geochemical environment enhanced the survival of fine-grained magnetite in the erosion–transportation system. This resulted in a sedimentary fabric indicator for the very fine-grained portion of a rock that would not normally be so well developed in younger rocks where detritus formed under a more evolved atmosphere.

Magnetic anisotropies

AMS records the directional variation of induced magnetization in all minerals. This encompasses paramagnetic minerals such as micas, clays, and mafic silicates with susceptibilities $<2000\,\mu SI$. Higher values are due to inclusions of ferromagnetic impurities. In contrast, the remainder of the rock matrix may be composed of diamagnetic minerals such as quartz, calcite and feldspar with very feeble anisotropies and a small diamagnetic susceptibility of approximately $-14\,\mu SI$. Finally, remanence-bearing minerals occur as traces in low concentrations, but they are highly susceptible; e.g. magnetite has a mean susceptibility of $2.8 \times 10^6\,\mu SI$.

The anisotropies are weak so that current directions may be very poorly determined because the slightest contamination by a few poorly oriented grains of higher susceptibility causes spurious anisotropy directions. Moreover, calcite, because of the parallelism of its c-axis and minimum susceptibility, could cause an inverse AMS with maximum susceptibility perpendicular to bedding (Rochette et al. 1992). The susceptibility anisotropy of quartz has been reported as both inverse and normal, depending on the origin of the samples (Hrouda 1986). Unfortunately, the size and shape of the samples was not given and studies of natural crystals are unlikely to be helpful, as the slightest contamination by inclusions can mask the intrinsic AMS of the quartz lattice. Hrouda (1986) quoted anisotropies of k_{ab}/k_c at 1.014 and 1.028 for synthetic quartz, which are probably more accurate reflections of the anisotropy of the pure lattice. Thus, synthetic quartz has a susceptibility of $-14.8\,\mu SI$ along the c-axis and $-14.5\,\mu SI$ in orthogonal directions. The susceptibility of remanence-bearing minerals, loosely called 'ferromagnetic' in this paper, is very high, up to $2.8 \times 10^6\,\mu SI$ for magnetite, so that microscopic traces of this mineral can swamp a conflicting but more significant palaeocurrent fabric of matrix silicates. Thus, it is very difficult to determine the matrix anisotropy (and infer palaeocurrent directions) where the matrix has low susceptibility. A further complication is that the most common remanence-bearing mineral, magnetite, has an anisotropy strongly influenced by 'magnetic grain size'. If the magnetite grain responds in a multidomainal (MD) manner, its long axis will produce the maximum low field susceptibility and its short axis will yield the smallest susceptibility. However, if the grain shows single domainal (SD) response, it is already saturated in its long axis so that this will appear as the minimum susceptibility direction (Stephenson et al. 1986; Jackson 1991). This generates inverse fabrics in which the minimum susceptibility defines the alignment of SD magnetite grains.

Separation of magnetic fabric contributions

A few simple techniques exist for isolating the anisotropy contributions of different minerals to the overall magnetic fabric of a rock. This approach is important to understand the alignment process and the particular mineral grains that it affected. The first, most simple, approach is the physical separation of minerals (Borradaile et al. 1986, 1990). Alternatively, the selective leaching of iron oxides may leave the matrix fabric intact (e.g. Borradaile et al. 1991; Jackson & Borradaile 1991) thus permitting its separate measurement. Finally, measurements of low field susceptibility at different frequencies can emphasize the response of different minerals according to their in-phase and out-of-phase response (Borradaile et al. 1991; Ellwood et al. 1993) or to different effective magnetic grain sizes of ferromagnetic grains (Thompson & Oldfield 1986; Jackson et al. 1989). However, these methods usually lack the precision needed to determine the anisotropy essential in estimating the palaeocurrent direction. Traditionally, AMS has been the preferred method, with the more recent addition of AARM. In this study, the AMS was measured in the usual way, at a low field (0.1 mT, 19 200 Hz). As a precaution, measurements were also made along the body diagonals of the sample and along the conventional 'principal plane' Girdler (1961) sample orientations. Borradaile & Stupavsky (1992) showed that this greatly increases sensitivity for practical reasons. On its own, the AMS combines the response of all minerals, paramagnetic, diamagnetic and ferromagnetic. Our samples are subarkoses so that the matrix should be characterized by a diamagnetic signal but the mean susceptibility was $+67 \mu S1$, indicating the presence of ferromagnetic contamination. (This was confirmed later by the samples' mean ARM susceptibility of 47.6 mA/m). The principal AMS directions are clearly related to sedimentary structure (Fig. 2a). However the results disagree with the commonly recognized pattern because the maximum susceptibility is perpendicular to bedding and the minimum susceptibility is parallel to the current direction. Although this has not been recorded previously in sediments, from the literature on metamorphic rocks, one recognizes this as an 'inverse fabric', caused by SD magnetite. To resolve this the AARM was measured by applying a laboratory anhysteretic remanence with a direct field of 0.1 mT while the alternating field (AF) decayed from 100 mT to zero. The anisotropy of remanence-bearing minerals is isolated in this way (Fig. 2b). The fabric is less well defined, because the fine-grained, sparse magnetites possess a weak fabric, but the inverse nature is confirmed, as most ARM maximum axes are steep.

AMS combines the magnetic response of all minerals whereas the AARM isolates the remanence-bearing ones, in this case SD magnetite. Thus, if the fabric of Fig. 2b is subtracted from that of Fig. 2a then any remaining fabric caused by the matrix grain alignment of the sediment can be visualized. To preserve the best accuracy we subtracted the double-precision, raw susceptibility measurements made in different directions through the sample. The alternative method, using the eigenvalues and eigendirections and working backwards, introduced too many calculation errors. The first approach was to estimate the volume of magnetite in each sample from the saturation remanence of the sample. This permitted an estimate to be made of the expected low-field susceptibility contribution to AMS by the SD magnetite. Unfortunately, this introduced so many inaccuracies that when the tensors were subtracted some components were negative and some were positive so that the resultant fabric ellipsoid was indeterminate. Thus, a simpler approach was adopted. First, the AMS and AARM tensors were normalized by their mean values, to yield dimensionless ellipsoids. It is known that the AARM fabric is a significant mask over the matrix AMS signal. Thus, increasing fractions of the AARM signal were subtracted from the AMS signal in incremental experiments. When this fraction exceeded 95% of the AMS signal, the remaining fabric was a slightly enhanced sedimentary fabric (Fig. 2c), now attributed to the quartz matrix.

The inverse fabric with maximum susceptibility (>0) perpendicular to bedding is still apparent because one now sees the dominant quartz grain fabric with quartz minimum susceptibility (-14.8) parallel to its long (c) axes, defining the current lineation. The alignment of quartz c-axes has been confirmed by observing the sign-elongation with a one-wavelength retardation plate. The quartz maxima (-14.5) are thus at a high angle to both major bedding and the avalanche slopes, being slightly imbricated in the avalanche (Figs 1 and 2a). Thus, both the quartz fabric and the much weaker magnetite fabric are inverse magnetic fabrics, which still faithfully record current flow by the alignment of their long grain axes. When the magnetite fabric is isolated by AARM (Fig. 2b) its alignment is rather weak, though discernibly inverse with maximum susceptibility steep and minimum susceptibility parallel to the movement direction. As noted above, this is understood in the sedimentation literature because some very fine grains settle onto the steep pore walls at the top of the avalanche surface during rainout. The SD, micron-size magnetite falls into this category, so that some samples have their long axes oriented steeply (horizontal maximum susceptibility, as the fabric is inverse; Fig. 2c). Confining magnetic

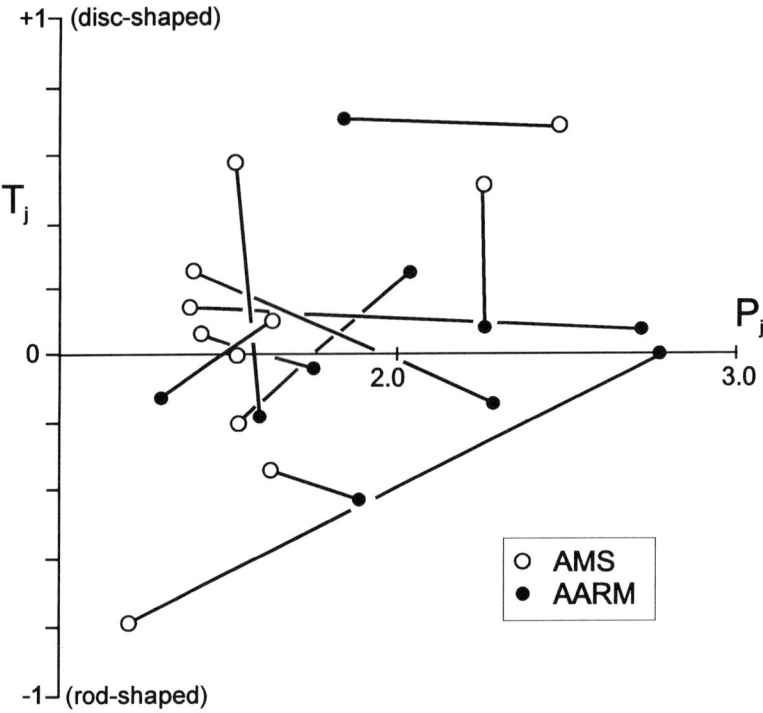

Fig. 3. The Jelinek (1981) fabric plot. This shows the symmetry or shape of the fabric ellipsoids (disc or planar fabrics, $T = +1$; rod shape or constricted, linear fabrics, $T = -1$) and their intensity ($P_j = 1$ = isotropic). Anisotropy of remanence (AARM) is generally more intense than AMS in the same sample and tends to have a more neutral shape ($T \approx 0$) than the AMS fabric. However, there is no consistent relationship between AMS and AARM for a sample, as shown by the crossing tie-lines.

fabric studies to AMS would probably have led to the conclusion that an inverse magnetite fabric caused the signal. However, by subtracting most of the magnetite signal it was possible to recognize an underlying, still stronger inverse fabric caused by a preferred alignment of diamagnetic quartz-grain framework with c-axes parallel to the current alignment direction.

In the analysis of current directions usually only the eigendirections of the anisotropy tensor are of interest, i.e. the grain alignment and foliation. However, it may also be of value to consider the strength and symmetry of the alignment in some circumstances. For this purpose, the Jelinek plot (Jelinek, 1981) provides a clear separation of symmetry (planar or constricted alignments) from intensity (isotropic v. extremely eccentric alignment distributions). The AARM and AMS for the samples are illustrated separately (Fig. 3). Generally AARM has higher eccentricities (P_j) due to the strong remanence anisotropy of magnetite, but the AARM and AMS ellipsoids are not systematically related for the samples. On the other hand, the quartz grains (millimetre size) would have greater inertia and there would have been less disturbance of their trajectories because of turbulence. Concerning symmetry, 8 of 11 samples have planar quartz fabrics ($T_j > 0$), emphasizing the bedding component of the current alignment. Six of 11 samples show a linear distribution ($T_j < 0$). This should not be taken to reflect the degree of current alignment because the intrinsical anisotropy of the grains must also be taken into account.

This work was funded by NSERC grants to G.B. and P.F.

References

BAGNOLD, R. A. 1954. Experiments on a gravity free dispersion of large solid spheres in a Newtonian fluid under shear. *Proceedings of the Royal Society of London, Series A*, **225**, 49–63.

BORRADAILE, G. J. 1987. Anisotropy of magnetic susceptibility: rock composition versus strain. *Tectonophysics*, **138**, 327–329.

—— & HENRY, B. 1997. Tectonic applications of magnetic susceptibility and its anisotropy. *Earth-Science Reviews*, **42**, 49–93.

—— & STUPAVSKY, M. 1992. Anisotropy of magnetic susceptibility: measurement schemes. *Geophysical Research Letters*, **22**, 1957–1960.

——, MACKENZIE, A. & JENSEN, E. 1990. Silicate versus trace mineral susceptibility in metamorphic rocks. *Journal of Geophysical Research*, **95**, 8447–8451.

——, —— & ——1991. A study of colour changes in purple–green slate by petrological and rock-magnetic methods. *Tectonophysics*, **200**, 157–172.

——, MOTHERSILL, J. S., TARLING, D. M. & ALFORD, C. 1986. Sources of magnetic susceptibility in a slate. *Earth and Planetary Science Letters*, **76**, 336–340.

CLOUD, P. 1973. Paleoecological significance of the banded iron-formation. *Economic Geology*, **68**, 1135–1143.

DUNLOP, D. J. & ÖZDEMIR, Ö. 1997. *Rock Magnetism: Fundamentals and Frontiers*. Cambridge Studies in Magnetism. Cambridge University Press, Cambridge.

ELLWOOD, B. B. & HOWARD, J. H., III 1981. Magnetic fabric development in an experimentally produced barchan dune. *Journal of Sedimentary Petrology*, **51**, 97–100.

——, TERRELL, G. E. & COOK, W. J. 1993. Frequency dependence and the electromagnetic susceptibility tensor in magnetic fabric studies. *Physics of the Earth and Planetary Interiors*, **80**, 65–74

FRALICK, P. W. & MIALL, A. D. 1989. Sedimentology of the Lower Huronian Supergroup (Early Proterozoic), Elliot Lake area, Ontario, Canada. *Sedimentary Geology*, **63**, 127–153.

GIRDLER, R. W. 1961. The measurement and computation of anisotropy of magnetic susceptibility in rocks. *Geophysical Journal of the Royal Astronomical Society*, **5**, 34–44.

HAMILTON, N., OWENS, W. H. & REES, A. I. 1968. Laboratory experiments on the production of grain orientation in shearing sand. *Journal of Geology*, **76**, 465–472.

HOLLAND, H. D. 1973. The oceans: a possible source of iron in iron-formations. *Economic Geology*, **68**, 1169–1172.

HROUDA, F. 1986. The effect of quartz on the magnetic anisotropy of quartzite. *Studia Geophysica et Geodaetica*, **30**, 39–45.

JACKSON, M. 1991. Anisotropy of magnetic remanence: a brief review of mineralogical sources, physical origins, and geological applications and comparison with susceptibility anisotropy. *Pure and Applied Geophysics*, **136**, 128.

—— & BORRADAILE, G. J. 1991. On the origin of the magnetic fabric in Cambrian Purple Slates of North Wales. *Tectonophysics*, **194**, 49–58.

—— & TAUXE, L. 1991. Anisotropy of magnetic susceptibility and remanence: developments in the characterization of tectonic, sedimentary and igneous fabric. *Reviews of Geophysics Supplement, US National Report to International Union of Geodesy and Geophysics*, 371–376.

——, SPROWIL, D. & ELLWOOD, B. 1989. Anisotropies of partial anhysteretic remanence and susceptibility in compacted black shales: grainsize- and composition-dependent magnetic fabric. *Geophysical Research Letters*, **16**, 1063–1066.

JELINEK, V. 1981. Characterization of the magnetic fabrics of rocks. *Tectonophysics*, **79**, T63–T67.

LOWE, D. R. 1976. Grain flow and grain flow deposits. *Journal of Sedimentary Petrology*, **46**, 188–199.

——1982. Sediment gravity flows: II. Depositional models with specific reference to the deposits of high-density turbidity currents. *Journal of Sedimentary Petrology*, **52**, 279–297.

POTTER, P. E. & MAST, R. F. 1963. Sedimentary structures, sand shape fabrics, and permeability. *Journal of Geology*, **71**, 441–471.

REES, A. I. 1968. The production of preferred orientation in a concentrated dispersion of elongated and flattened grains. *Journal of Geology*, **76**, 457–564.

ROCHETTE, P. 1988. Inverse magnetic fabrics in carbonate-bearing rocks. *Earth and Planetary Science Letters*, **90**, 229–237.

——, JACKSON, M. & AUBOURG, C. 1992. Rock magnetism and the interpretation of anisotropy of magnetic susceptibility. *Reviews of Geophysics*, **30**, 209–226.

STEPHENSON, A., SADIKUN, S. & POTTER, D. K. 1986. A theoretical and experimental comparison of the anisotropies of magnetic susceptibility and remanence in rocks and minerals. *Geophysical Journal of the Royal Astronomical Society*, **84**, 185–200.

TAIRA, A. & LIENERT, B. R. 1979. The comparative reliability of magnetic, photometric and microscopic methods of determining the orientations of sedimentary grains. *Journal of Sedimentary Petrology*, **49**, 759–772.

TARLING, D. H. & HROUDA, F. 1993. *The Magnetic Anisotropy of Rocks*. Chapman and Hall, London.

THOMPSON, R. & OLDFIELD, F. 1986. *Environmental Magnetism*. Allen and Unwin, London.

Diagenesis and remanence acquisition in the Cretaceous carbonate sediments of Monte Raggeto, southern Italy

B. D'ARGENIO,[1,2] V. FERRERI,[1] M. IORIO,[2] A. RASPINI[1] & D. H. TARLING[3]

[1] *Dipartimento di Scienze della Terra, Università 'Federico II', Largo San Marcellino, 10, 80138, Napoli, Italy*
[2] *Istituto di Ricerca, Geomare Sud, CNR, via A. Vespucci 9, 80142, Napoli, Italy*
[3] *Department of Geological Sciences, University of Plymouth, Plymouth PL4 8AA, UK*

Abstract: Lower Cretaceous limestones and dolomites drilled at Monte Raggeto, north of Naples, show a hierarchical organization in cycles, bundles and superbundles, similar to that found in other Cretaceous shallow-water carbonate rocks exposed in the Southern Apennines. These cycles have been interpreted as being caused by the influences of planetary orbital perturbations of the Earth (precession, obliquity), with the bundles and superbundles being considered to represent the short- and long-term eccentricity orbital cycles, respectively. The palaeomagnetic properties of an 88 m core (corresponding to a total interval of c. 3.2 Ma with a rock accumulation rate averaging 1 cm per 360 ± 16 years) were determined at 2 cm intervals. These show the same spectral periodicities as for the sedimentary properties determined at 1 cm intervals, but there are no correlations between these sedimentological and palaeomagnetic sets of data, when analysed on the scale of the bore-core. An example is given of a 15.50 m length of the bore-core, comprising a single superbundle corresponding to one of the long eccentricity terms, and including eight elementary cycles (each topped by a brief emersion episode) grouped in three bundles. These shallow marine carbonate sediments underwent very early diagenesis accompanied by marine cementation in the pore spaces at or immediately below the water–sediment interface in the same way as in modern-day tropical–subtropical environments. During brief emersions at the top of each cycle, meteoric early diagenesis also occurred. All non-carbonate grains that were present in the original sediment, including ferromagnetic ones, are likely to have been cemented in during these very early phases of diagenesis. It is concluded that the studied interval provides a model of the diagenetic evolution of the whole core, where the early diagenesis controlled the locking-in of a nearly continuous record of the geomagnetic field behaviour for the entire core. Thus depositional, diagenetic and palaeomagnetic properties can be used in conjunction to provide detailed information on the environmental variations and geomagnetic field behaviour during the evolution of this stratigraphic sequence at an ultra-fine resolution, as 1 cm of rock is equivalent, on average, to 360 ± 16 years.

Carbonate sediments deposited in shallow marine waters show very early cementation (see, e.g. Bathurst 1975; Tucker & Wright 1990), as evidenced by man-made objects and bones less than 50 years old found embedded in lagoonal environments (Purser 1980; see also Shinn 1969). Early cement precipitation develops in these environments if the pore fluids are supersaturated with respect to the calcium carbonate phase, as long as there are no kinetic factors inhibiting such precipitation; this process may also occur as a result of bacterial activity induced by the decay of organic matter. Other significant processes also occur during these early stages, including biogenic grain micritization and organic borings at the sediment–water interface, and burrowing (Bathurst 1975; James & Choquette 1983; Tucker & Bathurst 1990). If these deposits are exposed to meteoric waters, as occurs for large parts of the at present exposed late Pleistocene–Holocene shallow-water carbonate sediments, then they undergo strong diagenetic modifications. These involve karstic solution, recrystallization and cementation caused by interstitial waters whose composition is different from that of the original marine waters. Such early diagenetic (syn-diagenetic) modifications will thus affect very shallow marine sediments whenever there is a relative lowering of sea level shortly after their deposition.

D'ARGENIO, B., FERRERI, V., IORIO, M., RASPINI, A. & TARLING, D. H. 1999. Diagenesis and remanence acquisition in the Cretaceous carbonate sediments of Monte Raggeto, southern Italy. *In*: TARLING, D. H. & TURNER, P. (eds) *Palaeomagnetism and Diagenesis in Sediments*, Geological Society, London, Special Publications, **151**, 147–156.

In southern Italy, thick (>3000 m) Mesozoic carbonate sequences were originally deposited in shallow marine water within a subtropical–tropical environment. The sequences show a vertical organization of depositional characteristics and early diagenetic features which have a cyclical nature, suggesting repeated water depth and environmental oscillations. Very rapid cementation processes are, in general, recorded throughout the sequences. Only the uppermost part of the cycles shows narrow zones indicating early meteoric diagenetic episodes (karstic dissolution and/or pedogenesis). These are normally superimposed directly on carbonate sediments with typical subtidal depositional and diagenetic features, which form the vast majority of the sequence (Buonocunto et al. 1994; D'Argenio et al. 1997). The cyclic organization of such features can thus be explained by high-frequency sea-level fluctuations, with >90% of the deposits formed of shallow subtidal facies, only showing emersion-related features that are largely confined to narrow zones in their uppermost part of the cycle. Such a sedimentation pattern precludes an interpretation based on autocyclic tidal-flat progradation or episodic subsidence-controlled pulsations. Moreover, the stacking pattern of elementary cycles shows a hierarchy of at least three orders that correspond, with good reliability, to specific periodicities of the Earth's orbit perturbations, particularly the Milankovitch cyclicities ranging between $c.20$ and $c.400$ ka (D'Argenio et al. 1997; Raspini 1998). Such an interpretation has been largely confirmed by spectral analyses of sedimentological data (Longo et al. 1994; Brescia et al. 1996; D'Argenio et al. 1997) from numerous outcrops, as well as from some 130 m of bore-cores drilled and analysed in lower Cretaceous shallow marine carbonate deposits at Monte Raggeto, southern Italy (Fig. 1). The bore-core sequence not only documents such sedimentological processes but also records the secular variations (SVs) of the magnetic field (Tarling et al. 1999). This means that the grains carrying the geomagnetic signal must have been cemented very rapidly into the shallow-water carbonate sequence at a very early stage and then effectively isolated from most later diagenetic changes. Spectral analyses of both the sedimentological and palaeomagnetic data within the bore-core sequences show high-frequency cyclicities that correlate almost exactly with perturbations in the Earth's orbital variables (Iorio et al. 1995, 1996; Raspini 1996). The sedimentological and palaeomagnetic intensity datasets, obtained at 1 and 2 cm scales, respectively, in the Monte Raggeto bore-core

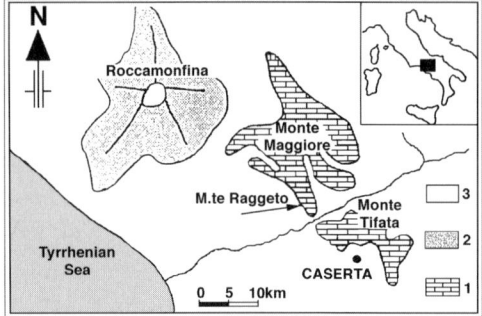

Fig. 1. The location of the area. 1, Mesozoic–Tertiary carbonate platform deposits; 2, Quaternary volcanic rocks; 3, Tertiary and Quatnary clastic deposits.

were examined to determine whether the sedimentological processes in some way influenced the cyclical oscillation of the palaeomagnetic field, clearly recorded during early diagenesis. In this paper the centimetre-scale analysis is illustrated with a 15.50 m section of well S1 of Monte Raggeto, which is entirely of Barremian age, but the data represent the preliminary results of a comparative study (sedimentological v. palaeomagnetic) carried out on the whole 88 m of Hauterivian–Barremian bore-core from Monte Raggeto.

The Monte Raggeto bore-cores

Sedimentology

The sequence examined comes from three wells drilled on the western side of Monte Raggeto (Monte Maggiore Group, Southern Apennines) located on small quarry terraces (S1, S2, S3 in Fig. 2). These bore-cores are 10 cm in diameter and were obtained with a 90% recovery rate, yielding a total thickness of about 130 m. Microstratigraphic (centimetre-scale) textural analysis on the whole core has shown several lithofacies, which can be grouped into four associations (Raspini 1996):

(a) Foralgal limestones. These comprise wackestone and wackestone–packestone with diversified thanatocoenoses. Benthic Foraminifera (textularids, verneulinids, miliolids, and lituolids) and green algae are often associated with large molluscan shells or small corals. Pervasive bioturbation is recorded locally. Intraclasts are absent and peloids seldom occur. The local bioturbation of the wackestone and wackestone–packestone, together with the thanatocoenosis diversity, suggest that the depositional environment was largely that of a well-oxygenated

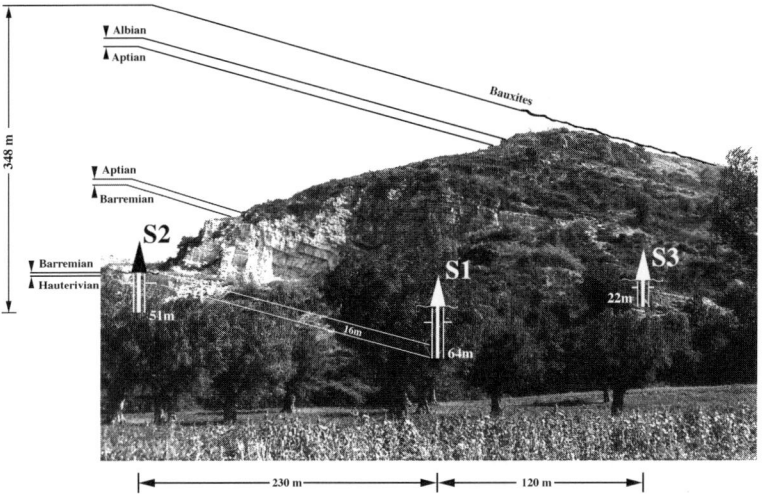

Fig. 2. The location of the wells and the stratigraphy of the section.

lagoonal environment of normal salinity ('open platform' *sensu* Wilson 1975; see also Buonocunto *et al.* 1994; D'Argenio *et al.* 1997).

(b) Biopeloidal limestones and dolomitic limestones. These have a grain-supported textures which include bioclasts and peloids, generally associated with small intraclasts. Thanatocoenosis encloses benthic Foraminifera, ostracods, molluscan shells (fragments) and green algae. This texture, locally showing weak laminations, suggests depositional environments characterized by relatively high-energy hydrodynamics (inner shoals) at the transition from open to restricted lagoon (Ferreri *et al.* 1997; Raspini 1998).

(c) Miliostralgal limestones and dolomitic limestones. Wackestone, wackestone–mudstone and bioturbated wackestone–packstone locally alternate with millimetre-size cryptalgal laminae. Small benthic Foraminifera (including miliolids) are associated with *Thaumatoporella*, ostracods and/or small gastropod shells. The mud-supported textures, with oligotypic thanatocoenoses characterized by a local abundance of ostracods, small miliolids and small gastropods, suggest depositional processes in subtidal environments characterized by restricted circulation ('restricted platform' *sensu* Wilson 1975; see also Buonocunto *et al.* 1994; D'Argenio *et al.* 1997).

(d) Laminate dolomitic limestones. These barren dolomites are characterized by millimetre-size cavities of variable shape often enlarged by karstic solution and occluded by mechanical geopetal fillings or by sparry calcite. Millimetre-thick laminae composed of peloidal packstones are also locally intercalated in these deposits, where they are generally considered to represent storm layers. These features suggest that deposition occurred within a largely tidal zone which episodically experienced storm events (Flügel 1982; Shinn 1986; D'Argenio *et al.* 1997).

Each of these lithofacies associations suggests an environment comprising relatively large sectors of a tropical carbonate platform within which several minor environmental variations can be recognized (lithofacies). Such variations, as seen in bore-core, can also be traced laterally in quarry outcrops, allowing relatively minor environmental variations to be recognized. Grain-supported deposits (limestones and dolomitic limestones formed by intraclasts, bioclasts, peloids and ooids) showing normal and reverse grading and erosional bases occur as episodic intercalations within the different lithofacies associations; such features suggest occasional high-energy deposits resulting from storm episodes affecting all the different tidal and subtidal settings (Ferreri *et al.* 1997). The vast majority of these carbonate sediments accumulated in a subtidal environment, interrupted occasionally by periods of emersion and discontinuities in deposition evidenced by karstic solution cavities and/or calcrete horizons occurring in the uppermost part of the lithofacies forming the top of the elementary cycles (see below). In such horizons, the complex network of cavities (from a few milimetres to centimetres in size) became filled by vadose silt (crystal silt *sensu* Dunham 1969) as well as by clayey deposits with a dominant greenish colour.

The variation of lithofacies along the sequence, as well as of early diagenesis (Fig. 3), extending from marine (micritized grains, micritic envelopes, rims of fibrous calcite around the grains)

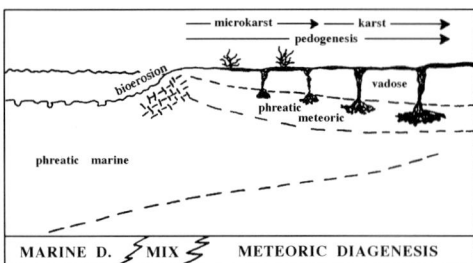

Fig. 3. Schematic model of the diagenetic modifications at cycle boundaries. (See text for further details.)

to meteoric (karstic cavities, geopetal crystal silt, calcrete horizons), allows typical sedimentary cycles to be recognized. Metre-scale cycles appear hierarchically organized; these elementary cycles form groups (bundles) which themselves form larger groups (superbundles). In the bore-core of Monte Raggeto the elementary cycles have an average thickness of 198 cm and are mostly formed of subtidal deposits showing, only in their upper parts, diagenetic features which testify to an early meteoric overprint. The bundles comprise from two to five elementary cycles and have an average thickness of 446 cm, and the superbundles are formed of two to four bundles and have an average thickness of 1154 cm. Bundles and superbundles may be recognized by the characteristics of their upper boundaries which, of course, correspond to tops of an elementary cycle and are marked by pervasive silt and/or karstic cavities of millimetre-size (Em1 type emersion surface, Fig. 4). The superbundle boundaries are marked by larger cavities which may penetrate downwards to cut through one or more elementary cycles and may also be characterized by thicker calcrete horizons (Em2 type emersion surface, Fig. 4). Subaerial erosional surfaces are marked by meteoric early diagenetic features which are directly superimposed on subtidal deposits. These, as well as the hierarchical cyclic organization of depositional and diagenetic features (elementary cycles, bundles and superbundles), suggest high-frequency eustatic oscillations controlled by planetary orbital perturbations (Milankovitch cyclicities), as have been widely proposed for other Mesozoic carbonate platform strata (Goldhammer et al. 1990; Buonocunto et al. 1994; Longo et al. 1994; Strasser 1994; Read et al. 1995; Grötsch 1996; D'Argenio et al. 1997; Raspini 1998).

Palaeomagnetism

Hand samples from quarry outcrops where the boreholes were later drilled had weak magnetic remanences but these were measurable on the JR4 spinner magnetometer at the Eötvös Lorand Geophysical Institute, Budapest, and at the University of Plymouth. The samples were subjected to either thermal or alternating field

Fig. 4. The vertical organization of lithofacies, early diagenesis and dolomitization intensity in 15.50 m of bore-core in the upper part of S1 well (Barremian pp). The lithofacies are defined (Raspini 1996) as: *Lithofacies associations*: A, foralgal limestones; B, biopeloidal limestones and dolomitic limestones; C, miliostralgal limestones and dolomitic limestones; D, laminate dolomitic limestones. *Lithofacies*: A1, wackestone and wackestone/packstone with benthonic foraminifera (among which Lituolids, Verneulinids, Textularids, Miliolids), green algae (among which *Salpingoporella melitae* and *Salpingoporella muhelbergii*), Thaumatoporella, ostracodes, peloids (see inset b); B1, intraclast peloidal g/p, packstone and grainstone with benthic foraminifera, rare green algae, cortoids and large pelecypod shell fragments; B2, peloidal g/p, and pack stone with benthic foraminifera, small intraclasts, Thaumatoporella and rare small pelecypod shell fragments; B3, peloidal packstone. C1, ostracode *Thaumatoporella* wackestone and m/w with small Miliolids and rare green algae (tempestites are shown by a triangle in this column); C2, *Salpingoporella* mudstone and mudstone/wackestone, locally dolomitized, with rare small benthic foraminifera (see inset a); C3, dolomitized mudstone with small gastropod moulds; C4, barren dolomites locally alternating with millimetre size cryptalgal laminae (see inset c). D1, Stromatolitic and loferitic dolomites. Em 1, Elementary cycle boundaries evidenced by pervasive dissolution and mm-size karstic cavities (see insets d and f). Em 2, Elementary cycle boundaries marked by mm-to-cm size karstic cavities (see inset e). Dolom. is the degree of dolomitization corresponding to the three intervals of Fig. 5. Horizontal dashed lines indicate cycle boundaries clearly observed in the corresponding quarry outcrop. Vertical dashed lines (oblique in the dolomitization log) indicate interruptions in the bore-core due to fragmentation of the limestones during the coring. In the interval illustrated here, the lithofacies suggests predominantly widespread lagoonal conditions typical of an inner area of a carbonate platform characterized by a relatively restricted circulation. Note that lithofacies B1, B2, B3, C3 and D1 are absent in this interval. Whatever the timing of any diagenetic overprint, it is notable some agreement remains between the cycle boundaries, the lithofacies (suggesting relatively more restricted environments) and the degree of dolomitization, although strong dolomitization may have hidden or even obliterated some elementary cycle boundaries by recrystallization or amalgamation processes.

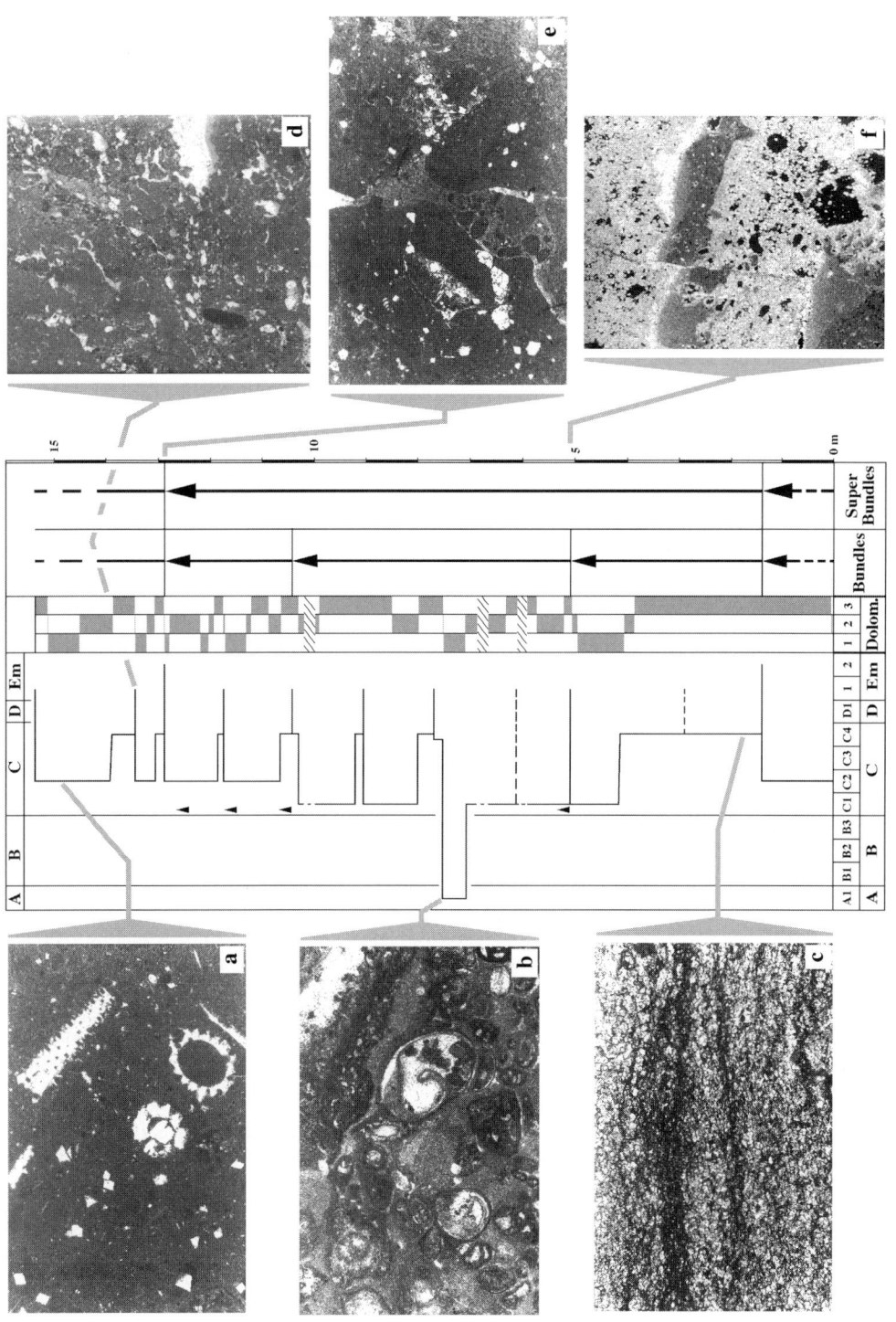

(AF) partial demagnetization. AF demagnetization revealed largely single-component remanences above 10 mT peak field and the same single-component remanences were also identified during heating to $c.\,550°C$ (Iorio et al. 1996). On the basis of these hand samples, the entire 130 m of bore-core was subsequently subjected to incremental AF demagnetization and measured using the 2 G long-core cryogenic magnetometer at the University of Southampton. (The third well has not yet been analysed.) The initial intensity of remanence averaged $47\,\mu A/m$, two orders of magnitude greater than the noise measured when no core was present $(0.9-1.0\,\mu A/m)$. This very weak intensity had the advantage that each magnetometer reading sensed the average remanence of some 5–7 cm thickness of core, corresponding to some 2000 years (Tarling et al. 1999). The initial intensity for most of the individual 1.5 m core lengths decreased gradually by 40–80% after demagnetization to 40 mT and, in the vast majority of cores, it was decided that the vectors isolated at 35 and 40 mT represented the characteristic magnetization of the bore-cores. The evidence confirming high magnetic stability, with a dominant single component above 20 mT, and commonly extending to 45–55 mT, has been discussed by Iorio et al. (1996, 1998b). The low-field susceptibility of the bore-cores, also measured at 2 cm intervals, was also very weak, with 95% of the core being dominantly diamagnetic (Tarling et al. 1999). Spectral analyses of the characteristic remanence directions (declination and inclination), the initial intensity and low-field susceptibility all showed essentially identical peak wavelengths, and these also corresponded with those derived from analyses of the lithological and depositional features. Furthermore, the normalized peak wavelengths (cm) in both sedimentological and magnetic parameters correlated almost exactly (Iorio et al. 1996) with the normalized peak periodicities (ka) predicted for Cretaceous time by Berger et al. (1992).

Strong correlations between variations in the Earth's orbital variables and high-frequency periodicities in depositional and palaeontological characteristics of carbonate lithologies have been identified in many other sequences (Fischer et al. 1990; Fischer & Bottjer 1991; Schwarzacher 1993). Cyclical variations in the magnetic properties of pelagic carbonate sediments have been reported mainly in low-field susceptibility and intensities (i.e. Robinson 1986; Bloemendal et al. 1988; Tarduno et al. 1991), where they have mostly been attributed to changes in the concentrations of terrigenous material and post-depositional dissolution of magnetite during diagenesis.

Napoleone & Ripepe (1989) have additionally reported cyclical variations in remanence directions in pelagic limestones, but only for very short stratigraphic intervals. In contrast to these earlier pelagic limestones studies, the carbonate sediments discussed here have lithologies that are highly responsive to climatic and sea-level oscillations and have very little or no terrigenous input but high sedimentation rates (Longo et al. 1994) accompanied by very early cementation. Consequently, the processes of remanence acquisition and the time when it becomes locked-in in these carbonate rocks is very different from those of pelagic carbonate depositional environments. In these bore-core sediments the characteristic remanences appear to be single vectors and have unblocking temperatures that are strongly suggestive of biogenetic magnetite (Tarling et al. 1999), as identified in modern shallow-water marine carbonate deposits (see discussion by McNeil (1990)). In areas of sediment accumulation away from a terrigenous input of magnetic particles, such as in these carbonate platform environments, biogenic magnetic particles are most likely to be the primary source for the magnetic minerals within the sediment (Vali et al. 1987; McNeil 1990), with a substantial contribution attributable to green algae and cyanophyta (Torres de Araujo et al. 1986; Chang et al. 1989; Stolz et al. 1989). These organisms produce chains of small grains of iron oxide ($c.\,0.3\,\mu m$ single-domain (SD) cubes of magnetite), which account for most of their body weight. The remains of such magnetotactic organisms clearly include these biogenic magnetite particles unless they are subsequently oxidized. Bacterially derived magnetites are extremely stable magnetically because of their SD grain-size range (Kirschvink & Gould 1981) and are thus able to retain their magnetizations unchanged for millions of years. In the absence of significant external sources of iron oxides or hydroxides, as in this case, biogenically produced magnetites are likely to have formed the majority of the sparse iron oxides in the bottom sediments. The bacterially active zone, where these grains will have been produced, lies at and immediately below the sediment–water interface. The sediment itself usually is rapidly lithified by very early marine cements before further burial. Up to, and immediately before this early cementation, the magnetite grains remain free to rotate within the carbonate sediments, but at the time of cementation their alignment is locked in the sediment, so the magnetic record in the bore-cores allows determination of the direction and strength of the geomagnetic field at the time of cementation rather than of deposition. However, cementation must also have occurred rapidly to preserve the observed details of short-term geomagnetic variations (Tarling et al. 1999) and to account for the very high precision with which the peak frequencies are defined. Such observations suggest that the magnetic grains within any 2 cm of sediment must have been locked in at a rate comparable with, but probably less than, the c. 2000 year smoothing represented by each individual magnetometer reading (Iorio et al. 1998b; Tarling et al. 1998).

Comparison of the properties

The magnetic properties of NRM remanence for the total core (Table 1) show clear cyclicities (Iorio et al. 1996), which were all present at all demagnetization levels up to 40 mT, except for the initial magnetic intensity (Iorio 1995). As the observed peak wavelength and those of the stratigraphic features, both expressed in centimetres, are remarkably similar (Brescia et al. 1996; Iorio et al. 1996), this immediately raised the question of whether the palaeomagnetic variables are themselves controlled by the sedimentological features, which are, in turn, controlled by changes in insolation in response to orbital variations. This seems unlikely, as previous studies have shown that the combined core, representing just over 3 Ma, comprises a nearly continuous record of predominantly normal polarity with several reversed intervals which correlate with those in other pelagic Lower Cretaceous magnetostratigraphic sequences (Iorio et al. 1998b). More importantly, from a diagenetic viewpoint, the core contains a smoothed record of geomagnetic SVs in which the average arithmetic change in direction between consecutive 2 cm levels is $18.6 \pm 19.1°$ per 2 cm (Tarling et al. 1999). As 1 cm, on average, corresponds to 360 ± 18 years, the median value, $12.7°$ per 2 cm, corresponds to an average annual rate of change in direction of almost $2°$ per century; almost exactly the same as recorded for modern-day geomagnetic SV (Tarling 1988). Tarling et al. (1999) also found that there were no linear correlations between the palaeomagnetic variables and the depositional properties, although the initial magnetic intensity showed a very weak correlation, <0.4, which raised the question of a some relationship for this factor. In this paper, the upper 15.50 m section of well S1, which is entirely of Barremian age, is used as an example. Within this section, 11 elementary cycles can be readily identified, eight of which are organized into three bundles, in turn forming a single superbundle, which can be related directly with the orbital long eccentricity cycle (Fig. 4). It was found that the initial intensity of magnetization through the studied core length oscillates at or below some $50 \mu A/m$ within each of the lithofacies associations, irrespective of depositional as well as early diagenetic (marine or meteoric) characteristics (Figs 4 and 5). However, there are some ten intervals of 10–50 cm thickness within which the initial intensity ranges between 60 and $200 \mu A/m$ (with the exception of two peaks with higher values at 632 and $435 \mu A/m$ corresponding respectively to intervals affected by recent karstic cavities with evident detrital infilling, and to late, very coarse dolomite). All these peaks occur immediately above each elementary cycle limit. It appears here that the peaks correlate clearly with the transgressive part of the cycle irrespective of the specific type of sediment.

Final comments and remarks

The lower Cretaceou sediments carbonates drilled at Monte Raggeto, in the Southern Apennines of Italy, show evident cyclic organization and hierarchical groupings into bundles of cycles and their superbundles. Very early diagenetic processes have affected the sequence, with good evidence of both marine diagenesis, which predominates throughout the sequence, and, at

Table 1. *Peak wavelength in the palaeomagnetic variables and peak orbital periodicities for Early Cretaceous time*

Intensity (cm)	Inclination (cm)	Declination (cm)	Astron. (years)
52	54	–	18 350
65	69	61	22 230
103	118	–	38 200
144	142	–	48 750
258	260	268	95 800
–	1073	982	403 800
Normalized			
1.0	1.0	–	1.0
1.2	1.3	–	1.2
2.0	2.2	–	2.0
2.7	2.6	–	2.6
4.9	4.8	–	5.2
–	19.8	–	22.0

The values refer to the entire bore-core. The palaeomagnetic properties are of the initial characteristic component of remanence (spectral peaks are present in the directions after partial demagnetization in all fields up to at least 40 mT). Astron: astronomical periodicities for Hauterivian and Barremian time as predicted by Berger et al. (1989, 1992).

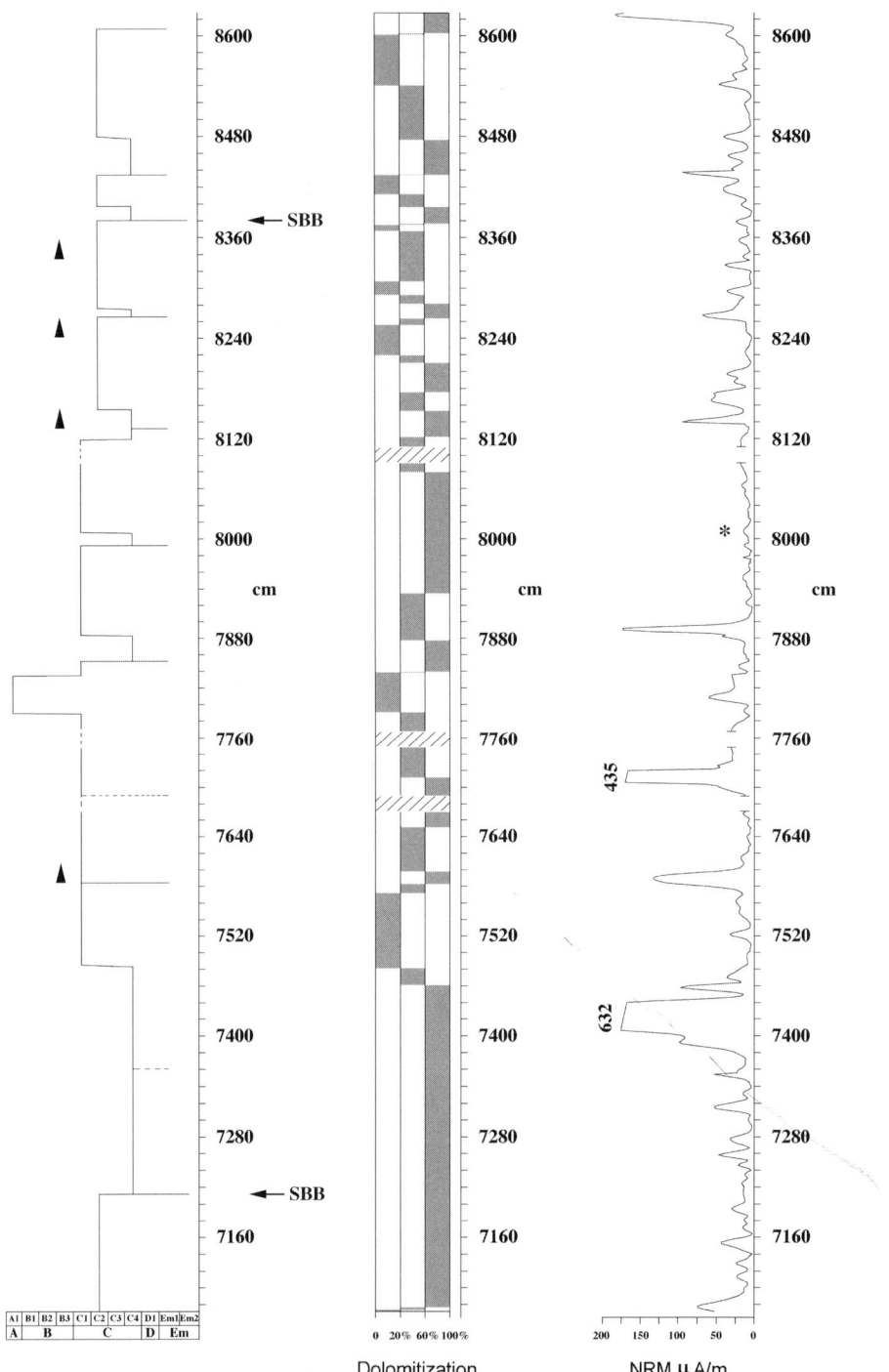

Fig. 5. Graphic log of the elementary cycles compared with the intensity of dolomitization (see also Fig. 4) and the intensity of remanence. Left: elementary cycles as shown in Fig. 4. Centre: dolomitization intensity, based on visual estimate. Right: initial magnetic intensity. (Interruptions in the log correspond to gaps in the bore-core; see Fig. 4.) It should be noted that immediately above cycle boundaries the magnetic intensity reaches values exceeding 50 μA/m, with the exception of the lower and upper superbundle boundaries (SBB) when the values tend to oscillate at or below 50 μA/m. The asterisk marks a break in the core and hence in the continuity of the intensity measurements, which could have hidden an intensity peak exceeding 50 μA/m.

each cycle top, early meteoric modification of the sediments during short periodic emersions which led to karstic type solution, i.e. pervasive solution up to true karst and, locally, soil formation (Iorio 1995; Raspini 1996). Such diagenetic features, and their associated textures and composition, are clearly strongly correlated with the Earth's orbital periodicities calculated for Early Cretaceous time (Longo et al. 1994; Iorio et al. 1995; Brescia et al. 1996). A climatic and eustatic control of such periodicities may be deduced, suggesting that the elementary cycles have a connection with planetary precession or inclination (or a combination of both) and that the bundles and superbundles relate to the short and long orbital eccentricity terms, respectively (D'Argenio et al. 1997). The palaeomagnetic variables show exactly the same periodicities and have similar very high correlations with the Earth's orbital variables, yet there is no correlation between the palaeomagnetic variables and the sedimentological and textural features. Consequently, the palaeomagnetic features are not controlled, overall, by the local depositional environment and are mostly likely to be caused by orbitally induced changes in the geomagnetic field itself. This is a similar conclusion to that obtained from analysis of marine magnetic anomalies in the Atlantic (D'Argenio et al. 1996) and the Pacific Oceans (Iorio et al. 1998a). The fact that fine-scale features of geomagnetic field behaviour are recorded by the sediments requires that cementation of the magnetic grains occurred extremely rapidly. This also means that it is possible to consider the use of past geomagnetic variations as another tool for evaluating the timing of early diagenetic processes. At present, this is being attempted in studies of the remaining core and further along the outcrop, on the basis of correlations that are now possible at a 1 cm level corresponding to an ultra-fine magnetic stratigraphy on a scale in which 1 cm of rock equates, on average, to $c.\ 360 \pm 18$ years.

This work is part of a research programme on high-resolution stratigraphy carried on at the Geomare sud Research Institute, National Research Council, Naples, and at the Department of Scienze della Terra, Federico II University, Naples, with the co-operation of the Department of Geological Sciences of Plymouth University. The sedimentological analysis came from the doctoral dissertation of A.R., and the palaeomagnetic analysis from the PhD thesis of M.I. The framework of the present paper arose from specific reasoning and discussions which involved all of the authors equally. D.H.T. acknowledges a Short Term Mobility Grant of CNR, which provided funds for his stay in Naples during 1997.

References

BATHURST, R. G. C. 1975. *Carbonate Sediments and their Diagenesis*. Elsevier, Amsterdam.

BERGER, A., LOUTRE, M. F. & DEHANT, V. 1989. Astronomical frequencies for pre-Quaternary palaeoclimate studies. *Terra Nova*, 1, 474–479.

——, —— & LASKYR, J. 1992. Stability of the astronomical frequencies over the Earth's history for palaeoclimate studies. *Science*, 255, 560–566.

BLOEMENDAL, J., TAUXE, L., VALET, J. P. & SHIPBOARD SCIENTIFIC PARTY 1988. High resolution, whole-core magnetic susceptibility logs from leg 108. In: RUDDIMAN, W., SARNTHEIN, M. & BALDAUF, J., et al. (eds) *Proceedings of the Ocean Drilling Program, Initial Reports*, 108, Ocean Drilling Program, College Station, TX, 1005–1013.

BRESCIA, M., D'ARGENIO, B., FERRERI, V., LONGO, G., PELOSI, N., RAMPONE, S. & TAGLIAFERRI, R. 1996. Neural net aided detection of astronomical periodicities in geologic records. *Earth Planetary Science Letters*, 139, 33–45.

BUONOCUNTO, F. P., D'ARGENIO, B., FERRERI, V. & RASPINI, A. 1994. Microstratigraphy of highly organized carbonate platform deposits of Cretaceous age. The case of Serra Sbregavitelli, Matese (Central Apennines). *Giornale di Geologia*, 56(2), 179–192.

CHANG, S.-B. R., STOLZ, J. F., KIRSCHVINK, J. L. & AWRAMIK, S. M. 1989. Biogenic magnetite in stromatolites. II. Occurrence in ancient sedimentary environments. *Precambrian Research*, 43, 305–315.

D'ARGENIO, B., FERRERI, V., AMODIO, S. & PELOSI, N. 1997. Hierarchy of high frequency orbital cycles in Cretaceous carbonate platform strata. *Sedimentary Geology*, 113, 169–193.

——, IORIO, M. & TARLING, D. H. 1996. Periodicity of magnetic intensities in magnetic anomaly profiles: the Cenozoic of the South Atlantic. *Geophysical Journal International*, 127, 141–155.

DUNHAM, R. J. 1969. Early vadose silt in Townsend mound (reef), New Mexico. In: FRIEDMAN, G. M. (ed.) *Depositional Environments in Carbonate Rocks*. Society of Economic Paleontologists and Mineralogists, Special Publication, 14, 139–181.

FERRERI, V., WEISSERT, H., D'ARGENIO, B. & BUONOCUNTO, F. P. 1997. Carbon-isotope stratigraphy. A tool for basin to carbonate platform correlation. *Terra Nova*, 9, 57–61.

FISCHER, A. G. & BOTTJER, D. 1991. Orbital forcing and sedimentary sequences. *Journal of Sedimentary Petrology*, 61(7), 1063–1069.

——, DE BOER, P. L. & PREMOLI SILVA, I. 1990. Cyclostratigraphy. In: GINSBURG, R. N. & BEAUDOIN, B. (eds) *Cretaceous Resources, Rhythms and Events*. NATO – Advanced Studies Institute, Series C, 304. D. Reidel, Dordrecht, 139–172.

FLÜGEL, E. 1982. *Microfacies Analysis of Limestones*. Springer-Verlag, Berlin.

GOLDHAMMER, R. K., DUNN, P. A. & HARDIE, L. A. 1990. Depositional cycles, composite sea level changes, cycle stacking pattern and the hierarchy of stratigraphic forcing. Examples from Alpine Triassic platform carbonates. *Geological Society of America Bulletin*, 102, 535–562.

GRÖTSCH, J. 1996. Cycle stacking and long-term sea-level history in the Lower Cretaceous (Gavrovo platform, NW Greece). *Journal of Sedimentary Research*, **66**, 723–736.

IORIO, M. 1995. *High resolution palaeomagnetic analysis of Cretaceous shallow water carbonates*. PhD thesis, University of Plymouth.

——, TARLING, D. H. & D'ARGENIO, B. 1998a. Periodicity of magnetic intensities in magnetic anomaly profiles: the Cretaceous of the Central Pacific. *Geophysical Journal International*, **133**, 233–244.

——, —— & ——1998b. The magnetic polarity stratigraphy of a Hauterivian/Barremian carbonate sequence from Southern Italy. *Geophysical Journal International*, **134**, 13–24.

——, ——, —— & NARDI, G. 1996. Ultra-fine magnetostratigraphy of Cretaceous shallow water carbonates, Monte Raggeto, Southern Italy. *In*: MORRIS, A. & TARLING, D. H. (eds) *Paleomagnetism and Tectonics of the Mediterranean Region*. Geological Society, London, Special Publication, **105**, 195–203.

——, ——, ——, —— & HAILWOOD, A. E. 1995. Milankovitch cyclicity of magnetic directions in Cretaceous shallow-water carbonate rocks, southern Italy. *Bollettino di Geofisica Teorica ed Applicata*, **37**, 109–118.

JAMES, N. P. & CHOQUETTE, P. W. 1983. Diagenesis, 6. Limestones – the sea floor diagenetic environment. *Geoscience Canada*, **10**, 162–179.

KIRSCHVINK, J. L. & GOULD, J. L. 1981. Biogenic magnetite as a basis for magnetic field detection in animals. *Biosystems*, **13**, 181–201.

LONGO, G., D'ARGENIO, B., FERRERI, V. & IORIO, M. 1994. Fourier evidence for high-frequency astronomical cycles recorded in Lower Cretaceous carbonate platform strata. Monte Maggiore, Southern Apennines, Italy. *In*: DE BOER, P. L. & SMITH, D. G. (eds) *Orbital Forcing and Cyclic Sequences*. International Association of Sedimentologists, Special Publication, **19**, 77–85.

MCNEIL, D. F. 1990. Biogenic magnetite from surface Holocene carbonate sediments, Great Bahama Bank. *Journal of Geophysical Research*, **95**, 4363–4371.

NAPOLEONE, G. & RIPEPE, M. 1989. Cyclic geomagnetic changes in Mid-Cretaceous rhythmite, Italy. *Terra Nova*, **1**, 437–442.

PURSER, B. H. 1980. *Sédimentation et diagenèse des carbonates néritiques récents, Tome 1*. Technip, Paris.

RASPINI, A. 1996. *Sedimentologia e ciclostratigrafia del Cretacico inferiore in facies di Piattaforma Carbonatica dell'Appennino centro-meridionale*. Doctoral dissertation, University of Bologna.

——1997. Microfacies analysis of shallow water carbonates and evidence of hierarchically organized cycles. Aptian of Monte Tobenna, Southern Apennines, Italy. *Cretaceous Research*, **19**, 197–223.

READ, J. F., KERANS, C., WEBER, L. J., SARG, J. F. & WRIGHT, F. M. 1995. *Milankovitch sea-level changes, cycles, and reservoirs on carbonate platforms in greenhouse and icehouse worlds*. Society of Economic Paleontologists and Mineralogists Short Course, **35**.

ROBINSON, S. G. 1986. The Late Pleistocene palaeoclimatic record of North Atlantic deep-sea sediments revealed by mineral-magnetic measurements. *Physics of the Earth and Planetary Interiors*, **44**, 22–47.

SCHWARZACHER, W. 1993. *Cyclostratigraphy and the Milankovitch Theory*. Elsevier, Amsterdam.

SHINN, E. A. 1969. Submarine lithification of Holocene carbonate sediments in the Persian Gulf. *Sedimentology*, **12**, 109–144.

——1986. Modern carbonate tidal flats: their diagnostic features. *Colorado School of Mines Quarterly*, **1**(81), 7–33.

STOLZ, J. F., CHANG, S.-B. R. & KIRSCHVINK, J. L. 1989. Biogenic magnetite in stromatolites. Occurrence in modern sedimentary environments. *Precambrian Research*, **43**, 295–305.

STRASSER, A. 1994. Milankovitch cyclicity and high-resolution sequence stratigraphy in lagoonal–peritidal carbonates (Upper Tithonian–Lower Berriasian, French Jura Mountains). *In*: DE BOER, P. L. & SMITH, D. G. (eds) *Orbital Forcing and Cyclic Sequences*. International Association of Sedimentologists, Special Publication, **19**, 285–301.

TARDUNO, J. A., MAYER, L. A., MUSGRAVE, R. & SHIPBOARD SCIENTIFIC PARTY 1991. High resolution, whole-core magnetic susceptibility logs from leg 130, Ontong Java plateau. *In*: KROENKE, L. W., BERGER, W. H., JANECEK, T. R., *et al*. (eds) *Proceedings of the Ocean Drilling Program, Initial Reports, 130*, Ocean Drilling Program, College Station, TX, 541–548.

TARLING, D. H. 1988. Secular variations of the geomagnetic field – the archaeomagnetic record. *In*: STEPHENSON, F. R. &, WOLFENDALE, A. W. (eds) *Secular Solar and Geomagnetic Variations in the last 10,000 Years*. Kluwer Academic, Dordrecht, 349–365.

——, IORIO, M. & D'ARGENIO, B. 1999. Geomagnetic secular variations and polarity transitions in Italian Lower Cretaceous shallow-water carbonates *Geophysical Journal International*, in press.

TORRES DE ARAUJO, F. F., PIRES, M. A., FRANKEL, R. B. & BICUDO, C. E. M. 1986. Magnetite and magnetotaxis in algae. *Biophysical Journal*, **50**, 375–378.

TUCKER, M. E. & BATHURST, R. G. C. (eds) 1990. *Carbonate Diagenesis*. Reprint Series, Vol. 1 of the International Association of Sedimentologists. Blackwell Scientific, Oxford.

—— & WRIGHT, V. P. 1990. *Carbonate Sedimentology*. Blackwell Scientific, Oxford.

VALI, H., FORSTER, O., AMARANTIDIS, G. & PETERSEN, N. 1987. Magnetotactic bacteria and their magnetofossils in sediments. *Earth and Planetary Science Letters*, **86**, 389–400.

WILSON, J. L. 1975. *Carbonate Facies in Geologic History*. Springer-Verlag, Berlin.

Diagenesis in platform carbonate rocks: a palaeomagnetic study of an upper Triassic–lower Jurassic section, Tata (Hungary)

EMŐ MÁRTON

Eötvös Loránd Geophysical Institute of Hungary, Paleomagnetic Laboratory, H-1145 Budapest, Columbus utca 17–23, Hungary

Abstract: The upper Triassic platform carbonate rocks at Tata are penetrated by three generations of neptunean dykes and cut by strike-slip faults connected to the platform's break-up in early Jurassic time. These platform carbonate rocks are conformably overlain by lower Jurassic pelagic carbonate rocks, with a clear change in sedimentation type at the Triassic–Jurassic boundary. Openings within the Triassic platform, including bedding-conformable cavities and *Megalodon* shells, are filled with Jurassic material showing late 'diagenetic' features with respect to the Triassic platform facies. The Triassic platform carbonate rocks were examined in a 10 m high outcrop and in a 200 m core, which commenced 40 m below the surface. The Triassic rock commonly has a weak initial natural remanent magnetization (NRM) and diamagnetic susceptibilities, in both outcrop and within the core, and has the same kind of magnetic mineralogy as in the 'diagenetic' material and stratified Jurassic rock, but the magnetic mineral content is higher in the Jurassic materials. The core has mixed polarity, whereas the Triassic outcrop and Jurassic materials have only normal polarity. The overall mean palaeomagnetic direction for the core is similar to those in two large quarries of similar age in the same area (the Gerecse Hills), but differs significantly from the overall mean directions of both the Jurassic 'diagenetic' material and the Jurassic beds, which are practically identical to each other. Some sites and individual samples in the Triassic outcrop carbonate rocks show Jurassic directions, but others show remagnetization circle distributions between this Jurassic direction and the present-day Earth's magnetic field. These observations suggest that the remanence of the latest Triassic time was reset during the early Jurassic diagenesis, but this was not associated with chemical changes as the mineralogies in all groups are the same.

Tata Kálvária Hill is one of the key Mesozoic sections in Hungary (Fig. 1) and consequently has been well studied geologically (e.g. Fülöp 1975; Haas 1995). The lower part of the section consists of platform carbonate rocks of late Triassic age, the end of which is marked by an erosive break, shown by *Megalodons* shells cut in half at the top of the Triassic deposits, and a sudden change from the platform to early Jurassic limestone pelagic facies. Higher in the section Ammonitico Rosso occurs, followed by Bathonian–Callovian chert and upper Jurassic–lower Cretaceous neritic limestones. An interesting feature of the outcrop is the occurrence of three generations of neptunean dykes, which penetrated the disintegrating platform in an irregular manner. Other infill materials occur in (1) palaeokarst features, some of which are bedding conformable, (2) *Megalodon* shells in which the shells themselves are often diagenetically replaced by calcite and filled with light red calcareous mud, and (3) gaps of a strike-slip character. The dyke materials and infills all comprise material derived from the light red lower Jurassic pelagic carbonate rocks, which contrast with the white upper Triassic platform facies. The thickness of the late Triassic, in outcrop, is only 10 m, but a 200 m borecore (T5) was drilled in the floor of an abandoned quarry, now a natural conservation area, which penetrated 180 m of the Triassic (upper Norian–Rhaetian) platform carbonate rocks. The material from the borecore differs from that of the outcrop as it consists of alternating supratidal, intertidal and subtidal horizons, whereas no supratidal members (often containing clay) are observed in the outcrop. Post-depositional materials, such as those of the neptunean dykes, also seem rare in the borecore, although this may be simply due to the limited representation of the section within the core. The abundance of obviously post-depositional material in the upper part of the Triassic carbonate platform, the presence of the lower Jurassic rocks

MÁRTON, E. 1999. Diagenesis in platform carbonate rocks: a palaeomagnetic study of an upper Triassic–lower Jurassic section, Tata (Hungary). *In*: TARLING, D. H. & TURNER, P. (eds) *Palaeomagnetism and Diagenesis in Sediments*, Geological Society, London, Special Publications, **151**, 157–165.

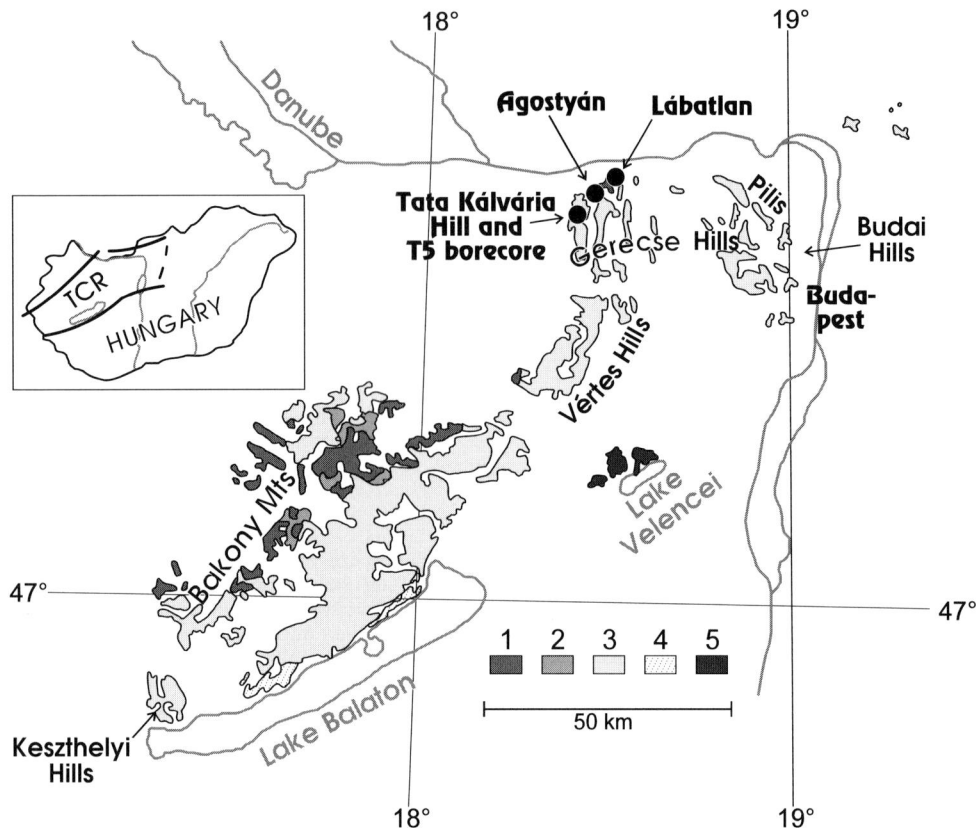

Fig. 1. The location of Tata Kálvária Hill and related sections in the Transdanubian Central Range. The Transdanubian Central Range (TCR) comprises the Keszthelyi Hills, the Bakony Mts, the Vértes Hills, the Gerecse Hills, and the Pilis and Budai Hills. Comparisons are made between the palaeomagnetic features of the Tata Kálvária Hill samples and those for identical aged sections at Lábatlan and Agostyán in the Gerecse Mts, i.e. in the same tectonic block. 1, Cretaceous; 2, Jurassic; 3, Triassic; 4, Palaeozoic sediments; 5, Palaeozoic granite.

(which supplied the material of the infillings) as a well-bedded sequence, as well as the possibility of reorienting the less affected borecore material renders Tata Kálvária Hill (Fig. 1) particularly suitable for studying the influence of diagenesis on the remanence of the Triassic platform carbonate rocks.

Sampling

Several of the regular beds, exposed in the 10 m thick outcrop of Triassic platform carbonate rocks, were drilled in the field to provide 4–8 magnetically oriented cores for each bed. The neptunean dykes, replacement material in the *Megalodon* shells, and the palaeokarst and strike-slip infill materials were also cored and oriented *in situ*. The lower Jurassic rocks, which outcrop only in the nature conservation area, were previously studied palaeomagnetically by Márton & Márton (1981). Several segments of the borecore were cut perpendicular to the bedding strike, starting at 40 m depth where

Table 1. *The palaeomagnetic mean directions in the borecore*

	N	D	I	k	α_{95}
Normal polarity	27	289	+53	28	6
Reversed polarity	6	120	−52	27	13
All samples	33	290	+53	30	5

The angle of separation between the normal and reversed mean directions is 6.4°, less than the critical angle of 13.2°, indicating a positive reversal test (McFadden & McElhinny 1990).

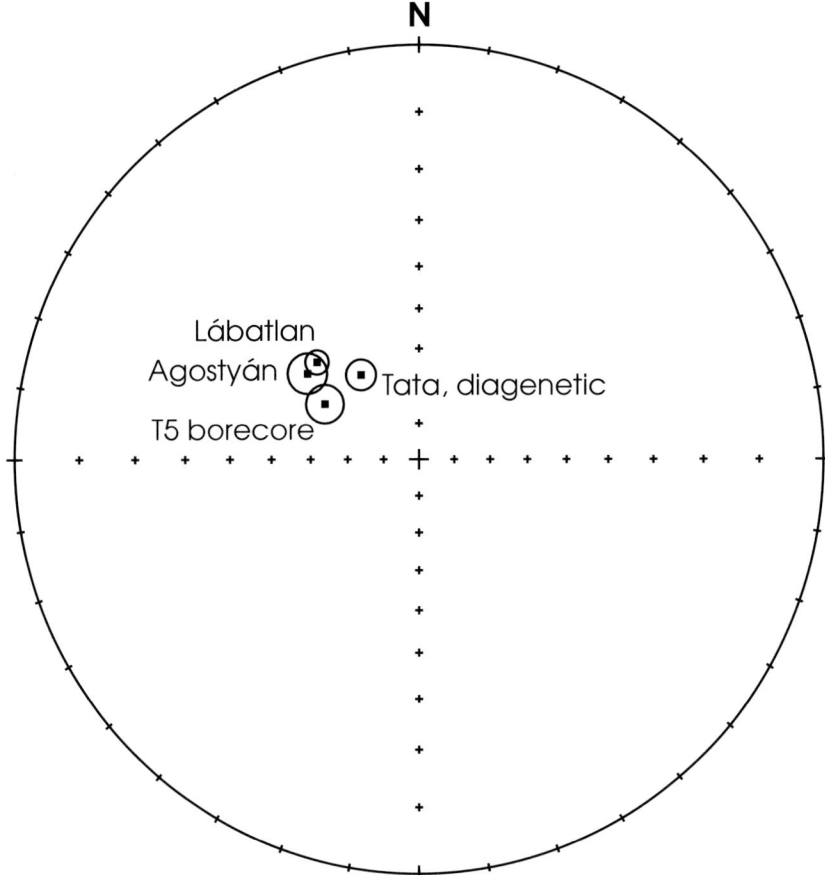

Fig. 2. Palaeomagnetic mean directions and precisions. T5 is the borecore carbonate rocks, Tata 'diagenetic' is the infill material, and Lábatlan and Agostyán are comparable upper Triassic platform carbonate rocks from the same tectonic block.

the stratification was clearly visible. From each segment, 2–5 palaeomagnetic cores were drilled and oriented with respect to the plane of the cut, so that the consistency of the palaeomagnetic signal for each segment could be checked. Finally, the orientation of the core segments, with respect to the north and the horizontal, was determined by matching the stratification

Table 2. *Magnetic polarities, initial NRM intensity and susceptibility ranges*

	Polarity	Intensity (A/m)	Susceptibility (SI)
Platform facies, outcrop	N	10^{-5} (10^{-4})	Diamag. 10^{-5}
Platform facies, borecore (T5)	N and R	10^{-5} (10^{-4})	Diamag. 10^{-5}
Early diagenetic bedding-conformable cavity fillings, fossil pseudomorphs	N	10^{-4}–10^{-3}	Diamag. 10^{-4}–10^{-3}
Neptunian dykes	N	10^{-4}–10^{-3}	Diamag. 10^{-4}
Late diagenetic (strike-slip)	N	10^{-4}–10^{-3}	10^{-5}–10^{-4}

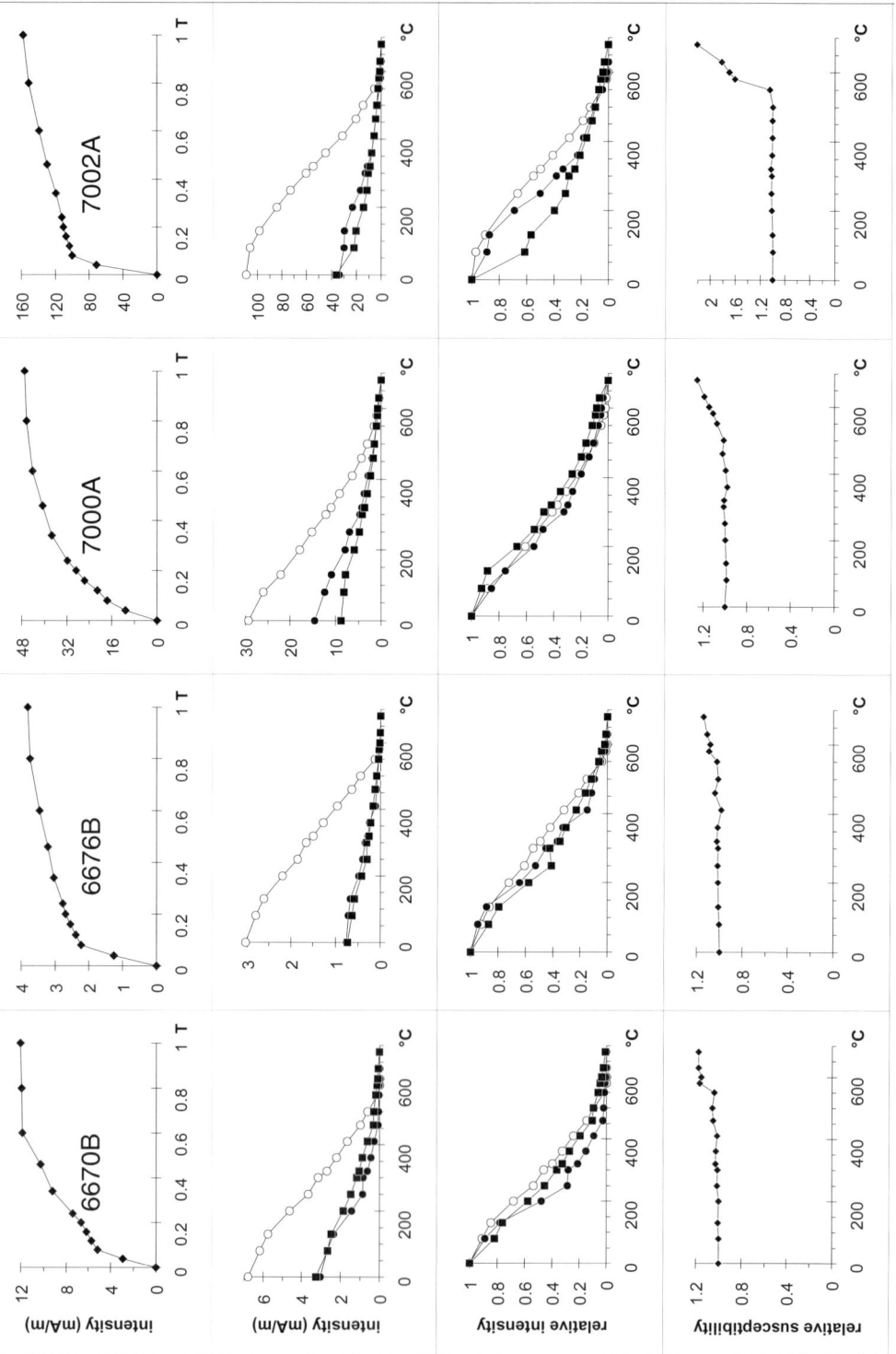

Fig. 3. Four examples of the IRM identification of the magnetic mineralogy. From top to bottom: IRM acquisition curves in applied fields up to 1 T; thermal demagnetization of the three IRM components (○, soft, 0.2 T; ●, medium, 0.36 T; ■, hard, 1.0 T). Samples 6670B and 6676B are Triassic platform carbonate rocks from

observed in the core with that measured in the outcrop at the top of the drill core. Although the tectonic tilt was shallow, the remarkable consistency in palaeomagnetic directions between each core segment and those obtained from rocks of the same age elsewhere within the same tectonic unit (Table 1 and Fig. 2) demonstrated that such reorientation had been highly successful.

Magnetic properties and magnetic minerals

The platform facies in the outcrop, as well as in the borecore, were characterized by diamagnetic susceptibilities and very weak initial natural remanent magnetization (NRM) intensities (Table 2). The intensity of the NRM was one or two orders of magnitude higher in the 'diagenetic' material and magnetic susceptibilities were

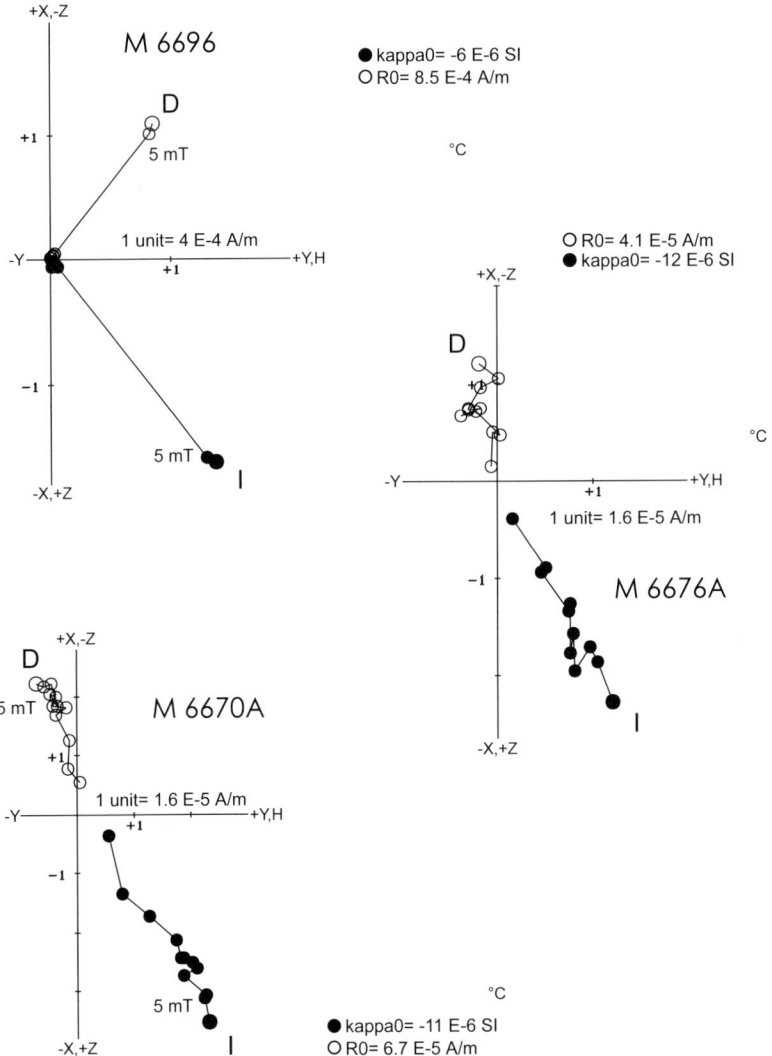

Fig. 4. Typical demagnetization curves for the Triassic outcrop platform carbonate rocks. The scales differ on the modified Zijderveld diagrams showing decrease with AF treatment; 1 unit is 400 μA/m, 16 μA/m, and 16 μA/m for M6696, M6676A and M6670A, respectively. ○, Declinations ($D = x$ v. y); ●, inclinations ($I = h$ v. z). ○, Normalized NRM intensity change with temperature, with initial values of 850 μA/m, 41 μA/m, and 67 μA/m; ●, normalized susceptibility v. temperature, with initial susceptibilities of -6 μSI, -12 μSI and -11 μSI for M6696, M6676A and M6670A, respectively. In sample M6696, the NRM intensity drops 90% between 5 mT and 150°C and suggests the unusual presence of goethite.

typically positive, although diamagnetic susceptibilities occasionally occurred in the neptunean dykes and very rarely in the oldest cavity infillings. Isothermal remanent magnetization (IRM) acquisition curves showed that both the platform facies and the 'diagenetic' material contain both soft and hard magnetic minerals, but the IRM intensities are much higher in the latter (Fig. 3). Stepwise thermal demagnetization (Fig. 4) of the three-component IRM (see Lowrie 1990), suggested that the soft component was magnetite (unblocking by 580°C and a fairly stable susceptibility up to 500°C or higher). The hardest component gradually decayed on heating and was completely demagnetized at 680°C, i.e. hematite was identified in the regular beds as well as in the infills. Goethite, though detected in some samples (e.g. specimen M6696, Fig. 4), was not common and none of the three-component IRM diagrams suggested that goethite contributed to the hardest component. The medium–hard component was completely demagnetized by 580°C, suggesting that very fine-grained magnetite also occurred in the samples. Hematite was considered not to be an actual carrier of the NRM as all specimens were fully demagnetized

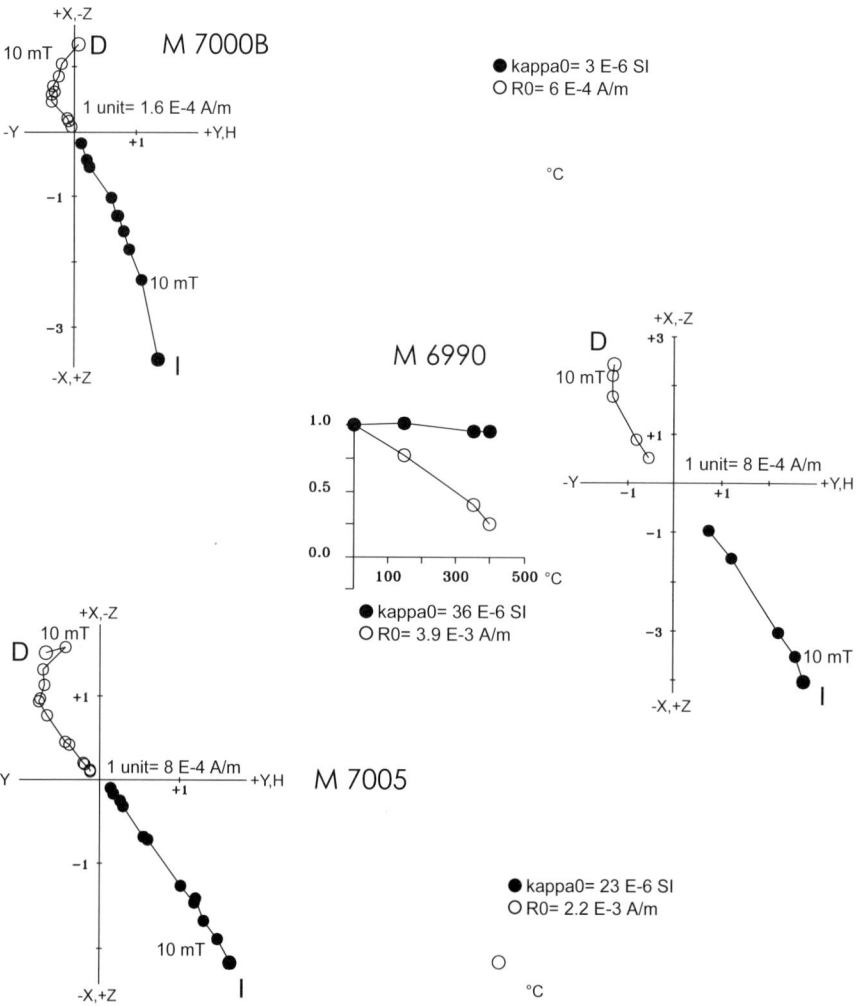

Fig. 5. Typical demagnetization curves for the 'diagenetic' material. Symbols as for Fig 4. M7000B, infilling in a *Megalodon* shell (oldest diagenetic material); M6990, neptunean dyke; M7005, infilling in a strike-slip opening (latest diagenetic material).

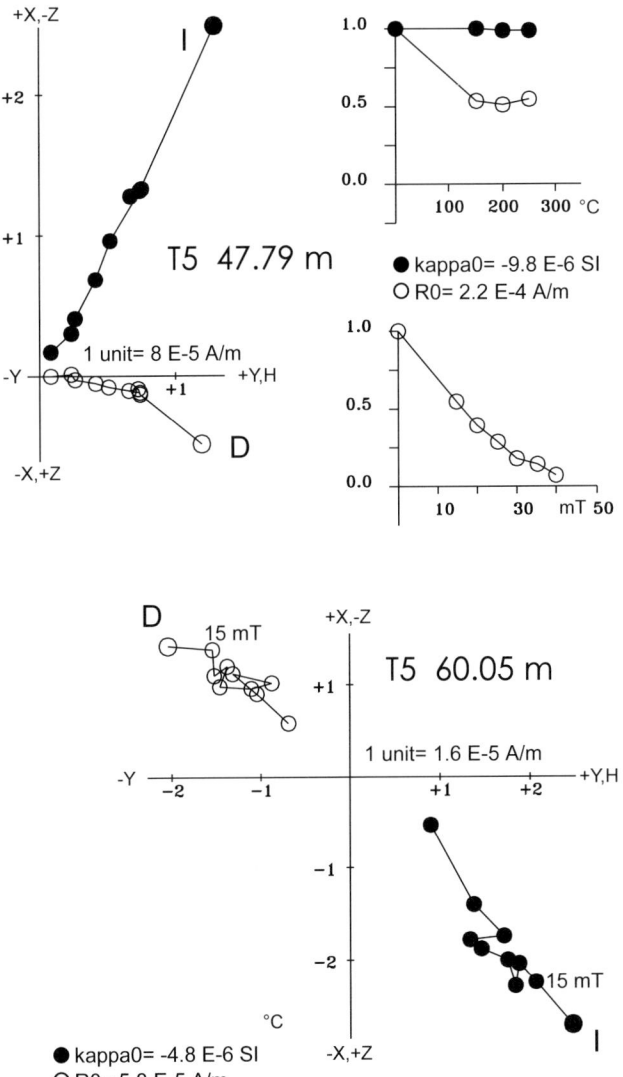

Fig. 6. Typical demagnetization curves for the borecore material. Symbols as for Fig. 4. Specimen T5 47.79 m was demagnetized first by the thermal, then by the AF method.

by the Curie-point of the magnetite or by 40 mT alternating field (AF) (Figs 4–6).

Palaeomagnetic directions and polarities

The majority of the Triassic platform carbonate samples from the borecore yielded good palaeomagnetic signals. In contrast, only four of the 16 sampled outcrop sites yielded consistent mean directions and these differed from the overall mean direction for the borecore T5 (Fig. 7). Individual sample directions from the outcrop carbonate rocks (Fig. 7b) plotted between the direction of the 'diagenetic' material, which clustered around the early Jurassic overall mean direction (Fig. 7a), and that of the local present-day geomagnetic field. A further difference between the borecore and the outcrop samples was that both normal and reversed polarity were found in the former, with a positive reversal test (Table 1), whereas the polarity was exclusively normal in the latter.

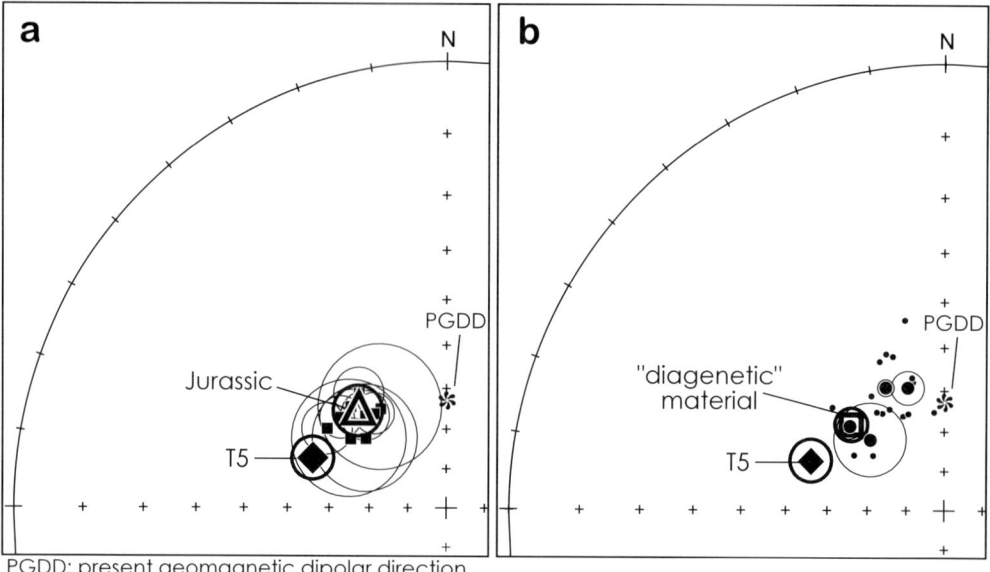

Fig. 7. Comparison between the mean palaeomagnetic directions. The circles represent the cones of confidence at a 95% level (α_{95}). (**a**) T5, overall borecore; Jurassic, overall mean; ■, site mean directions for the 'diagenetic' material. (**b**) T5, overall borecore mean; 'diagenetic' mean: larger dots, with α_{95}, are the outcrop site means for the top 10 m of the Triassic deposits; smaller dots, without α_{95}, are individual sample characteristic magnetizations from the top 10 m of the Triassic rocks.

Discussion and conclusions

The magnetic properties of the Triassic platform carbonate rocks in the outcrop are similar to those of the borecore material. Both have a very weak NRM and predominantly diamagnetic bulk susceptibilities. In the 'diagenetic' materials, the intensity of the NRM and the susceptibility are higher than in the platform carbonate rocks. However, the compositions of magnetic minerals and proportion of the soft and hard magnetic minerals seem to be similar in both facies. The most important differences between the borecore and the outcrop platform carbonate rocks of the outcrop are in their directions and polarity. The outcrop platform carbonate rocks have similar polarity and directions to those of the 'diagenetic' and dyke materials as well as the bedded Jurassic pelagic limestones. It is clear, therefore, that the samples from the top 10 m of the Triassic deposits in the outcrop must have been overprinted in Jurassic time and, to some extent, also in the present field (Fig. 7b). This means that the NRM of the uppermost part of the Triassic platform carbonate rocks must have been diagenetically influenced during the erosion and the break-up of the platform. However, the material of the borecore, being at some distance from the surface (or below the water–sediment interface) preserved the same early diagenetic remanence as found in samples of the same age and within the same tectonic block at Lábatlan and Agostyán. The identity of the magnetic mineralogy in all facies considered also means that this remagnetization at the top of this Triassic platform must have occurred by a realignment of the magnetic domains into the direction of the early Jurassic magnetic field and was not related to diagenetic changes in the magnetic mineralogy.

Thanks are due to J. Haas, for helping to reorient the T5 cores, to A. Mindszenty for sorting out the age relationship of the 'diagenetic' material, and to G. Imre for technical assistance. This work was partially supported by OTKA (Hungarian National Science Foundation) Research Grant T007368.

References

FÜLÖP, J. 1975. *Mesozoic Horsts at Tata* (in Hungarian). Geologica Hungarica, Series Geologica, Special Publications of Geological Institute of Hungary, Budapest, **16**, 1–225.

HAAS, J. 1995. Upper Triassic platform carbonates of the northern Gerecse Mts (in Hungarian, with English abstract). *Földtani Közlöny*, **125**(1–2), 27–64.

LOWRIE, W. 1990. Identification of ferromagnetic minerals in a rock by coercivity and unblocking temperature properties. *Geophysical Research Letters*, **17**, 159–162.

MÁRTON, E. & MÁRTON, P. 1981. Mesozoic paleomagnetism of the Transdanubian Central Mountains and its tectonic implications. *Tectonophysics*, **72**, 129–140.

MCFADDEN, P. L. & MCELHINNY, M. W. 1990. Classification of the reversal test in paleomagnetism. *Geophysical Journal International*, **103**, 725–729.

The influence of dolomitization on the magnetic properties of Lower Palaeozoic carbonate rocks in Estonia

ALLA SHOGENOVA

Institute of Geology at Tallinn Technical University, 7 Estonia Avenue, Tallinn, Estonia
(e-mail: alla@gi.ee)

Abstract: The low-field magnetic susceptibility and saturation isothermal remanent magnetization of Estonian Lower Palaeozoic carbonate core and outcrop samples were studied, together with their chemical composition. Dolomites of different age and genesis were found to be of different chemical composition and different magnetic properties. The dolomitization of the Lower and Middle Ordovician sequences is diagenetic and is represented by widespread dolomitized layers as well as dolomitization associated with tectonic disturbances. All studied dolomites have an increasing total iron content. In Upper Ordovician and Silurian dolomites this increase is not enough to change magnetic properties significantly. In the rocks, sampled from a fracture zone and from a dolomite layer of the Väo Formation in the Middle Ordovician sequence and another widespread, several metres thick dolomite layer of the Volkhov stage from the Lower Ordovician sequence, the magnetic susceptibility of dolomites was found to be several times higher than that of limestones of the same age. Changes in magnetic susceptibility were found to be associated with substitution of Fe^{2+} for Mg^{2+} in the crystalline lattice of dolomites.

There have been few previous studies of the magnetic properties of sedimentary rocks in Estonia. In the carbonate rocks studied here, the magnetic properties are dominated by the iron minerals that are mostly found in their more argillaceous parts. Most of the pure limestones are characterized by a zero or negative magnetic susceptibility (diamagnetism), whereas in others, the susceptibility increases to paramagnetic or ferromagnetic with increasing content of clay, iron oolites and glauconite impurities. There are also layered formations which show varying degree of dolomitization, and some fracture zones show that dolomitization was accompanied by remagnetization. This variability in the magnetic properties of the Ordovician and Silurian carbonate rocks has allowed comparative studies to be undertaken, using regression and factor analyses, to improve the understanding of the processes of dolomitization and to assess their effects on the magnetic properties of the carbonate rocks in the Estonian sedimentary basin.

Geology

Estonia is on the southern slope of the Baltic Shield, which is a strongly metamorphosed Precambrian crystalline basement. This is overlain unconformably by a condensed (<800 m) Upper Vendian and Lower Palaeozoic sedimentary cover (Suveizdis 1979). The dominant structure in the basin is the Estonian Homocline (Puura & Vaher 1997), which dips gently southwards at some 2.9–4.4 m/km (Shogenova 1996), decreasing in thickness from c. 100 m in northern Estonia to some 800 m in the south, e.g. 784 m in the Ruhnu borehole. Both major and minor linear disturbances and fracture zones complicate the homocline, but most of the flexures have low amplitudes (<30 m) and correspond to faults in the crystalline basement. The basement faults control the occurrence of most of the secondary alterations to the carbonate rocks, which include karst solution, leaching, dolomitization, red pigmentation by iron oxides and hydroxides, and sulphide mineralization (Vingisaar & Taalman 1974; Pichugin *et al.* 1976).

The Ordovician rocks are predominantly epicontinental marine limestones which show varying degrees of dolomitization (Põlma 1982). The clay content tends to be highest in the central part of the syncline, and fine, dispersed pyrites occur throughout the sequence (Põlma 1982). Glauconite and hematite or goethite impurities occur in the Lower (≤20 m thick) and Upper (≤90 m thick) Ordovician sequences. Bituminous layers occur throughout the sequence, especially

Table 1. Physical properties and chemical composition of different origin limestones and dolomites

Rock type	Silurian			Upper Ordovician			Vão Formation Middle Ordovician			Volkhov stage Lower Ordovician			Fracture zone Middle Ordovician		
	Avg	STD	Num	Avg	STD	Num	Avg	STD	Num	Avg	STD	Num	Avg	STD	Num
Magnetic properties															
Magnetic susceptibility κ ($\times 10^{-5}$ SI)															
Limestones	**1.09**	2.46	36	**1.73**	2.44	25	**2.07**	1.54	69	**7.08**	3.22	13	**3.23**	3.61	27
Argil. limst.	**3.42**	2.91	27	**5.27**	3.9	27	**2.28**	0.47	8	**7.68**	2.58	12	**3.94**	2.33	32
Marls	**9.39**	3.55	13	**9.54**	4.72	16	–	–	–	**9.08**	1.88	4	**7.36**	1.69	7
Dolomites	*1.68*	*1.14*	*30*	*3.24*	*1.02*	*12*	*19.4*	*1.76*	*4*	*19.6*	*12*	*10*	*7.33*	*4.53*	*21*
Argil. dolom.	*4.39*	*2.04*	*16*	*5.29*	*2.12*	*9*	*20.4*	*1.56*	*5*	*16.5*	*11.7*	*8*	*9.15*	*3.8*	*14*
Dolom. marl	*7.61*	*2.31*	*12*	*9.74*	*6.53*	*4*	–	–	–	–	–	–	–	–	–
Saturation isothermal remanent magnetiz., SIRM ($\times 10^{-3}$ A/M)															
Limestones	**10.2**	9.21	35	**10.5**	8.85	25	**12.1**	7.83	69	**456.8**	615	13	**31.5**	38.5	28
Argil. limst.	**21.7**	10.6	27	**205.5**	600	26	**22.1**	19	8	**821**	1091	12	**38.3**	21.1	32
Marls	**41.23**	18.4	12	**1429**	2026	16	–	–	–	**1663**	1064	4	**60.7**	10	7
Dolomites	*16.4*	*12.2*	*13*	*13.6*	*12.2*	*12*	*6.5*	*3*	*4*	*358*	*394*	*10*	*22.2*	*9.86*	*21*
Argil. dolom.	*30*	*25.3*	*16*	*256*	*708*	*9*	*8.2*	*3.56*	*5*	*1160*	*1071*	*8*	*36.5*	*23.4*	*14*
Dolom. marl	*39.1*	*20.2*	*12*	*1146*	*1856*	*4*	–	–	–	–	–	–	–	–	–
Density properties															
Grain density, σ_g (g/cm³)															
Limestones	**2.72**	0.02	36	**2.72**	0.03	27	**2.71**	0.01	69	**2.73**	0.02	13	**2.7**	0.06	28
Argil. limst.	**2.72**	0.04	28	**2.74**	0.03	27	**2.71**	0.01	8	**2.74**	0.02	12	**2.7**	0.08	30
Marls	**2.74**	0.02	13	**2.74**	0.03	16	–	–	–	**2.75**	0.01	4	**2.69**	0.08	7
Dolomites	*2.79*	*0.04*	*13*	*2.8*	*0.05*	*12*	*2.84*	*0.01*	*4*	*2.86*	*0.02*	*10*	*2.78*	*0.04*	*19*
Argil. dolom.	*2.76*	*0.11*	*16*	*2.79*	*0.05*	*9*	*2.84*	*0.01*	*5*	*2.86*	*0.02*	*8*	*2.79*	*0.05*	*13*
Dolom. marl	*2.78*	*0.03*	*12*	*2.77*	*0.01*	*4*	–	–	–	–	–	–	–	–	–
Bulk density, σ_w (g/m³)															
Limestones	**2.65**	0.06	36	**2.65**	0.06	25	**2.66**	0.02	69	**2.68**	0.02	13	**2.62**	0.09	28
Argil. limst.	**2.61**	0.07	28	**2.54**	0.07	27	**2.65**	0.02	8	**2.63**	0.04	12	**2.57**	0.1	30
Marls	**2.52**	0.06	13	**2.56**	0.04	16	–	–	–	**2.58**	0.02	4	**2.48**	0.06	7
Dolomites	*2.66*	*0.07*	*13*	*2.67*	*0.14*	*12*	*2.81*	*0.01*	*4*	*2.76*	*0.07*	*10*	*2.66*	*0.1*	*19*
Argil. dolom.	*2.56*	*0.14*	*16*	*2.62*	*0.09*	*9*	*2.8*	*0.03*	*5*	*2.77*	*0.03*	*8*	*2.63*	*0.12*	*13*
Dolom. marl	*2.54*	*0.1*	*12*	*2.61*	*0.12*	*5*	–	–	–	–	–	–	–	–	–
Chemical composition															
Dolomite (%)															
Limestones	**7.71**	7.46	37	**9.63**	8.62	25	**2.42**	3.16	75	**4.86**	4.79	13	**10.3**	7.17	28
Argil. limst.	**12.9**	6.41	28	**21.1**	11.96	27	**3.08**	1.66	9	**9.11**	5.85	12	**12.5**	6.26	32
Marls	**16.8**	4.26	13	**28.97**	11.06	16	–	–	–	**9.38**	1.73	4	**15.8**	7.9	8
Dolomites	*82.7*	*4.87*	*13*	*86.4*	*6.17*	*12*	*70.9*	*1.54*	*4*	*66.8*	*7.7*	*10*	*79.5*	*9.06*	*21*
Argil. dolom.	*67.7*	*10.7*	*16*	*83.5*	*12.7*	*9*	*63.8*	*4.59*	*5*	*84*	*6.6*	*8*	*71.2*	*8.3*	*14*
Dolom. marl	*45.2*	*10.3*	*13*	*69.2*	*16.7*	*5*	–	–	–	–	–	–	–	–	–

	Avg	STD	Num	Avg	STD	Num	Avg	STD	Num	Avg	STD	Num	Avg	STD	Num
Limestones	**4.75**	2.72	37	**4.59**	3.14	25	**4.33**	2.47	75	**4.91**	2.36	13	**7.14**	1.73	28
Argil. limst.	17.4	4.01	28	17.2	4.5	27	11.9	2.32	7	17.2	4.51	12	15.6	4.9	32
Marls	36.5	9.55	13	30.5	7.08	16	–	–	–	26.8	2.14	4	31.5	3.15	8
Dolomites	4.87	2.49	13	4.8	1.81	12	7.13	0.52	4	7.63	1.43	10	7.06	1.56	21
Argil. dolom.	16.9	4.09	16	17.3	5.53	9	14.5	3.36	5	14.8	2.95	8	14.6	3.43	14
Dolom. marl	34.3	9.9	13	36.9	7.91	5	–	–	–	–	–	–	–	–	–
FeO (%)															
Limestones	**0.13**	0.1	36	**0.2**	0.17	24	**0.11**	0.08	73	**0.09**	0.1	13	**0.19**	0.24	28
Argil. limst.	0.19	0.12	27	0.27	0.28	23	0.09	0.05	9	0.26	0.32	12	0.29	0.27	31
Marls	0.46	0.31	13	0.31	0.4	16	–	–	–	0.36	0.25	4	0.53	0.46	8
Dolomites	0.33	0.2	12	0.32	0.21	12	0.85	0.54	4	0.66	1.04	10	0.42	0.4	19
Argil. dolom.	0.29	0.26	15	0.35	0.35	9	0.83	0.27	5	1.02	1.46	8	1.02	0.61	14
Dolom. marl	0.42	0.28	13	0.63	0.95	5	–	–	–	–	–	–	–	–	–
Fe_2O_3 (%)															
Limestones	**0.13**	0.1	35	**0.32**	0.16	24	**0.52**	0.27	73	**1.67**	1.09	13	**0.49**	0.35	27
Argil. limst.	0.68	0.3	27	0.74	0.43	23	0.71	0.46	9	1.31	0.92	12	0.79	0.73	31
Marls	1.47	0.72	13	1.34	0.98	16	–	–	–	1.08	0.69	4	0.92	0.82	8
Dolomites	0.28	0.16	12	0.98	1.24	12	0.85	0.54	4	2.23	1.33	10	0.79	0.61	19
Argil. dolom.	0.74	0.5	15	0.88	0.65	9	0.83	0.27	5	1.5	1.04	8	0.67	1.16	14
Dolom. marl	1.29	0.44	13	1.51	1.37	5	–	–	–	–	–	–	–	–	–
$Fe_2O_{3\ total}$ (%)															
Limestones	**0.3**	0.22	35	**0.54**	0.25	24	**0.64**	0.3	73	**1.77**	1.07	13	**0.7**	0.28	27
Argil. limst.	0.88	0.36	27	1.03	0.33	23	0.82	0.48	9	1.6	0.87	12	1.1	0.71	31
Marls	1.98	0.8	13	1.69	0.85	16	–	–	–	1.48	0.68	4	1.49	0.82	8
Dolomites	0.64	0.24	12	1.33	1.34	12	2.63	0.18	4	2.96	1.23	10	1.24	0.39	19
Argil. dolom.	1.07	0.43	15	1.26	0.68	9	2.71	0.08	5	2.62	1.11	8	1.77	0.5	14
Dolom. marl	1.75	0.42	13	2.2	1.06	5	–	–	–	–	–	–	–	0.85	–
MnO (%)															
Limestones	**0.02**	0.06	37	**0.06**	0.11	25	**0.07**	0.08	75	**0.22**	0.07	13	**0.1**	0.05	21
Argil. limst.	0.02	0.03	28	0.12	0.18	27	0.19	0.38	9	0.17	0.05	12	0.07	0.01	20
Marls	0.04	0.04	13	0.15	0.1	16	–	–	–	0.17	0.04	4	0.07	0.01	4
Dolomites	0.06	0.04	13	0.1	0.09	12	0.28	0.01	4	0.28	0.2	10	0.18	0.07	14
Argil. dolom.	0.04	0.02	16	0.13	0.12	9	0.27	0.02	5	0.3	0.19	8	0.16	0.04	4
Dolom. marl	0.06	0.06	13	0.12	0.07	5	–	–	–	–	–	–	–	–	–
Cl (%)															
Limestones	**0.03**	0.02	37	**0.02**	0.02	25	**0.01**	0.02	75	**0.03**	0.02	13	**0.03**	0.02	21
Argil. limst.	0.05	0.03	28	0.03	0.04	27	0.01	0.02	9	0.03	0	12	0.03	0.03	20
Marls	0.09	0.07	13	0.09	0.08	16	–	–	–	0.02	0.01	4	0.02	0.01	4
Dolomites	0.11	0.08	13	0.1	0.11	12	0.16	0.08	4	0.07	0.03	10	0.07	0.04	14
Argil. dolom.	0.07	0.05	16	0.06	0.03	9	0.1	0.01	5	0.07	0.04	8	0.07	0.01	4
Dolom. marl	0.1	0.06	13	0.06	0.08	5	–	–	–	–	–	–	–	–	–

Avg is average, STD is standard deviation, Num is number.

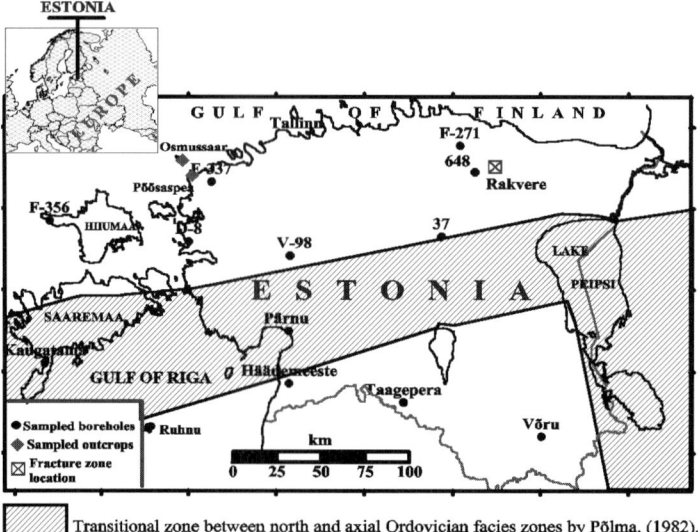

Fig. 1. Map of Estonia showing the sampling locations and main structural zones.

along the axial zone (Fig. 1), with commercial oil shales occurring in the Middle Ordovician units (≤80 m) in northeastern Estonia. Oolites are sometimes formed of goethite or, less frequently, francolite, with phosphatic nodules commonly having pyrite crystals within them or on their surfaces. The dolomitization of the Ordovician carbonate rocks is independent of the facies and is mainly associated with tectonic zones, being more common in the north and in the transition zone, and comparatively rare further south, i.e. within the axial zone (Põlma 1982). The widespread dolomitization in the Ordovician succession is mostly considered to be secondary (Vingisaar & Taalman 1974; Pichugin et al. 1976; Põlma 1982) but it is unclear whether it is associated with early (Põlma 1982) or late diagenesis (Pichugin et al. 1976). One of the most dolomitized layers in the Middle Ordovician sequence is the Pae member (c. 0.2–0.6 m thick) of the Väo Formation (Table 1); this has been ascribed to diagenetic processes by Orviku (1940), but to late dolomitization by Kiipli (1983a, b). However, the topmost part of the Upper Ordovician sequence is represented by shallow-water nearshore sediments which include primary dolomites deposited in high-salinity conditions (Põlma 1982). The overlying Silurian rocks comprise marine carbonates, marls and shales that were deposited in the same continental marginal basin, but there is a clear facies arrangement into five belts. Dolomitization occurs throughout the section and is mostly of early and late diagenetic origin, the latter being marked by its high, often cavernous, porosity. There are few Silurian primary

Table 2. Factor loadings for Upper Ordovician and Silurian data sets

	Silurian		Upper Ordovician	
	Factor 1	Factor 2	Factor 1	Factor 2
κ	**0.80**	**0.31**	**0.81**	**0.38**
SIRM	−0.04	**0.30**	**0.65**	0.18
σ_w	**−0.68**	0.14	**−0.61**	0.28
σ_g	0.04	**0.70**	−0.05	**0.88**
SiO_2	**0.95**	0.07	**0.76**	0.07
TiO_2	**0.94**	0.05	**0.94**	0.16
Al_2O_3	**0.96**	0.09	**0.94**	0.19
FeO	**0.44**	**0.46**	0.15	0.11
Fe_2O_3	**0.69**	0.09	**0.54**	**0.33**
MnO	0.07	**0.67**	0.30	0.27
MgO	0.01	**0.70**	−0.29	**0.86**
CaO	**−0.70**	**−0.54**	**−0.47**	**−0.75**
Na_2O	**0.63**	0.19	**0.53**	0.16
K_2O	**0.93**	0.04	**0.90**	0.17
P_2O_5	**0.43**	0.16	0.10	0.13
SO_3	**0.48**	−0.02	−0.10	0.01
Cl	0.24	**0.62**	0.17	**0.47**
Expl. var.	6.73	2.61	5.65	2.87
Prp. Total	0.40	0.15	0.33	0.17
Eigenvalue	7.19	2.16	6.15	2.37
Variance (%)	49.27	12.72	36.2	13.93

Extraction: principal components. Varimax normalized rotation of factors. Significant loadings given in bold type.
Expl. Var., explored variance calculated as sum of loadings of all factors. Prp. Total, total proportion calculated as Expl. Var. value divided by number of variables.
κ, magnetic susceptibility; SIRM, saturation isothermal remanent magnetization; σ_w, bulk density of wet samples; σ_g, grain density.

dolomites. Iron is present in the form of sulphides (pyrite and marcasite), oxides (hematite and magnetite), hydroxides (hydrohematite and hydrogoethite) and silicates (glauconite). The iron is mainly of terrigenous origin and its distribution is controlled by the facies pattern. The highest content, up to 5% in the marls and shales and up to 7% in red-coloured limestones and shales, is in the deep-water facies. Secondary dolomites can have an iron content up to 1.7% whereas nearshore lime-stones contain <0.5% iron (Jürgenson 1988).

Sampling and analytical methods

Some 500 samples were collected from outcrops and boreholes in both Ordovician and Silurian rocks (Fig. 1, Table 1), with particular attention being paid to their immediate background and tectonic setting. Chemical analyses of altered rocks from some localities have been described previously (Bityukova et al. 1996; Shogenova & Puura, 1997), as have the petrophysical properties and chemical composition of the Ordovician carbonate rocks from six wells (P2279–P2284) near Rakvere (Fig. 1). These had been drilled across a low-resistivity anomaly zone (Shogenova & Tuuling 1990; Shogenova

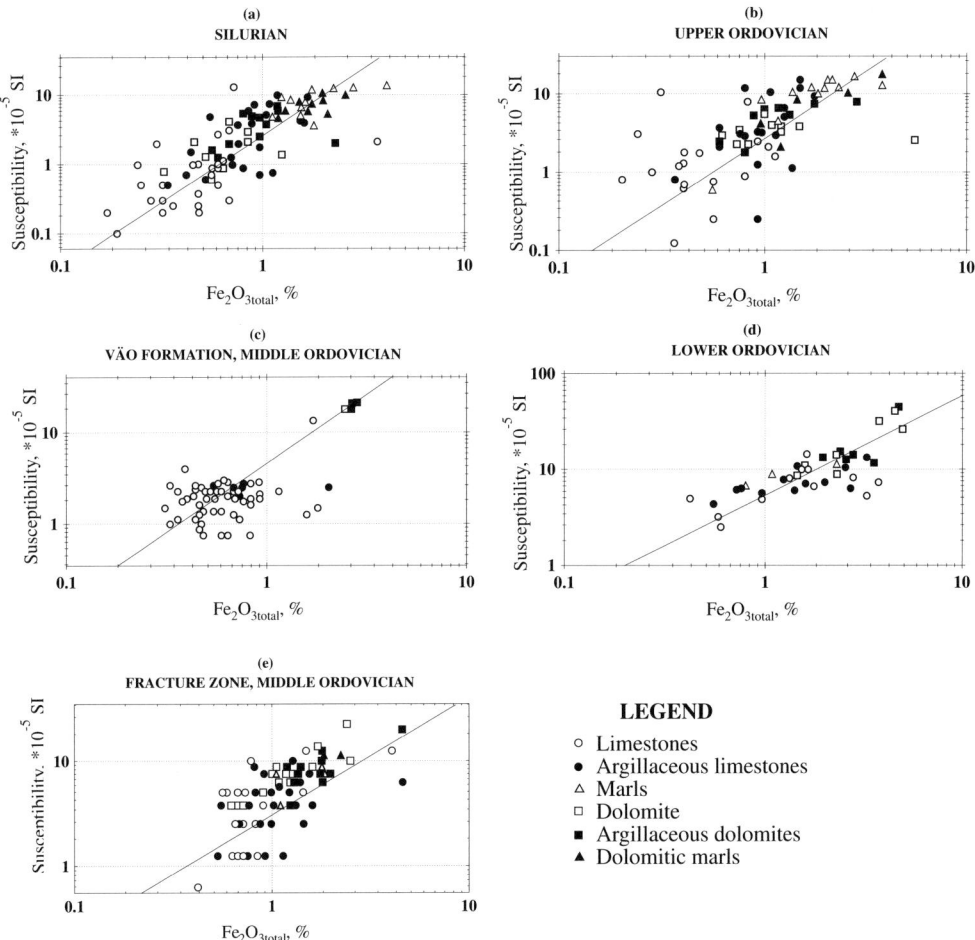

Fig. 2. Magnetic susceptibility v. Fe_2O_3 total content (both on a log-decimal scale). (**a**) Silurian rocks; (**b**) Upper Ordovician rocks; (**c**) the Middle Ordovician Väo Formation, sampled near fissures; (**d**) Lower Ordovician rocks; (**e**) Middle Ordovician fracture zone. Using an insoluble residue (IR) of $SiO_2 + TiO_2 + Al_2O_3 + Na_2O + K_2O$, calcite content of $[CaO - 30.4 \times (MgO/21.8) \times (1 + (44/56)]$ and dolomite content of $MgO \times 4.6/[(MgO \times 4.6) + \text{calcite content}]$, the lithologies are defined as follows: limestones, IR < 10% and dolomite <50%; argillaceous limestones, 10% < IR < 25% and dolomite <50%; marls, IR > 25% and dolomite <50%; dolomites, IR < 10% and calcite <50%; argillaceous dolomites, 10% < IR < 25% and calcite >50%; dolomite marls, IR > 25% and calcite <50%.

& Puura 1997) associated with an oil shale and phosphorite deposit that was interpreted as being a low-amplitude fracture zone. This low-resistivity anomaly zone, identified using the dipole electrode configuration (Vaher 1986), is situated near Rakvere township (Fig. 1). It was interpreted as a fracture zone with small displacement amplitudes. Within this zone there was a clearly defined MgO anomaly associated with an increase in FeO, NaO, Al_2O_3 and SO_3, and a decrease in Fe_2O_3. Within this anomaly, the carbonate rocks were affected by secondary alterations such as dolomitization, red colouring, oxidation, solution (karstification) and leaching. From these rocks, 101 Middle Ordovician samples are considered here. These data, from exposures and the bore-cores, have been expanded by determining the magnetic and density properties and chemical compositions of all other rock samples. The Middle Ordovician Väo Formation dataset comprises about 100 samples from a quarry along the Laagna road near Tallinn, where both pure limestones, typical of the Väo Formation and the Pae dolomite bed, could also be sampled in clear structural relationship to each other.

Chemical analyses comprised the measurement of the CaO, MgO, SiO_2, TiO_2, Al_2O_3, FeO, Fe_2O_3, Fe_2O_{3total}, MnO, Na_2O, K_2O, P_2O_5, SO_3, CO_2, and Cl contents. The methods for measuring the bulk (wet sample) and grain densities have been described by Shogenova & Puura (1997). The magnetic susceptibilities and saturation isothermal remanent magnetizations (SIRMs) of 24 mm side cubic samples were determined, following the methods described by Prijatkin & Poljakov (1983). The susceptibilities were measured with an IMV-2 susceptibility meter using a field of 0.08 mT, and the SIRM was measured with an ION-1 instrument (rock generator type) after the samples had been magnetized in a magnetic field of 30 mT. The technique and some results of complex

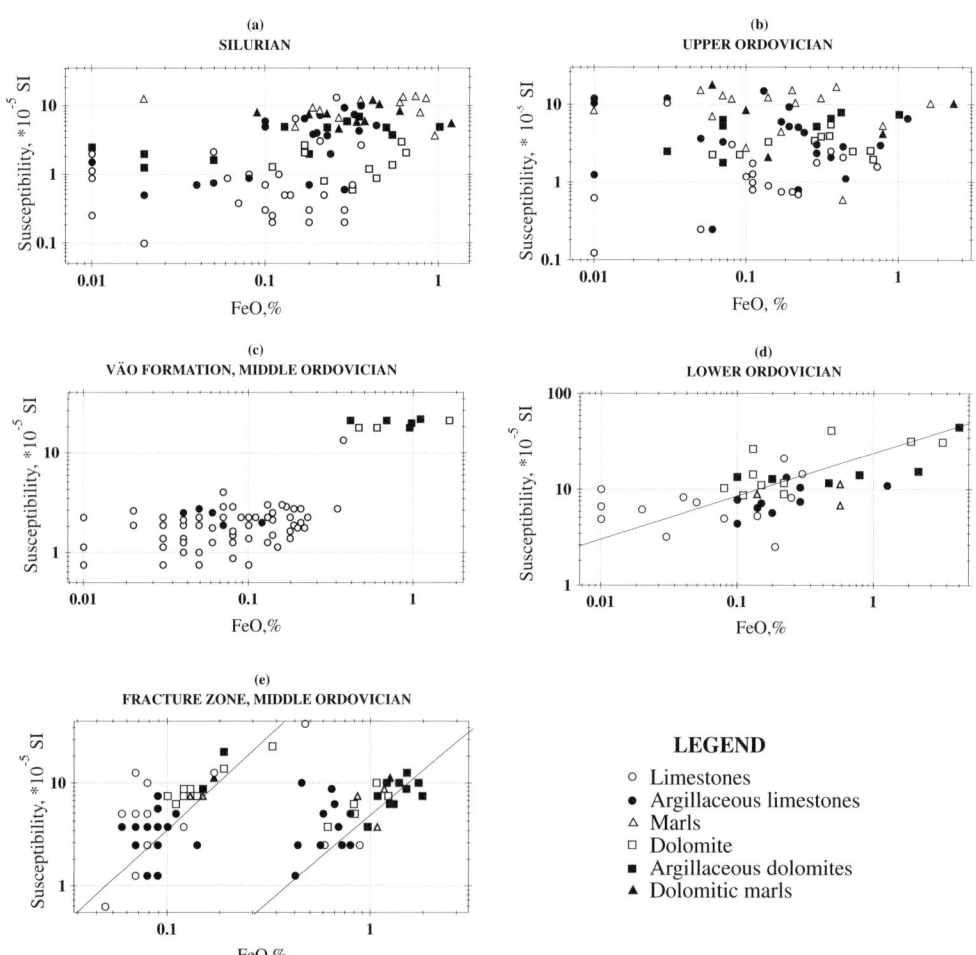

Fig. 3. Magnetic susceptibility v. FeO content (on a log-decimal scale). Description as for Fig. 2.

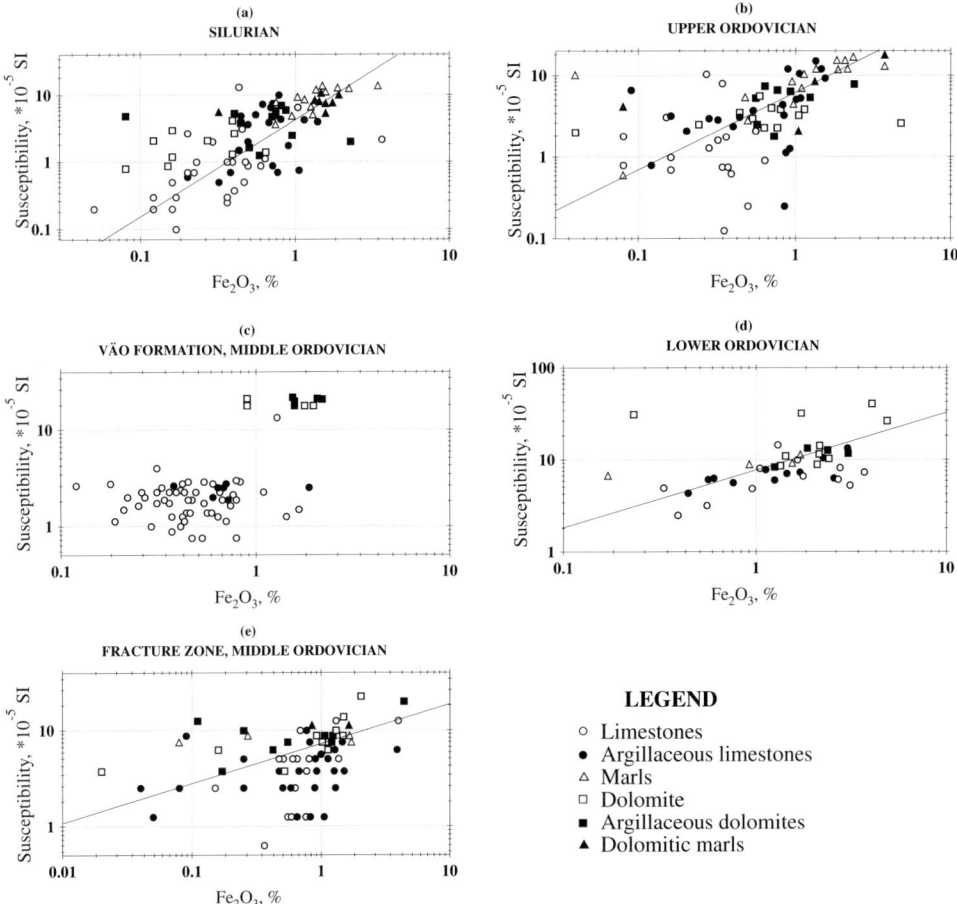

Fig. 4. Magnetic susceptibility v. Fe_2O_3 content (both on a log-decimal scale). Description as for Fig. 2.

analysis of petrophysical and chemical variables of these rock samples were partly described earlier (Shogenova et al. 1998).

Results

Susceptibility and the form of iron oxide and its total content

The magnetic susceptibility of six lithologies (limestones, argillaceous limestones, marls, dolomites, argillaceous dolomites and dolomitic marls) were compared with their total iron oxides content (Fe_2O_{3total}) and their form (Table 1, Figs 2–4). In general, the susceptibility correlated well with the total iron content (Fig. 2). The Silurian carbonate rocks had average magnetic susceptibilities between 1.09×10^{-5} SI in limestones and 9.39×10^{-5} SI in marls, and from 1.68×10^{-5} SI in the dolomites to 7.61×10^{-5} SI in dolomitic marls (Table 1, Fig. 2a). The average total iron content in Silurian carbonates increased from 0.3% in limestones to 1.98% in marls, and from 0.6% in dolomites to 1.8% in dolomitic marls (Table 1, Fig. 2a). The average FeO content increased from 0.13% in limestones to 0.46% in marls, and from 0.3% in dolomites to 0.4% in dolomitic marls (Fig. 3a), whereas the mean Fe_2O_3 increased from 0.13% in limestones to 1.47% in marls, and from 0.3% in dolomites to 1.3 % in dolomitic marls (Fig. 4a).

In the Upper Ordovician sequence the average magnetic susceptibility increased from 1.73×10^{-5} SI in limestones to 9.54×10^{-5} SI in marls, and from 3.24×10^{-5} SI in dolomites to 9.74×10^{-5} SI in dolomitic marls. The mean total iron content increased from 0.54% in limestones to 1.69% in marls, and from 1.33% in dolomites to 2.2% in dolomitic marls (Fig. 2b). The average FeO content increased from 0.2% in limestones to 0.31% in marls, and from 0.32% in dolomites to 0.63% in dolomitic marls (Fig. 3b). The average Fe_2O_3 content increased from 0.32% in limestones to 1.34% in marls, and from 0.98% in dolomites to 1.51% in dolomitic marls (Fig. 4b). In the Väo Formation, the average total iron content increased from 0.64% in limestones to 2.6% in the dolomite layer, and was matched by an increase in average susceptibility from 2.07×10^{-5} SI in limestones to 19×10^{-5} SI in dolomites (Table 1, Fig. 2c). The average FeO content increased from 0.11% in limestones to 0.9% in dolomites (Fig. 3c). The mean Fe_2O_3 content increased less: from 0.52% in limestones to 0.9% in dolomites (Fig. 4c).

The Lower Ordovician rocks show a total iron increase from 1.77% in limestones to 2.96% in dolomites, which correlates with an increase in average susceptibility from 7.08×10^{-5} in limestones to 19.6×10^{-5} SI in dolomites (Fig. 2d). The average FeO content increased from 0.09% in limestones to 0.66% in dolomites (Fig. 3c). The relatively high mean of Fe_2O_3 content in limestones (1.67%) increased to 2.23% in Volkhov dolomites (Fig. 4c).

In the fracture zone crossing the Middle Ordovician sequence the magnetic susceptibility increased on average from 3.23×10^{-5} SI in limestones to 7.36×10^{-5} SI in marls, 7.3×10^{-5} SI in dolomites and 9.2×10^{-5} SI in argillaceous dolomites (Table 1, Fig. 2e). This was caused by an increase in the average total iron content from 0.7% in limestones to 1.49% in marls, 1.2% in dolomites and 1.8% in argillaceous dolomites. An increase in both iron forms was determined. The mean FeO increased from 0.19% in limestones to 0.53% in marls, 0.4% in

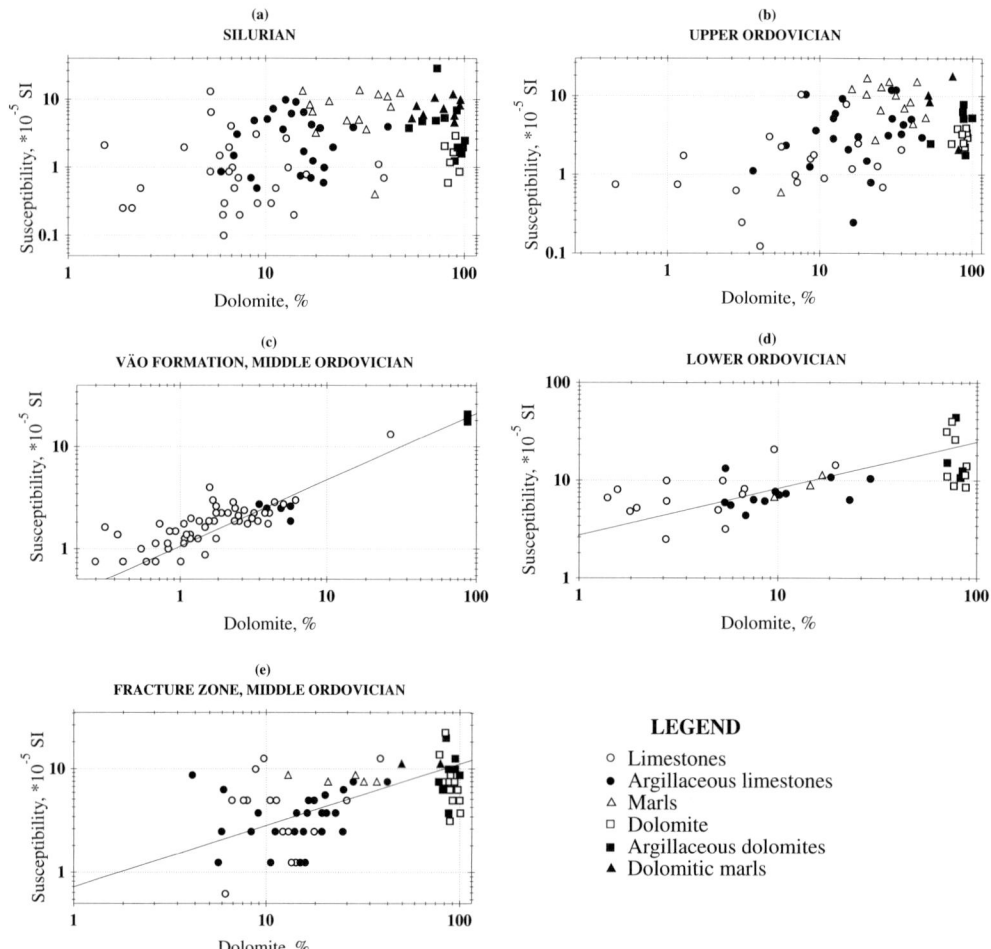

Fig. 5. Magnetic susceptibility v. dolomite content (on a log-decimal scale). Description as for Fig. 2.

dolomites and 1% in argillaceous dolomites (Fig. 3e). The average Fe_2O_3 content increased from 0.49% in limestones to 0.92% in marls, 0.8% in dolomites and 0.7% in argillaceous dolomites (Fig. 4e).

Susceptibility and the degree of dolomitization and clay content

To study the relationships of susceptibility with the degree of dolomitization (Fig. 5) and the clay content of the dolomites, discussed above, the insoluble residue content in samples was also calculated (Fig. 6). The best correlation between susceptibility and dolomitization was in rocks of the Middle Ordovician Väo Formation (Fig. 5c). This correlation also existed but with more scattered samples in the Lower Ordovician sequence (Fig. 5d) and in the fracture zone (Middle Ordovician rocks) (Fig. 6e). The correlation is very poor and very scattered in the Silurian (Fig. 5a) and Upper Ordovician datasets (Fig. 5b). The best correlation between magnetic susceptibility and clay content (insoluble residue) is in the Silurian rocks (Fig. 6a) and, to a lesser extent, in the Upper Ordovician rocks (Fig. 6b). In the Väo Formation (Fig. 6c), there are clearly two susceptibility groups, although these samples had a very low insoluble residue content (<20%), whereas there is no correlation between these variables in the Lower Ordovician samples (Fig. 6d). In the fracture zone (Fig. 6e), there is a clear distinction between the pure carbonate rocks and the argillaceous rocks, but neither variables shows any correlation.

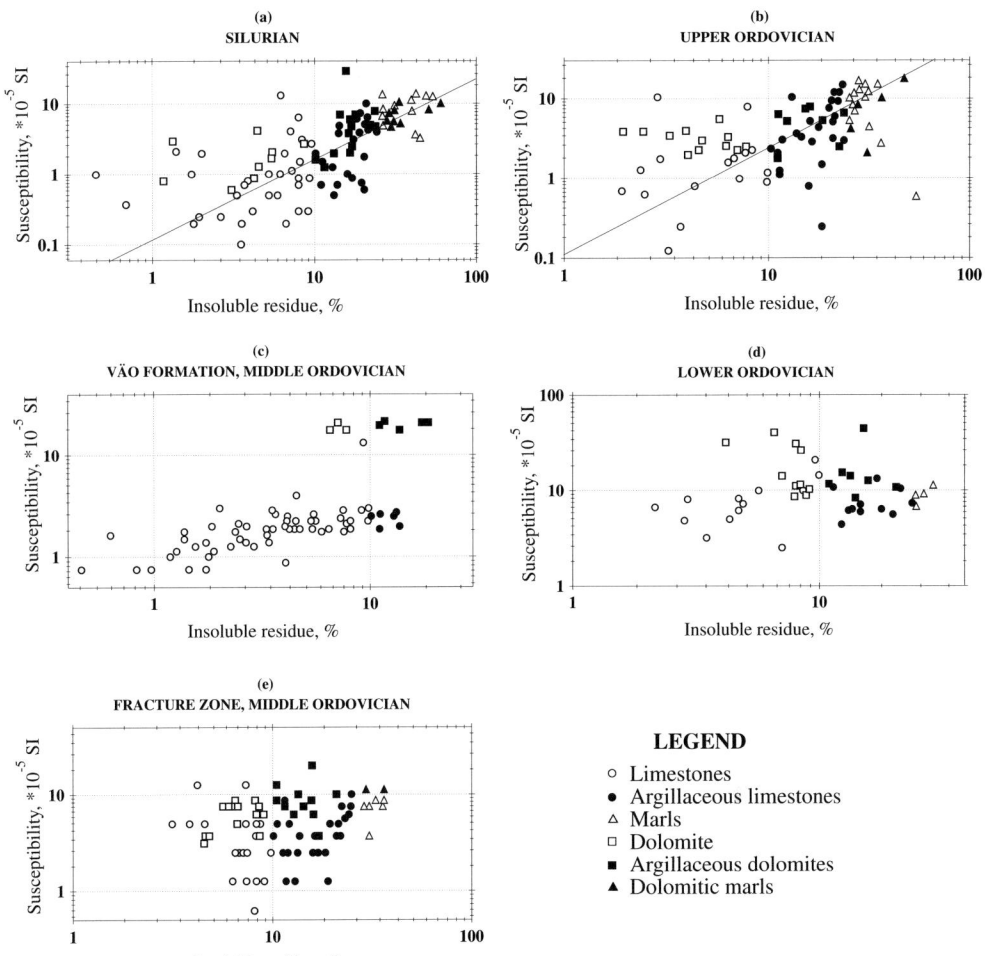

Fig. 6. Magnetic susceptibility v. insoluble residue content (on a log-decimal scale). Description as for Fig. 2.

Isothermal remanence and total iron contents

The variations of SIRM in all of these Lower Palaeozoic rocks clearly depend on Fe_2O_{3total} (Fig. 7), but in a more complicated way than the susceptibility (Figs 4–6). The lowest SIRM values occur in the purer limestones, which are of Silurian age (Fig. 7a), where there is also a clear correlation. However, in all other groups (Fig. 7b–e), there is no clear correlation, probably because the SIRM depends not only on the total amount of iron present but also on the form and grain size. The highest SIRM values are associated with the Upper Ordovician marls (Fig. 7b) and the Lower Ordovician secondary dolomites, which contain glauconite and goethite.

Changes of magnetic properties in the dolomitized fracture zone

The 101 samples from the six boreholes drilled across a resistivity anomaly provided examples of alteration by secondary dolomitization, reddening, solution and leaching. The zone itself shows progressive increases in the effective porosity, induced polarization, grain density and magnetic properties; a decrease in the bulk densities and the velocities of elastic waves; and changes in the apparent resistivity and dielectric constants. These are associated with (1) increases in effective (open) porosity, (2) increased secondary dolomitization of the carbonate rocks, and (3) changes in reduction–oxidation potential as reflected in the forms of iron. Shogenova & Puura (1997) have shown that the MgO anomaly

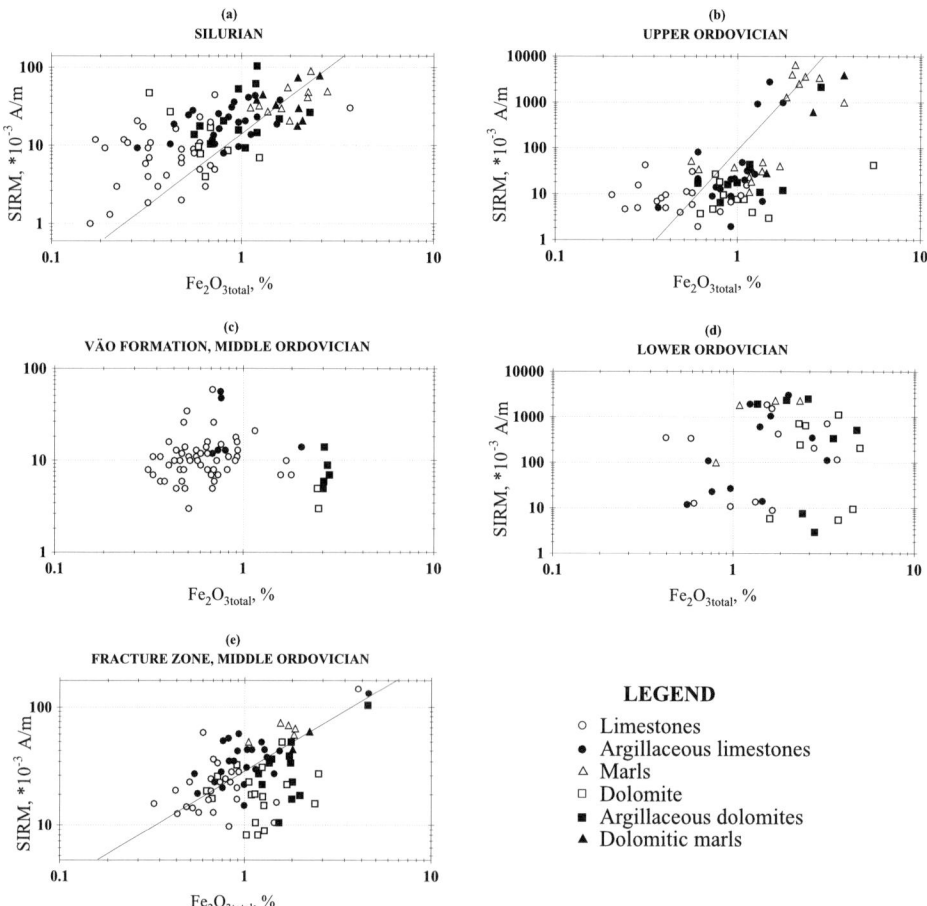

Fig. 7. Saturation isothermal remanent magnetization v. Fe_2O_{3total} content (on a log-decimal scale). Description as for Fig. 2.

in the fracture zone was associated with an increase of the total iron oxide (Fe_2O_{3total}) and FeO content, and a simultaneous decrease in Fe_2O_3 content. This sequence was accompanied by increases of Na_2O, Al_2O_3 and SO_3 content, but there was no change in the clay content. Both the magnetic susceptibility and SIRM increase in the fracture zone for all rock types involved (Figs 2e, 3d, 4d, 5e, 6e and 7e) in all formations.

Changes in the dolomitization in the Väo Formation

The limestones of the Väo Formation are relatively pure and therefore their properties can be compared with those of the 0.7 m thick Pae dolomite member. Previous studies (Bitjukova et al. 1995) had shown that the MgO concentration in the dolomite layer in the Pae member shows a positive correlation with FeO content, whereas de-dolomitized areas close to some fissures (Kiipli 1983a) showed a simultaneous decrease in FeO content. The present study showed that the magnetic susceptibility of rocks from the dolomite layer had a susceptibility some ten times higher than in the surrounding limestones, and the total iron content increased some four times.

R-mode factor analyses

Seventeen variables, related to magnetic susceptibility, SIRM, bulk and grain densities, and chemical composition, were then examined statistically using regression and R-mode factor analyses to investigate relationships between them. Such analyses first require the different variables to be normalized, in this case by determining the mean and standard deviation for each variable. The mean value was then subtracted from each measurement and the balance was divided by the standard deviation. The computation of the correlation matrices was then simplified by replacing the large number of variables by a smaller number of factors. This was done by using the R-mode factor analysis, in which the relationships between m variables can be regarded as reflecting the variations between p mutually uncorrelated underlying factors, i.e. each factor corresponds to a linear combination of the original variables, so that little or no information is lost (Reiment & Jöreskog 1993). The covariance matrices were then calculated, from which the communalities and eigenvalues could be determined.

Silurian and Upper Ordovician rocks

In Silurian datasets the first factor accounted for about 42% of parameter variations (Table 2). It had high loadings on those oxides which entered into the argillaceous part of the rocks; SiO_2, TiO_2, Al_2O_3, K_2O, Na_2O, SO_3 and P_2O_5, which had positive loadings of 0.95, 0.94, 0.96, 0.93, 0.63, 0.48 and 0.43, respectively. Both of the iron forms, Fe_2O_3 and FeO, also had high loadings in these rocks (0.69 and 0.44). This dominant factor also controls the magnetic susceptibility, which has high loading (0.80), whereas the bulk density had a large negative loading (-0.68). The second factor accounted for only 12.7% of parameter variations and has high loadings on MgO, MnO, Cl and grain density (0.7, 0.67, 0.62 and 0.70, respectively), whereas FeO had a lower loading (0.46). The second factor also controlled the magnetic susceptibility, but with less loading than the first factor (0.31), and controlled SIRM with loading 0.3. In the Upper Ordovician dataset (Table 2) the first dominant factor accounts for 36.2% of parameter variations, with high loadings for SiO_2, TiO_2, Al_2O_3, Na_2O and K_2O (0.76, 0.94, 0.94, 0.53 and 0.90,

Table 3. *Factor loadings for Middle Ordovician and Lower Ordovician rocks*

	Väo Formation Middle Orovician		Lower Ordovician	
	Factor 1	Factor 2	Factor 1	Factor 2
κ	**0.97**	0.11	**0.81**	0.05
SIRM	−0.31	**0.54**	0.00	**−0.69**
σ_w	**0.96**	−0.13	**0.77**	**0.51**
σ_g	**0.97**	0.05	**0.94**	0.08
SiO_2	**0.54**	**0.81**	0.00	**−0.97**
TiO_2	0.32	**0.89**	0.21	**−0.83**
Al_2O_3	**0.56**	**0.79**	−0.01	**−0.97**
FeO	**0.89**	0.06	**0.70**	−0.03
Fe_2O_3	**0.82**	0.19	0.04	0.09
MnO	**0.49**	0.18	**0.68**	0.25
MgO	**0.98**	0.07	**0.83**	0.10
CaO	**−0.91**	−0.33	**−0.81**	0.39
Na_2O	0.02	**0.84**	**0.40**	−0.23
K_2O	**0.42**	**0.87**	−0.02	**−0.96**
P_2O_5	−0.19	**0.44**	0.09	0.33
SO_3	**0.60**	**0.42**	0.19	0.19
Cl	**0.84**	−0.04	**0.63**	0.10
Expl. var.	8.35	4.43	5.07	4.68
Prp. Total	0.49	0.26	0.30	0.28
Eigenvalue	9.35	3.44	5.24	4.51
Variance (%)	55.02	20.21	30.84	26.55

Extraction: principal components. Varimax normalized rotation of factors. Significant loadings given in bold type.

respectively); only Fe_2O_3 had a significant loading on the first factor (0.54). The second factor for this rocks had high loadings on MgO, grain density and Cl (0.86, 0.88 and 0.47, respectively) and accounted for 13.9% of the variations in the parameters. This factor also controlled the magnetic susceptibility, but with lower loading (0.38).

The Middle Ordovician Väo Formation

The first dominant factor in this dataset had high loadings on MgO, FeO, Fe_2O_3, Cl, SiO_2 and Mn (0.98, 0.89, 0.82, 0.84, 0.54 and 0.49, respectively) (Table 3). This factor controlled magnetic susceptibility, grain and bulk densities with high loadings (0.97, 0.97 and 0.96, respectively). The first factor accounted for 55% of the parameter variations. The second factor, which accounted for 20% of parameter variations, controlled SiO_2, TiO_2, Al_2O_3, K_2O, Na_2O, SO_3, P_2O_5, and SIRM (0.81, 0.89, 0.79, 0.87, 0.84, 0.42, 0.44 and 0.54, respectively).

The Lower Ordovician rocks

The first factor accounted for 30.8% of parameter variations and controlled MgO, FeO, MnO and Cl with loadings 0.83, 0.7, 0.68 and 0.63, respectively, and the magnetic susceptibility and the wet and grain densities (0.81, 0.77, 0.94, respectively). (Table 4). The second factor accounted for 26.6% of the parameter variations and controlled changes in chemical parameters which entered into the argillaceous component of the rocks, SiO_2, TiO_2, Al_2O_3 and K_2O, with high negative factor loadings (−0.97, −0.83, −0.97 and −0.96, respectively). This factor also controlled SIRM, with a loading of −0.69, and bulk density, with a positive loading (0.51).

Fracture zone in Middle Ordovician rocks

Here the first dominant factor, accounting for 30.8% of parameter variations, had high positive loadings for MgO, FeO, Fe_2O_3, MnO and Cl (0.79, 0.81, 0.53, 0.88 and 0.36, respectively) (Table 5). This factor also controlled variations in the susceptibility (0.90), grain density (0.76), and bulk density (0.59). The second factor accounted for 28.4% of the parameter variations and had high loadings on SiO_2, TiO_2, Al_2O_3, K_2O, Fe_2O_3 and SO_3 with negative loadings −0.96, −0.96, −0.96, −0.56, −0.54 and −0.61, and, correspondingly, positive loading on SIRM (0.58) and negative loading on bulk density (−0.56).

Table 4. *Factor loadings for Lower Ordovician and Middle Ordovician fracture zone*

	Factor 1	Factor 2
κ	**0.90**	0.21
SIRM	0.06	**0.58**
σ_w	**0.59**	**−0.56**
σ_g	**0.76**	−0.18
SiO_2	0.00	**0.96**
TiO_2	0.07	**0.96**
Al_2O_3	0.07	**0.96**
FeO	**0.81**	0.23
Fe_2O_3	**0.53**	**0.54**
MnO	**0.88**	−0.23
MgO	**0.79**	−0.15
CaO	**−0.69**	**−0.49**
Na_2O	−0.07	−0.05
K_2O	−0.16	**0.56**
P_2O_5	**0.31**	**−0.33**
SO_3	**−0.51**	**0.61**
Cl	**0.36**	−0.21
Expl. var.	5.07	4.99
Prp. Total	0.30	0.29
Eigenvalue	5.24	4.83
Variance (%)	30.79	28.39

Extraction: principal components. Varimax normalized rotation of factors. Significant loadings given in bold type.

Comments and interpretation

The increase in susceptibility in the Väo Formation is clearly correlated with the dolomite content and with an increase in the content of both iron forms (Figs 2c, 3c, 4c and 5c), but all other comparisons of susceptibility and the degree of dolomitization indicate groupings of different lithologies rather than any correlation (Fig. 5). This correlation is revealed more easily by the factor analyses, related to the dominant factor only in Lower and Middle Ordovician datasets (Table 3). In the Silurian and Upper Ordovician datasets the first factor, which has high loadings on those oxides, particularly the iron oxides, is associated with the argillaceous part of the rocks, and can thus be interpreted as a 'clay content' factor, which will have been largely determined during primary sedimentation. In contrast, the second factor in the same sequences has high loadings on MgO and can thus be interpreted as a diagenetic dolomitization factor arising subsequent to the initial deposition. The second factor in the Silurian and Upper Ordovician datasets has high loadings on MgO and was interpreted as a dolomitization factor. In all datasets, dolomitization also increased the grain density. There is some evidence for a difference in chemical composition between the Silurian and

Upper Ordovician dolomites, as their geochemical loadings are different; both show increased dolomitization being associated with increased Cl content, but with MnO and FeO only in the Silurian rocks (loadings 0.67 and 0.46), and with Fe_2O_3 in the Upper Ordovician rocks (0.33), but there is no significant change in their magnetic variables. In contrast, the first factor is interpreted as the dolomitization factor in the Väo Formation. Here the dolomitization caused a particular increase in both iron forms, magnetic susceptibility, and grain and bulk density. However, the significant loading of some of the insoluble residue components is probably due to about half of the dolomite samples being somewhat argillaceous (Fig. 5c), with insoluble residue contents of 10-20%. Dolomitization is also associated with increase in Cl, iron sulphides (correlation in iron oxides and SO_3) and MnO. The second factor can be interpreted as the clay content factor associated only with increases in SIRM. The Lower Ordovician dataset includes widespread dolomites of Volkhov stage but the role of dolomitization in parameter variations here is less clear than in the Väo Formation, and, although dolomitization in these rocks is the first factor, it accounts for only 30.8% of parameter variation, close to the second factor (26.6%). It is associated with increases in magnetic susceptibility, grain and wet densities, and FeO. The second factor was interpreted as a clay content factor and is marked by increases in SIRM and decreases in bulk density. In the fracture zone in the Middle Ordovician rocks, the second, clay content, factor causes an increase in SIRM, because of the increase in Fe_2O_3, and a decrease in wet density. The first factor is again interpreted as indicating dolomitization with associated increases in iron oxides, magnetic properties, grain and wet densities, and in Cl and Mn contents. These diagenetic changes are mostly a result of late diagenetic dolomitization of the rocks associated with tectonic disturbances, which may be explained by replacement of Mg^{2+} by Fe^{2+} in the crystalline lattice of the late diagenetic dolomites (Goldsmith 1983; Kiipli 1983b; Morse & Mackenzie 1990), and probably with the presence of ankerite with the dolomite (Pichugin et al. 1976).

Discussion and conclusions

Five geochemical types of dolomite can be recognized in Estonia, each with different magnetic properties. These are: (1) the Silurian very weakly magnetic dolomites with significant increases in MnO and Cl content; (2) the Upper Ordovician very weakly magnetic dolomites with significant increase in Cl content; (3) remagnetized dolomite in the Pae member of the Väo Formation with increases in FeO, Fe_2O_3, MnO and Cl and, magnetically, a marked increase in magnetic susceptibility; (4) remagnetized Lower Ordovician dolomite layers with increases in FeO and Cl; (5) dolomites in the fracture zone, which have been remagnetized by migrating fluids and marked by significant increases in FeO, Fe_2O_3, MnO and Cl.

The regression and factor analysis of these weakly magnetic carbonate rocks shows that their properties are influenced by two main factors: variations in clay content and dolomitization. Dolomitization appears to have had least affect on the magnetic properties of the Silurian and Upper Ordovician rocks, but is particularly marked in the Lower and Middle Ordovician dolomites. Such statistical analyses have not been a traditional approach, but the results are similar to those obtained in other regions of the world using dolomites of different ages (Purser et al. 1994; Tucker & Wright 1994) and the results are consistent with those of studies of remagnetization and demagnetization of carbonate rocks in other sedimentary basins (e.g. Elmore et al. 1993). They therefore appear to provide a way in which primary and secondary features can be more readily identified, thereby leading to a far greater understanding of the complex processes by which dolomitization occurs.

The rock samples were described by T. Saadre and K. Suuroja from the Geological Survey of Estonia, and T. Kiipli from the Institute of Geology. Magnetic properties of rocks were measured in the Petrophysical laboratory of the Research Institute of Earth Crust of Saint Petersburg University. Bulk chemical analyses were carried out in the Ukraine State Institute of Mineral Resources. The forms of iron were determined in the Institute of Geology, Estonian Academy of Sciences, by Õ. Roos and T. Linkova. The author is grateful to V. Puura, R. Vaher and T. Kiipli, and to two anonymous referees for helpful comments and suggestions. This work was funded in part by a grant from the Volkswagen Foundation under the joint German–Estonia project 'Analysis and integrated interpretation of Estonian Palaeozoic sedimentary basin structures' (No. 1/70 410) and Grant 2191 from the Estonian Science Foundation.

References

BITYUKOVA, L., SHOGENOVA, A., PUURA, V., SAADRE, T., SUUROJA, K. 1996. Geochemistry of major elements in middle Ordovician carbonate rocks: comparative analysis of alteration zones, north Estonia. *Proceedings of the Estonian Academy of Sciences*, **45**(2), 78–91.

ELMORE, R. D., LONDON, D., BAGLEY, D. & FRUIT, D. 1993. Remagnetization by basinal fluids: testing the hypothesis in the Viola limestone, Southern Oklahoma. *Journal of Geophysical Research*, **98**, 6237–6254.

GOLDSMITH, J. R. 1983. Phase relations of rhombohedral carbonates. *In*: REEDER, R. J. (ed.) *Carbonates: Mineralogy and Chemistry*. Reviews in Mineralogy, Mineralogical Society of America, **11**, 49–96.

KIIPLI, T. 1983a. Dolomites of the Estonian Middle Ordovician Väo Formation (in Russian with English summary). *Proceedings of the Academy of Sciences of the Estonian SSR*, **32**(2), 60–68.

—— 1983b. On the genesis of Ordovician and Silurian dolomites at the contact with Devonian deposits (in Russian with English summary). *Proceedings of the Academy of Sciences of the Estonian SSR*, **32**(3), 110–117.

JÜRGENSON, E. 1988. *Deposition of the Silurian Beds in the Baltic* (in Russian with English summary). Valgus, Tallinn.

MORSE, J. W. & MACKENZIE, F. T. 1990. *Geochemistry of Sedimentary Carbonates*. Elsevier, Amsterdam.

ORVIKU, K. 1940. Lithologie der Tallinn-Serie (Ordovizium, Estland) 1. *Acta et Comm. Univ. Tartuensis, A*, **36**, 1–216.

PICHUGIN, M. S., PUURA, V. A., VINGISSAAR, P. A. & ERISALU, E. K. 1976. Regional metasomatic dolomitization associated with tectonic disturbances in Lower Paleozoic of the Northern Baltic region. *International Geology Review*, **19**(8), 903–912.

PÕLMA, L. 1982. *Comparative Lithology of the Ordovician Carbonate rocks in the Northern and Middle East Baltic* (in Russian with English summary), Valgus, Tallin.

PRIJATKIN, A. A. & POLJAKOV, E. E. 1983. *Petrophysical Investigations of Rocks* (in Russian), Leningrad University.

PURSER, B., TUCKER, M. & ZENGER, D. (eds) 1994. *Dolomites: a volume in honour of Dolomieu*. Blackwell Scientific, Oxford.

PUURA, V. & VAHER, R. 1997. Cover structure. *In*: RAUKAS, A. & TEEDUMÄE, A. (eds) *Geology and Mineral Resources of Estonia*. Estonian Academy, Tallinn, 167–177.

REIMENT, R. A. & JÖRESKOG, K. G. 1993. *Applied Factor Analysis in the Natural Sciences*. Cambridge University Press.

SHOGENOVA, A. 1996. Use of the computer for the structural analyses of the Ordovician sedimentary basin in Estonia. *In*: FÖRSTER, A. & MERRIAM, D. F. (eds) *Geology Modeling and Mapping*. Plenum, New York, 199–214.

—— & PUURA, V. 1997. Petrophysical changes caused by dolomitization and leaching in fracture zones of lower Paleozoic carbonate rocks, North Estonia. *In*: MIDDLETON, M. (ed.) *Nordic Petroleum Technology Series: One, Second Nordic Symposium on Petrophysics, Fractured Reservoirs*. Nordisk Energi-Forskningsprogram, Saghellinga, Norway, 155–185.

——, TUULING, I. 1990. Physical properties of carbonate rocks in the fracture and dolomitization zone (North Estonia) (in Russian with English summary). *Proceedings of the Estonian Academy of Sciences*, **3**(2), 41–49.

——, BITYUKOVA, L. & GÖTZE, H.-J. 1998. Technique and results of complex analysis to study processes in sedimentary basin (in Estonia). *Physics and Chemistry of the Earth (and Solar System)*, **23**(3), 327–337.

SUVEIZDIS, P. I. (ed.) 1979. *Baltic Tectonics* (in Russian), Mosklas, Vilnius, p. 92.

TUCKER, M. E. & WRIGHT, V. P. 1994. *Carbonate Sedimentology*. Blackwell Scientific, Oxford.

VAHER, R. M. 1986. Application of the resistivity method for locating fracture zones in northeastern Estonia (in Russian with English summary). *Proceedings of the Academy of Sciences of the Estonian SSR, Geology*, **35**(4), 146–155.

VINGISAAR, P. & TAALMAN, V. 1974. Survey of the dolomitization of the lower Palaeozoic carbonate rocks of Estonia (in Russian with English summary). *Proceedings of the Academy of Sciences of the Estonian SSR, Chemistry, Geology*, **23**(3), 237–242.

Alteration of magnetic properties of Palaeozoic platform carbonate rocks during burial diagenesis (Lower Ordovician sequence, Texas, USA)

HERBERT HAUBOLD

Palaeomagnetic Laboratory 'Gams', Institut für Geophysik, Montanuniversität Leoben, Gams 45, A-8130 Frohnleiten, Austria (e-mail: herbert.haubold@unileoben.ac.at)

Abstract: Palaeomagnetic and sedimentological investigations of samples from two sections of correlative Iapetan platform carbonate rocks from Texas, USA, were made to test whether their magnetic properties reflect diagenetic alteration associated with regional and local tectonism. The Honeycut Formation (Llano Uplift area, central Texas), in close proximity to the late Palaeozoic Ouachita orogenic belt, exhibits a distinct correlation between magnetization intensity, magnetization age (direction) and lithofacies. Mudstones preserve their weak primary Early Ordovician magnetization, whereas dolo-grainstones carry a strong Pennsylvanian magnetization residing in authigenic magnetite. Fluid migration associated with the Ouachita Orogeny has been focused in lithofacies with high permeability and caused dolomite recrystallization and pervasive remagnetization. Magnetization intensity trends covary with fluid/rock ratios. However, aquitards were either not affected or less affected by these fluids. Unlike the Honeycut Formation, permeable rocks of the El Paso Group (Franklin Mountains, west Texas) carry only a non-pervasive Pennsylvanian magnetization. Therefore, a larger percentage of El Paso Group samples retain a primary Early Ordovician signature. This area is further removed from the Ouachita front, and, thus, the influence by Pennsylvanian orogenic fluids was less pronounced.

The aim of this study is to contribute to the understanding of the link between sedimentological properties and magnetic properties of Palaeozoic carbonate rocks affected by diagenetic processes. This problem is of significant interest because it provides a quantitative measure of the timing, intensity, and spatial distribution of diagenetic events identified by textural and chemical criteria. Although the directions of natural remanent magnetization (NRM) of sedimentary rocks have been the subject of many investigations, their NRM intensities have been less well studied. This investigation considers effects of diagenetic events on both NRM intensity and NRM direction. Discerning diagenetic alteration of NRM intensity yields important information on the nature of fluid–rock interaction events. Specifically, the stratigraphical distribution of NRM intensities and magnetization directions, in comparison with depositional lithofacies and geochemical properties, can often effectively elucidate ancient pathways of reactive fluids. Thus, integrating palaeomagnetic methods with petrographical and geochemical methods, palaeohydrological conditions can be reconstructed both on a local scale within a single outcrop and on a regional scale for long-distance fluid migration events. The Cambro-Ordovician platform carbonate rocks of Texas were chosen for investigation because they are exposed in two isolated areas across the platform, each of which underwent a distinctively different structural development. Regional and local tectonism gave rise to fluid migration events that affected the rocks. Thus these areas provide the opportunity to compare time- and facies-equivalent strata that were exposed to very different diagenetic modification processes. The details of major and minor diagenetic events recorded in these rocks and their relative timing have been well established by previous petrographical and geochemical work, so these rocks are very suitable for an integrated study involving palaeomagnetic and rock magnetic data with petrographical and geochemical data, to interpret the influence of diagenetic fluids, particularly their timing and pathways. The investigated sections constitute a thick sequence of peritidal carbonate rocks deposited on a vast gently dipping platform in the western Iapetus Ocean. All of the formation boundaries within the Ellenburger Group of the Llano Uplift area and in the El Paso Group of the Franklin Mountains

HAUBOLD, H. 1999. Alteration of magnetic properties of Palaeozoic platform carbonate rocks during burial diagenesis (Lower Ordovician sequence, Texas, USA). *In*: TARLING, D. H. & TURNER, P. (eds) *Palaeomagnetism and Diagenesis in Sediments*, Geological Society, London, Special Publications, **151**, 181–203.

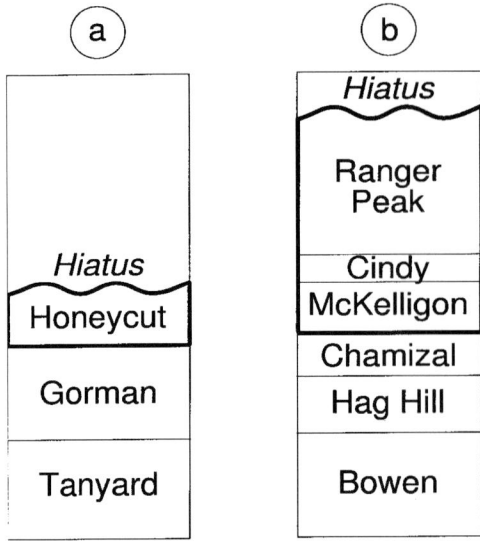

Fig. 1. Stratigraphical correlation of (**a**) the Ellenburger Group (Llano Uplift area) and (**b**) the El Paso Group (Franklin Mountains area). The formations sampled are more strongly outlined. (See also Fig. 2.)

(Fig. 1) were defined on the basis of genetic depositional units (Cloud & Barnes 1946; Lucia 1968). These strata have been correlated using biostratigraphy and lithostratigraphy (Cloud & Barnes 1946; LeMone 1968; Clemons 1989; Johnson 1991; Goldhammer et al. 1993). As the result of episodic tectonism during deposition (Rankey et al. 1994), the identification of depositional cyclicity within these carbonate rocks is somewhat controversial. High-order sequences are typically incomplete and vertical facies successions are rarely predictable (Kerans & Lucia 1989; Lindsay & Koskelin 1991; Goldhammer et al. 1993). Deposition terminated when platform uplift, or sea-level fall, caused significant karstification in mid-Ordovician time (Lucia 1969; Kerans 1988; Candelaria & Reed 1992). The Llano Uplift in central Texas (Fig. 2) is a broad dome that has never undergone major internal deformation in spite of its close proximity to the Ouachita orogenic belt (Flawn et al. 1961). During Palaeozoic time, the Llano block formed a structural promontory on the southwestern margin of the North American craton. A hypothesis to explain the internal stability of the Llano block calls upon southeastward displacement during the Ouachita Orogeny that caused the tectonic escape of this block (Algeo 1992). The Franklin Mountains are a fault block range in west Texas that was uplifted and deformed during the Rio Grande rifting or during Laramide thrusting in late Cretaceous–early Tertiary time (Shepard & Walper 1982). Structural features in the southern Franklin Mountains suggest that extensional tectonism probably accounts for the

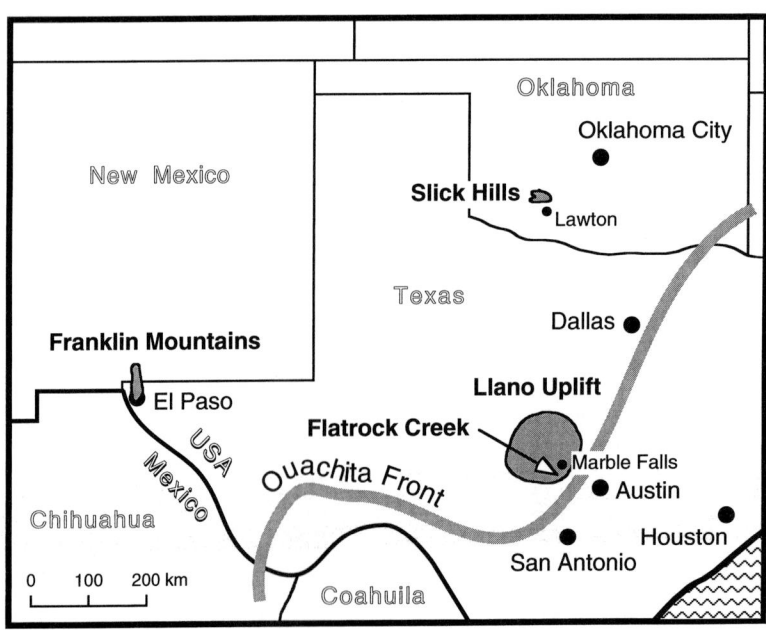

Fig. 2. Flatrock Creek (Llano Uplift area) and Franklin Mountains sampling localities.

observed deformation. At the west boundary fault zone there is considerable displacement within Pleistocene strata (Stacey et al. 1992).

Previous studies

The Ellenburger Group carbonates contain several generations of dolomite (Lee & Friedman 1987; Kupecz 1989; Kupecz & Land 1991; Amthor & Friedman 1992). Early replicative dolomite formed in a subtidal setting. Three generations of coarse-grained, brightly orange luminescent, ^{18}O-depleted late-stage dolomites formed during burial by multiple events of dissolution and re-precipitation of pre-existing dolomite. There are both massive late-stage dolomites and late-stage dolomite cements. The distribution of late-stage dolomite is related to porosity and permeability of the rocks. Replacement of precursor dolomites was caused by hot, reactive fluids that were expelled from basinal shales, probably during the Ouachita Orogeny (Kupecz 1989; Kupecz & Land 1991). In the El Paso Group, early dolomite is associated with tidal-flat facies and typically occurs when regression of sea level was similar to subsidence rates. During Silurian or Devonian time the rocks were affected by a major karst event involving cavern development and subsequent collapse brecciation and, therefore, generated fluid conduits (Lucia 1969, 1995). Some El Paso Group limestones contain trace amounts of ^{18}O-depleted late-stage replacive dolomite, which indicates some interaction of the rocks with warm fluids or ^{18}O-depleted ground water during burial diagenesis at an unknown time (Stepanek 1987). The remagnetization of Palaeozoic sedimentary rocks in a Pennsylvanian or Permian field over a large portion of the North American craton is well documented. Two remagnetization mechanisms, chemical remanent magnetization (CRM) and thermoviscous remanent magnetization (TVRM), have been widely accepted as the processes probably responsible for remagnetization of sedimentary rocks during late Palaeozoic time (McCabe & Elmore 1989). CRM acquisition can be diagenetic, i.e. pre-existing minerals transform into new minerals, or authigenic, i.e. new magnetic minerals precipitate as a response to a chemical change in their environment. An NRM is stable if its relaxation time exceeds the age of remanence acquisition, but prolonged exposure to elevated temperatures decreases the relaxation time. Consequently, the minerals may acquire a TVRM. Theoretical temperature–time functions for both multi-domain (MD) and interacting clusters of single-domain (SD) magnetite grains do not sufficiently clarify the problem of TVRM blocking and unblocking for complex natural mineral assemblages (Dunlop 1973, Pullaiah et al. 1975). Temperatures required to overprint magnetization components are typically higher than those predicted. There are no known published palaeomagnetic investigations of the Ellenburger Group. In the Franklin Mountains, the Precambrian igneous rocks carry a stable magnetization reflecting the time of emplacement (Spall 1971), whereas the El Paso Group carbonate rocks from the Florida Mountains, east of the Franklin Mountains, contain only a Holocene magnetization. However, the Bliss sandstone that underlies the El Paso Group was remagnetized during tectonism in late Cretaceous–early Tertiary time (Geissman et al. 1991). The top of the Ellenburger Group of central Texas was never buried deeper than $c. 280$ m (Kupecz & Land 1991). Vitrinite reflectance data from the overlying Mississippian Barnett Shale and the Barnett Limestone yield maximum temperature estimates of $<45°C$ (Wiggins 1982). However, fluid inclusions in megaquartz grains of the Ellenburger Group, cogenetic with the late-stage dolomite, homogenize between 76 and $102°C$. These values reflect the temperatures of warm orogenic fluids that penetrated these rocks (Kupecz & Land 1991). As far as is known, the maximum diagenetic or burial temperatures of the El Paso Group in the Franklin Mountains have not been investigated, but the maximum conodont alteration index (CAI) of shales from the overlying Devonian Woodford Formation in southwest New Mexico is 1.92 (Cromer 1991). Temperature–time estimates based on laboratory heating of Ordovician conodonts (Epstein et al. 1977) suggest that even if very short periods (2 ka) of high heat input are considered, the maximum temperatures indicated by CAI values <2 are below $150°C$. In the Franklin Mountains area El Paso Group rocks are less than 200 m deeper in the stratigraphical section than the Woodford Formation.

Methods

The sampling was designed to capture the spatial distribution and facies dependence of rock characteristics. Approximately 200 drill cores (2.5 cm diameter) were oriented magnetically, from which a standard 2.2 cm long plug was cut for magnetic analyses, and a smaller specimen was cut from the same core for chemical analyses. In the Franklin Mountains, oriented hand samples were also collected, with unoriented hand samples being taken from both areas for petrographical investigations. Lithofacies analysis was conducted on outcrops. After completion of the magnetic analyses, all of the drill plugs were cut in half and studied under a binocular microscope together with cut and polished hand samples. Acetate peels for light microscopy were produced from approximately a quarter of the cut drill plugs. Each specimen was categorized on the basis of its depositional lithofacies. Ten representative etched specimens were inspected using a scanning electron microscope, and a luminoscope was used to test chips made from all of the dolomitic samples for cathodoluminescence properties. The NRM intensities and directions were measured with a two-axis cryogenic magnetometer in a magnetically shielded room. Each sample was subjected to a detailed stepwise demagnetization procedure by using alternating field (AF) and/or thermal demagnetization. The isolation of the characteristic magnetization vectors was based on the collinearity of at least three vector endpoints, as determined by visual inspection of orthogonal projections (Zijderveld 1967). Vector components were calculated using principal component analysis (Kirschvink 1980), rejecting vector trajectories with mean angular deviations >12°. The magnetization ages were assessed by comparing the mean directions with the apparent polar wander path for the North American craton (Irving & Irving 1982; Van der Voo 1990). All bulk susceptibility values were diamagnetic and are not discussed further. Isothermal remanent magnetization (IRM) acquisition experiments were used to assist in the identification of the magnetic mineral content of 25 selected representative samples. Subsequently, ten of these samples were given a three-component IRM by applying successively weaker fields along the orthogonal axes. Stepwise thermal demagnetization of this composite IRM provided a method to identify different coercivity fractions of the rocks (Lowrie 1990). The carbonate mineralogical composition of half of the samples was determined by powder X-ray diffraction. For stable oxygen and carbon isotope analyses, powders from a third of the samples were digested in H_3PO_4. Evacuated sample powders of limestones and dolomites were reacted at 25 and 50°C, respectively. Evolved CO_2 was analysed relative to the PDB standard using a ratio mass-spectrometer. Dolomite oxygen isotope values were corrected to fractionation at 25°C by adding 1.22‰. The National Bureau of Standards (NBS) 19 and 20 standards were used to test the reproducibility of the results; the precision was ±0.1‰.

The honeycut formation, Llano Uplift area

Data are presented from a profile of the Honeycut Formation from the Llano Uplift area, central Texas, in the vicinity of Marble Falls (Figs 1 and 2). The investigated profile forms a part of the Flatrock Creek section (Cloud & Barnes 1946; Barnes 1982).

Depositional lithofacies

Cryptalgal laminites are abundant and formed in a muddy intertidal or supratidal environment. Although most of the cryptalgal laminites are

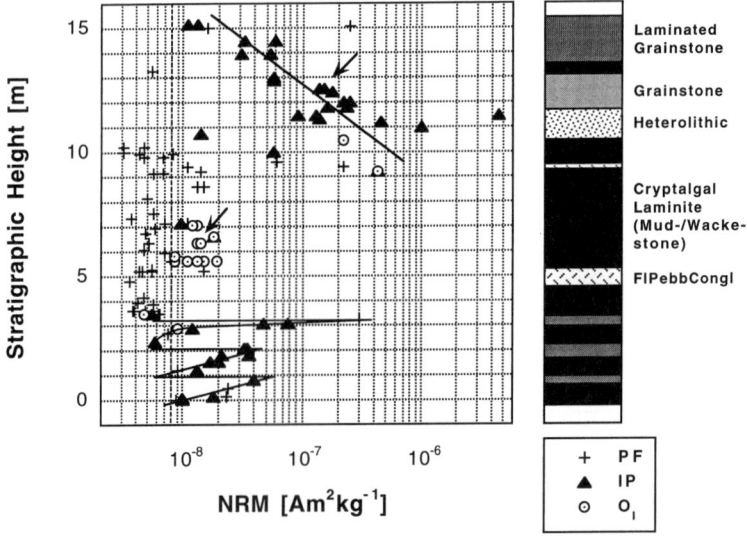

Fig. 3. NRM intensity profile, Honeycut Formation. O_l, Lower Ordovician; lP, Pennsylvanian; PF, samples which contain either only a pervasive present field magnetization or that were too weakly magnetized ($< 8 \times 10^{-9}$ A m^2/kg) to determine an ancient magnetization. Arrows indicate those samples depicted in Figs 7 and 8. Magnetization intensity and direction (age) correspond to lithology. The coarse-grained rocks carry a strong Pennsylvanian magnetization as a result of influence by orogenic fluids associated with the Ouachita Orogeny. Intensity trends (indicated with lines) reflect fluid/rock ratios. However, fine-grained rocks retain a depositional(?) weak Early Ordovician magnetization owing to their low permeability.

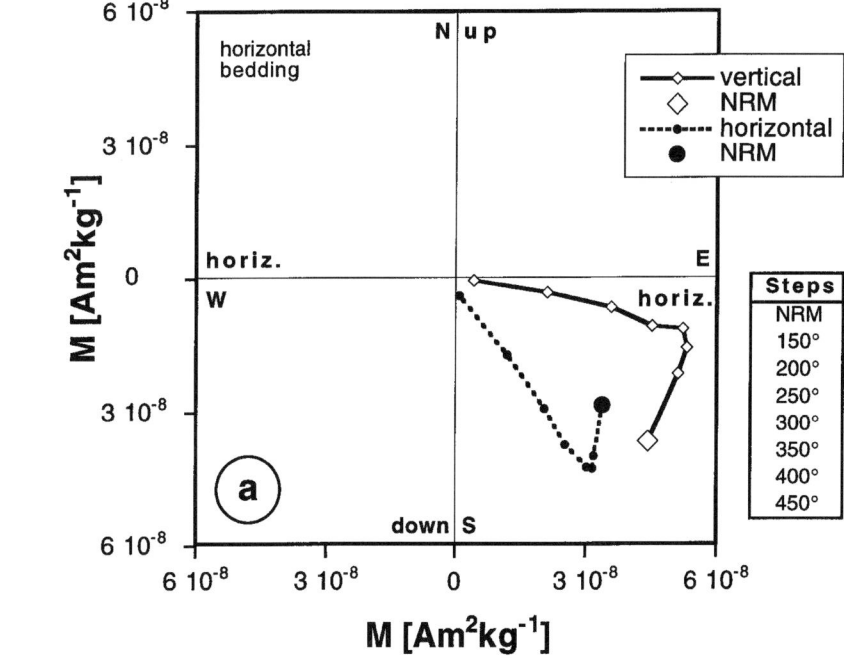

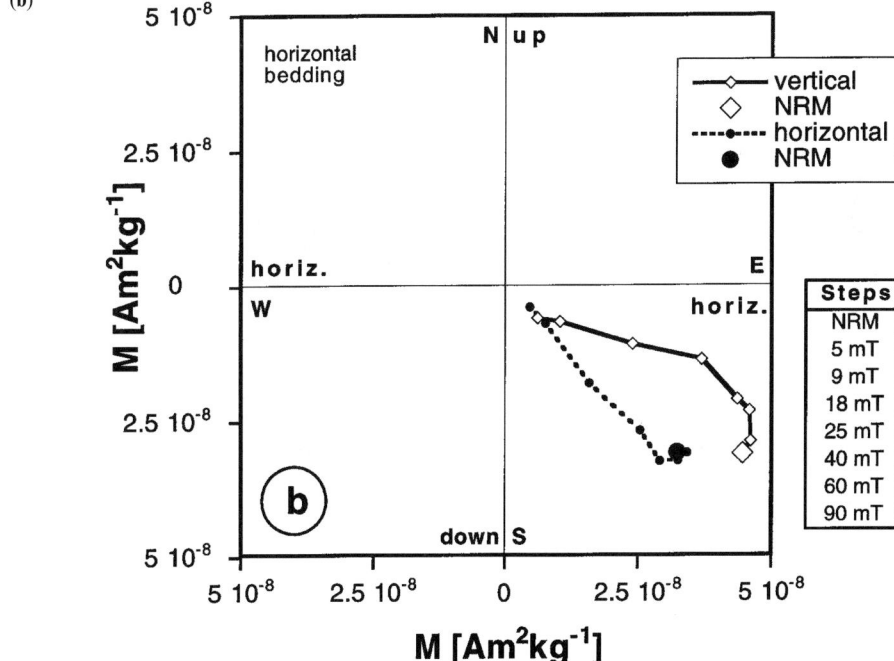

Fig. 4. Representative orthogonal projections, Honeycut Formation. A modified Zijderveld diagram is used. (a) Thermal cleaning reveals a stable shallow southeasterly direction that decays univectorially towards the origin. This is interpreted to be of Pennsylvanian age. (b) The same component was also isolated during AF demagnetization. (c) Another conspicuous characteristic magnetization component with a very shallow inclination and an easterly declination is interpreted to be of Early Ordovician age.

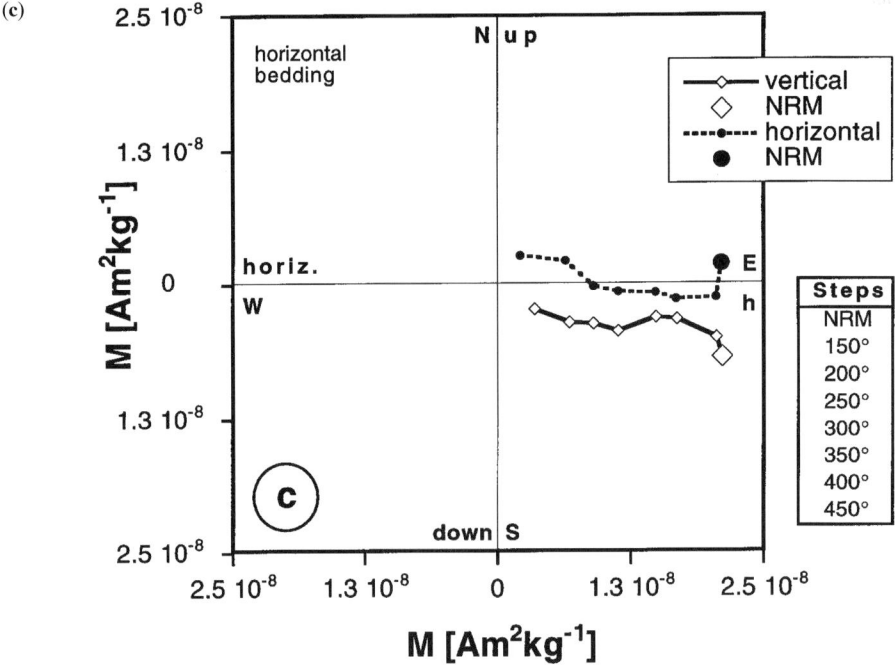

Fig. 4. (continued).

limestones, small dolomite crystals are typically associated with stylolites. Stylolite surfaces form a network throughout the laminite strata. Thin beds of flat-pebble conglomerates with small (up to 5 mm) flat intraclasts in a dense fine-grained matrix reflect early lithification on a tidal flat and subsequent reworking by storm activity. The sampled section contains several strata of current-laminated dolo-grainstones, all of which consist of coarse (up to 0.7 mm), orange luminescent late-stage replacement dolomite crystals. One stratum of dolo-grainstones does not exhibit any lamination, possibly because it was obliterated during dolomite recrystallization. Current-laminated grainstones are the result of alternating traction deposition of carbonate sand and surface veneering through mud settling. A succession of heterolithic beds consists of patches of dolo-grainstone and calcitic mudstones, and reflects the transition between depositional environments.

Intensities of NRM

The distribution of NRM intensities in this section shows a correlation with lithology (Fig. 3). Within the lowermost 3 m of the section three consecutive upward increasing intensity trends form a sawtooth pattern in which the peak values all occur within grainstone beds. The fine-grained rocks in the middle part of the section, between about 3 and 9.5 m are only weakly magnetized and do not display any intensity trends. At $c.$ 9.5 m the intensity increases by two orders of magnitude relative to underlying beds. The maximum intensities occur at 11.50 m followed by a progressive upward intensity decrease. This clear trend is associated with massive grainstones in the highest part of the section. In essence, coarse-grained lithofacies (grainstones) are characterized by high magnetization intensities, whereas fine-grained lithofacies (laminites, flat-pebble conglomerates) are weakly magnetized. (Such trends cannot be obtained using bulk susceptibility measurements because of the diamagnetic character of the rocks.)

Directions of NRM

Most samples of the Honeycut Formation carry an initial northerly magnetization with a steep inclination which is a modern magnetization overprint because of its similarity to the present Earth's magnetic field direction in central Texas. Typically, this direction is removed after heating the samples to between 150 and 250°C. After thermal cleaning, many samples exhibit a southeasterly magnetization component with a shallow inclination that is usually stable up to 450°C. The same magnetization direction also occurs in several AF demagnetized samples in which a peak field of 90 mT reduces the NRM intensity by c. 85%. Some other samples with a steep southeasterly NRM direction lost more than 90% of their NRM intensity after being exposed to only 5 mT. Fewer than 10% of the samples contain a shallow easterly magnetization component which, during thermal demagnetization, is usually stable up to 450°C. As the bedding is horizontal, no tilt correction was needed for this section (Fig. 4). The specimen directions show a strong modern overprint and two conspicuous characteristic magnetization directions (Fig. 5), which include two reference palaeo-directions and a large cluster of vectors close to, but not exactly corresponding to the present field direction because of incomplete coercivity separation during demagnetization. A large cluster of vectors lie close to the reversed polarity Pennsylvanian palaeo-direction, and six samples show corresponding normal polarity directions. Another cluster occurs close to the reverse Early Ordovician palaeo-direction, with two samples with a corresponding normal polarity magnetization. The Pennsylvanian magnetic palaeopole position for the North American craton is well defined, but not that of the Early Ordovician palaeopole position (Van der Voo 1990). However, primary magnetization directions observed in Lower Ordovician carbonate rocks from Arkansas confirm these positions (Farr et al. 1993). In addition to the clusters, there are wide streaks of magnetization vectors between the two ancient directions and the present field direction as a result of incomplete coercivity separation during demagnetization. Those vectors that form clusters corresponding to ancient magnetic reference directions were extracted from the dataset (Fig. 6). Twelve samples carry a mean magnetic direction

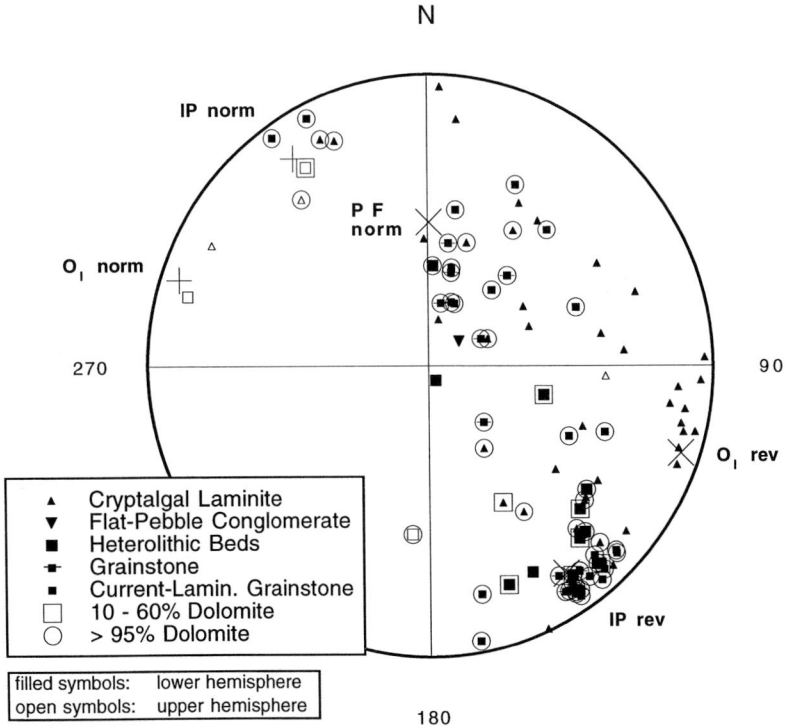

Fig. 5. Specimen magnetization components, Honeycut Formation. Crosses show expected magnetizations (palaeo-directions): ×, lower hemisphere; +, upper hemisphere. The vectors display distinct clusters close to the reverse polarity Pennsylvanian palaeo-direction and close to the reverse Lower Ordovician palaeo-direction. Some other specimens carry the same directions with normal magnetization polarity. Early Ordovician signals reside in fine-grained limestones, whereas Pennsylvanian signals predominantly reside in dolo-grainstones. The broad girdle-like streak of magnetization vectors between the ancient directions and the present field direction (PF) is the result of insufficient coercivity separation.

(102°, 10°; $\kappa = 74.0$, $\alpha_{95} = 4.9°$; VGP (virtual geomagnetic pole): 8°N, 162°E) interpreted as being of Early Ordovician age by comparison with the reference pole (14°N, 159°E) for the North American craton (Van der Voo 1990). Forty-four samples carry a magnetization interpreted as being of Pennsylvanian age (144°, 16°; $\kappa = 33.0$; $\alpha_{95} = 3.8°$; VGP: 39°N, 131°E) by comparison with the reference pole (40°N, 128°E). The magnetization direction is linked to depositional lithofacies in that the Early Ordovician magnetization components exclusively reside in fine-grained rocks, whereas all of the grainstones carry a Pennsylvanian magnetization (Fig. 3). All samples also contain some modern magnetization overprint. Many cryptalgal laminite samples were very weakly magnetized. After only one or two demagnetization steps the intensity of these samples fell below the noise level of the magnetometer and could not be analysed. In general, strong magnetizations were of Pennsylvanian age, whereas the weak magnetizations were of Early Ordovician age. Accordingly, there is a clear correlation between lithology and magnetization age. The Pennsylvanian magnetizations reside primarily in the grainstone facies, whereas the Early Ordovician magnetizations are restricted entirely to cryptalgal laminites. Most Early Ordovician magnetizations occur in the thick succession of cryptalgal laminites in the middle part (3–9.5 m) of the section (Fig. 3). Two cryptalgal laminite samples from this part carry a Pennsylvanian magnetization. These samples contain partly dolomitized stylolites. Other cryptalgal laminite successions in the lowest 2 m of the section were completely remagnetized during Pennsylvanian time.

Magnetic mineralogy

Rocks carrying a Pennsylvanian remagnetization have IRM acquisition that exhibits a steep initial remanence increase which flattens, but does not saturate in fields of 1 T. This indicates that although a low-coercivity (soft) magnetic mineral is the predominant magnetic carrier, there is also a high-coercivity (hard) mineral which contributes to the IRM. Unblocking spectra of three-component IRM demonstrate that the predominant magnetic carrier is a soft magnetic

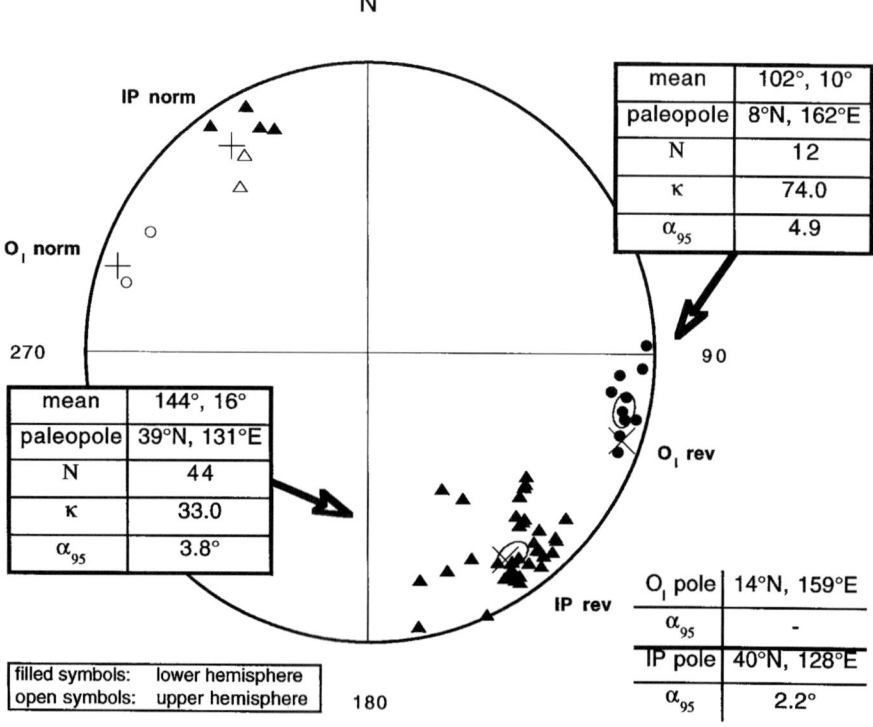

Fig. 6. Mean directions of components of remanence, Honeycut Formation, and Fisher (1953) statistics. $\bigcirc$, α_{95}. Normal polarity vectors were reversed and included in the calculations. The results show that these characteristic vectors reflect magnetization during early Ordovician time ($\bullet$) and Pennsylvanian time ($\blacktriangle$).

mineral, which begins to unblock at 450°C, and a medium magnetic carrier, which begins to unblock at 500°C (Fig. 7). A hard magnetic mineral carries the weakest IRM and does not unblock at 650°C, the highest temperature applied. This unblocking behaviour is interpreted as indicating the presence of coarse-grained magnetite (soft), fine-grained magnetite (soft to medium) and hematite (hard). Rocks with an Early Ordovician magnetization also have very steep IRM acquisition curves that do not flatten before more than 80% of the saturation magnetization is gained (Fig. 8), but these saturate in an ambient field of $c.\,0.8\,\text{T}$. A low-coercivity mineral, probably magnetite, is the predominant carrier of the IRM and the high-coercivity mineral is essentially insignificant.

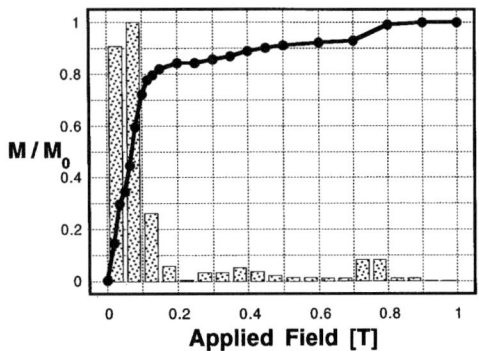

Fig. 8. Example of an IRM acquisition curve, Honeycut Formation with Early Ordovician magnetization. The steep initial slope indicates a soft mineral as the predominant carrier. These specimens typically saturate in an ambient field of less than 1 T, so the contribution of a hard magnetic mineral to the IRM is insignificant.

Stable isotopic composition and carbonate mineralogy

The limestones have $\delta^{18}O$ values mostly between -4 and $-6‰$ PDB, and the dolomites mostly between -2 and $-5‰$. The $\delta^{18}O$ values for both limestones and dolomites fall a narrow range between zero and $-2‰$. The stable isotopic composition of limestones that carry a Pennsylvanian magnetization is no different from that of limestones that carry an Early Ordovician magnetization (Figs 9 and 10), i.e. there is no correlation between the $\delta^{18}O$ and the age of magnetization. The dolomite content of the section covaries with grain size: grainstones are almost completely dolomitic; laminites are almost completely calcitic. Accordingly, the heterolithic facies exhibits a mixed mineralogy. Only in the lowest part of the section are two successions of laminites replaced by dolomite (Fig. 10). All of the dolomite in this section is coarsely crystalline, orange luminescent late-stage dolomite. In the dolomitic samples, there is a negative covariance between dolomite content and NRM intensity. Dolomitized cryptalgal laminites have the lowest magnetization intensity and the highest dolomite content (Fig. 11).

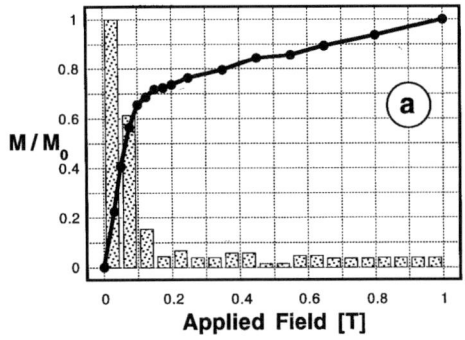
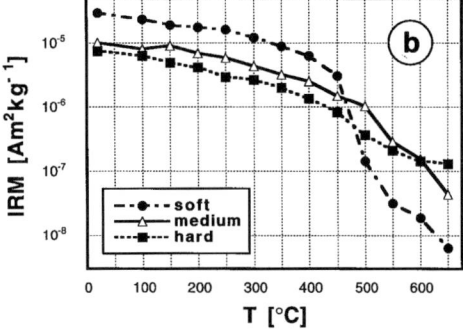

Fig. 7. Representative IRM acquisition and IRM thermal demagnetization curves, Honeycut Formation with Pennsylvanian magnetization. (**a**) IRM acquisition. The steep initial slope indicates a very soft (low-coercivity) magnetic mineral, but, at higher applied fields, the sample slowly continues to acquire remanence, reflecting the presence of a hard magnetic mineral. Bars indicate remanence change within the ambient field range. (**b**) Corresponding thermal unblocking spectrum of orthogonal component IRM (0.1, 0.4, 1 T). The unblocking temperatures indicate coarse-grained magnetite (soft), fine-grained magnetite (soft to medium) and hematite (hard).

The Upper El Paso Group, Franklin Mountains

The following section concerns the McKelligon Formation drill samples collected from McKelligon Canyon (Franklin Mountains, west Texas),

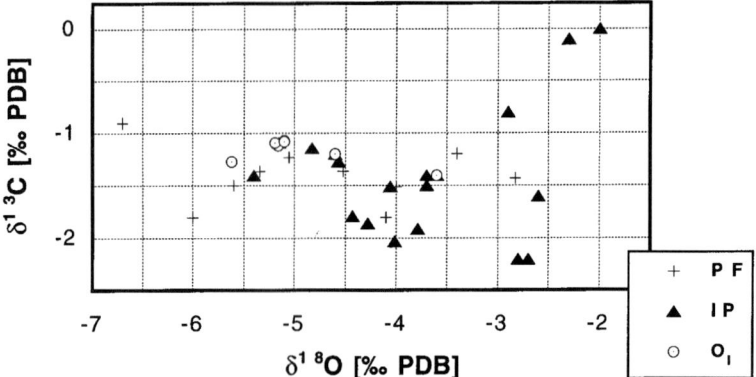

Fig. 9. Stable isotope cross-plot, Honeycut Formation. The range of $\delta^{18}O$ values is wider than that of $\delta^{13}C$ values (both axes have the same scale). Limestone samples that carry a Pennsylvanian magnetization show no correlation between stable isotopic signature and magnetization age (see Fig. 10).

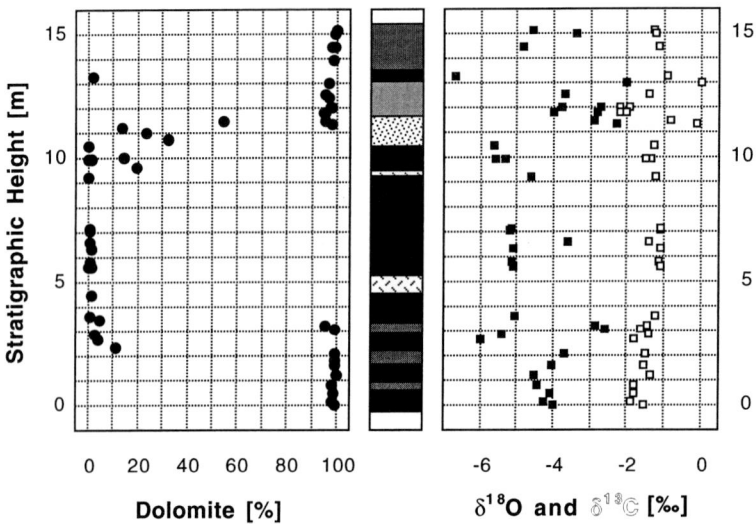

Fig. 10. Dolomite content and stable isotopic composition, Honeycut Formation. This figure corresponds to the profile in Fig. 3. The dolomite content covaries with grain size in that coarse-grained rocks are dolomitized, whereas fine-grained rocks are limestones. The pervasive Pennsylvanian magnetization of the dolo-grainstones records late-stage recrystallization during the Ouachita Orogeny. Carbon isotopes display a narrow range typical for Lower Ordovician carbonate rocks, whereas oxygen isotopes reflect diagenetic alteration. However, the $\delta^{18}O$ values for both limestones and dolomites are heavy in comparison with those for Lower Ordovician carbonate rocks elsewhere, suggesting that Pennsylvanian orogenic brines, at this particular site, were at only moderate temperatures (see Fig. 9). ■, $\delta^{18}O$; □, $\delta^{13}C$.

and presents results from oriented hand specimens of the McKelligon, Cindy, and Ranger Peak Formations collected on Sugarloaf Mountain and Comanche Peak in the southern Franklin Mountains (Figs 1 and 2).

Depositional lithofacies

Barrel-shaped thrombolitic algal–sponge mound complexes characterize the McKelligon Formation. They consist of skeletal wackestone that commonly grades into packstone. The micritic matrix indicates a muddy environment, such as that of a protected shelf lagoon (Klement & Toomey 1967). The tops of several bioherm complexes are composed of an intraclastic packstone grading into grainstone. Angular intraclasts within a micritic, partly dolomitized, matrix reflect high-energy transgressive phases in a predominantly muddy environment. The mounds are

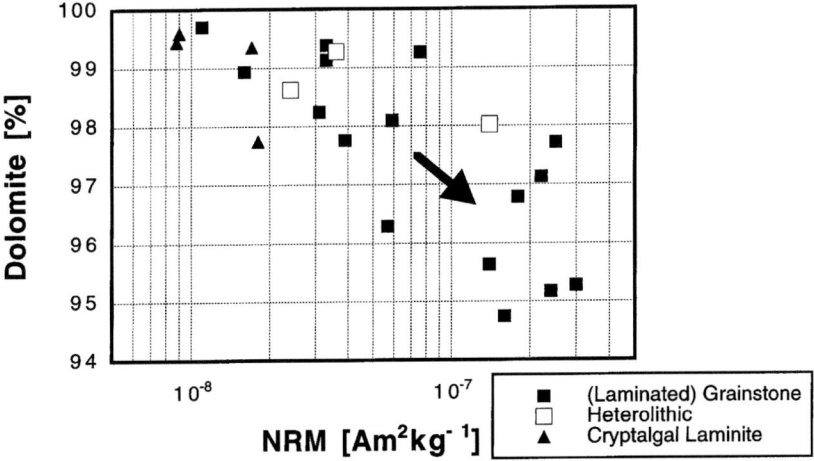

Fig. 11. Dolomite/NRM intensity ratios of dolomite samples, Honeycut Formation. These ratios suggest that, during water–rock interaction, the dolomite content of dolomitic rocks was slightly reduced as the magnetic intensity increased (arrow).

separated laterally by grainstone-filled channels. These grainstones are composed of mud-free arenitic debris of bioherm-derived material that was generated in place and reworked by storms. Mound horizons vertically alternate with lithoclastic beds. The burrowed skeletal wackestones–packstones, with angular intraclasts, indicate a high-energy subtidal shelf environment above wave-base (Toomey 1970). Lithoclastic beds are partly replaced by finely crystalline non-luminescent dolomite. Ribbon rocks are intercalated layers of calcitic grainstone and dolomitic mudstone with a characteristic flaser-bedding. These rocks formed on a tidal flat close to the subtidal

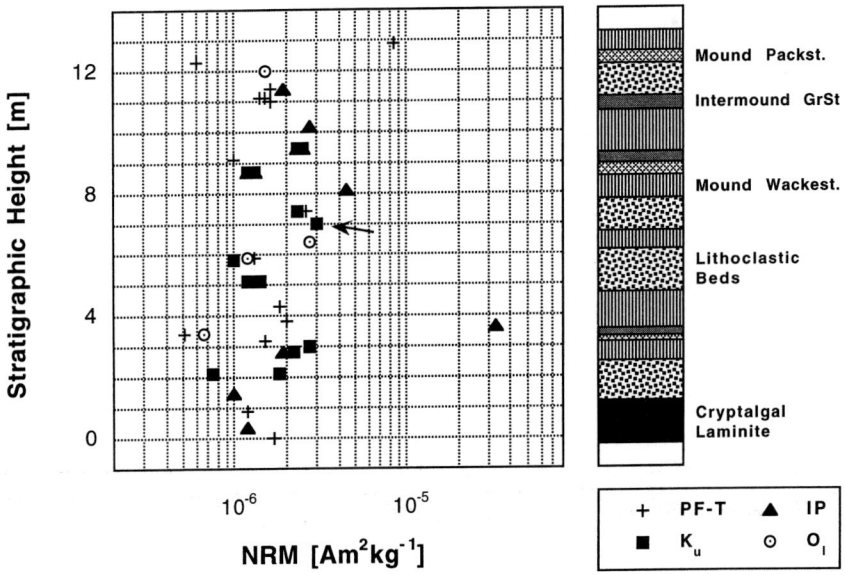

Fig. 12. NRM intensity profile, McKelligon Formation. O_l, Lower Ordovician; lP, Pennsylvanian; K_u, Upper Cretaceous; PF-T, present field or Tertiary. The arrows indicate the sample depicted in Fig. 18. The intensity has a narrow range with no trends. Neither the intensity nor direction (age) correspond to the lithology.

zone where mud could still settle. The dolomite is non-luminescent and fine grained. Cindy Formation rocks include well-sorted cross-bedded quartz arenites that reflect a foreshore environment. The quartz grains are embedded in a finely dolomitized muddy matrix. Thin beds of dolo-mudstone with desiccation features and small anhydrite crystals provide evidence for periods of exposure. Current-laminated silty dolo-mudstones are interpreted as a tidal-flat deposit in which the varying amounts of silt indicate the proximity to the shore during deposition (McDowell 1983). These rocks contain finely crystalline, non-luminescent early-stage dolomite (Stepanek 1987). The Ranger Peak Formation rocks contain somewhat nodular dark, thin-bedded mudstones that are typically finely dolomitized and were deposited in a tidal-flat setting. The intraclastic–peloidal packstone–grainstone zones are interpreted as tidal-channel deposits (Goldhammer et al. 1993). Flat-pebble conglomerates contain coarse, flat intraformational clasts enclosed in a predominantly muddy matrix.

Natural Remanent Magnetization

The magnetization intensity of the upper El Paso Group falls in a narrow range in spite of considerable differences in lithological composition, for the profile in McKelligon Canyon (Fig. 12) and also for the hand samples collected in the southern Franklin Mountains; there is no correlation between magnetization intensity and depositional lithofacies. The directions of remanence of all samples from the El Paso Group carry modern magnetization overprints. In most cases, high temperatures (250–350°C) are required to completely remove this magnetization direction, but it was removed at about 150°C in approximately a fifth of the samples, at which temperature most samples showed a noticeable drop in NRM. Some samples contain a shallow easterly magnetization component that is interpreted as being of Early Ordovician age. The maximum unblocking temperatures for these samples are 400–550°C. Other samples carry a thermally stable shallow southeasterly magnetization component that is interpreted as being of

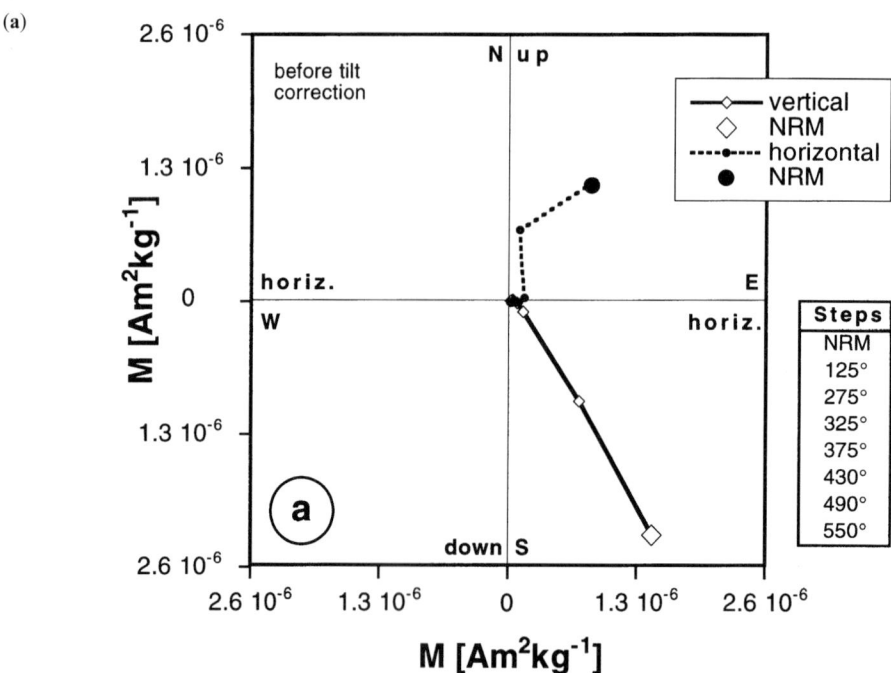

Fig. 13. Representative orthogonal projection, McKelligon Formation. A modified Zijderveld diagram is used. (**a**) This sample contains a random component, probably acquired during sample storage, and a strong modern overprint. (**b**) An enlargement illustrating that, after these components are thermally removed, a weak, but fairly stable magnetization component could be isolated, which has a very shallow inclination and an easterly declination indicating an Early Ordovician age. (**c**) Many samples carry a north-northwesterly component with a steep inclination interpreted as being of Late Cretaceous age.

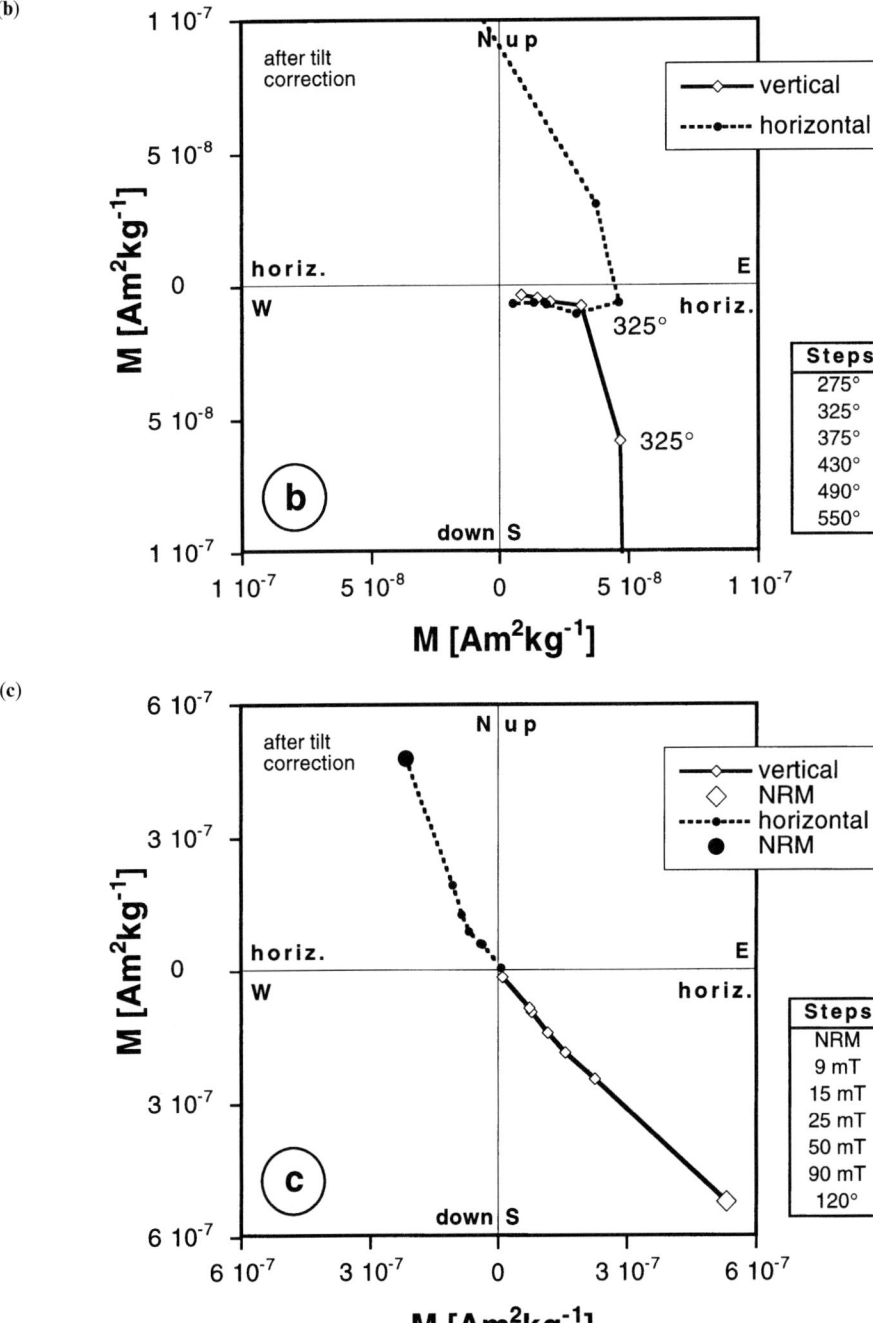

Fig. 13. (continued)

Pennsylvanian age. Many samples carry a steep north-northwesterly magnetization component (Fig. 13) with variable maximum unblocking (400–670°C). The characteristic magnetization components of the McKelligon Formation from McKelligon Canyon (Fig. 14) and the results from specimens from the McKelligon, Cindy and Ranger Peak Formations in the southern Franklin Mountains (Fig. 15) do not contain a pervasive Pennsylvanian remagnetization; approximately a third of the samples retain their Early Ordovician signature. Strata in the Franklin Mountains are tilted to the west with dip angles of about 35° so, after tilt correction, a north-northeasterly magnetization direction could be isolated in specimens from all of the investigated upper El Paso Group formations. These directions clustered close to the Late Cretaceous reference direction. However, whereas some samples exhibit this magnetization direction after their magnetization vectors are corrected for tilt, others show a similar direction before tilt correction (Fig. 16). No Silurian or Devonian characteristic remanent magnetizations were observed, even though some breccias were sampled that could be of this age. Twenty samples of the McKelligon Canyon and southern Franklin Mountains samples have a mean direction, after normal polarity vectors had been reversed (109°, 7°; $\kappa = 22.4$; $\alpha_{95} = 8.0°$; VGP: 14°N, 150°E) which was comparable with the Early Ordovician reference pole (14°N, 159°E) for the North American craton (Van der Voo 1990). Nineteen samples carry a mean direction of magnetization interpreted as being of Pennsylvanian age (145°, 14°; $\kappa = 22.6$; $\alpha_{95} = 7.0°$; VGP: 39°N, 120°E; reference pole: 40°N, 128°E). Seventeen samples carry a magnetization interpreted as being of Late Cretaceous age (325°, 54°; $\kappa = 67.7$; $\alpha_{95} = 4.4°$; VGP: 61°N, 178°E; reference pole: 67°N, 189°E). It was not possible to conduct a fold test or a conglomerate test (Fig. 17). There is no correlation between depositional lithofacies and magnetization directions, except for the Late Cretaceous magnetization, which seems to reside in the lithoclastic beds facies. In contrast to the Honeycut Formation, the Early Ordovician magnetizations in the El Paso Group are not restricted to fine-grained lithologies (Figs 12, 14 and 15).

Magnetic mineralogy

Samples that carry an Early Ordovician magnetization show steep initial IRM acquisition curves and typically acquire 65–85% of their saturation

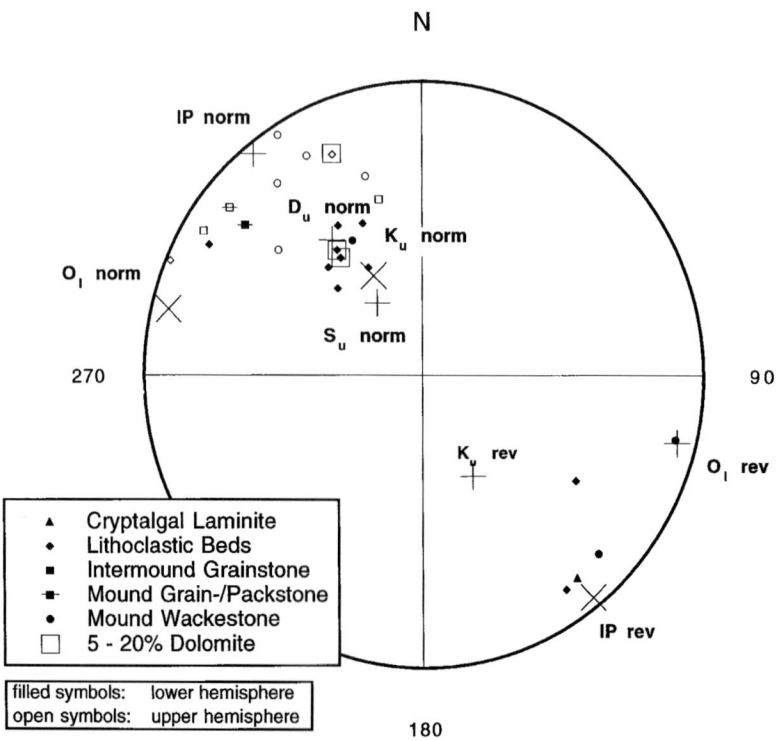

Fig. 14. Tilt-corrected magnetic components, McKelligon Formation (McKelligon canyon). Crosses show expected magnetizations (palaeo-directions): ×, lower hemisphere; +, upper hemisphere. These specimens carry characteristic magnetizations of Early Ordovician, Pennsylvanian, and Late Cretaceous age. Devonian or Silurian magnetization vectors were not observed, despite a major karst event occurring during this time. The characteristic directions do not correspond to particular lithofacies or to mineralogy.

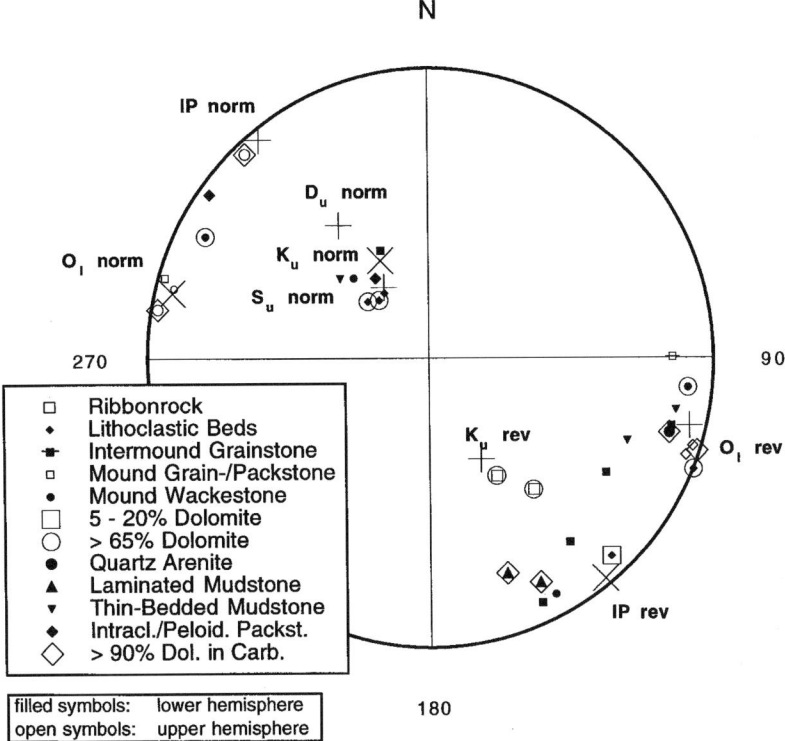

Fig. 15. Tilt-corrected magnetic components, upper El Paso Group (southern Franklin Mountains). The specimens show the same directions as for the McKelligon Formation (MeKelligon canyon). Again, there is no connection between the directions and lithologies. Massive early-stage dolomite (>90% dolomite in carbonate part of the rock) of the Cindy and Ranger Peak formations does not correspond to particular magnetization directions.

remanence in a field of 0.2 T. Only a few samples did not saturate until 1 T. Thermal demagnetization of three-component IRM demonstrated that the soft and medium remanences began to unblock at 400°C, characteristic of magnetite. The hard component typically showed a decay below 100°C but then began to unblock only at 600°C, suggesting the presence of both goethite and hematite. Samples carrying either the Pennsylvanian or Late Cretaceous magnetization did not show consistent IRM acquisition behaviour; some acquired c. 75% of their maximum IRM in an ambient field of 0.2 T but also continued to acquire remanence in higher fields and had not saturated in 1 T. Unblocking spectra of three-component IRM show unblocking of the soft and medium IRM beginning at 400°C, suggestive of magnetite, whereas the hard IRM decayed around 100°C, suggestive of goethite. In some cases, the hard IRM intensity increased at higher temperatures because of the transformation of goethite to hematite. At 600°C the hard IRM began to unblock, indicating hematite. However, the variable behaviour suggests that not all hematite was generated by the transformation of goethite, and that sometimes hematite was already present in the unheated samples (Fig. 18). Importantly, other samples that carry a Pennsylvanian or Late Cretaceous magnetization showed different behaviours during IRM acquisition, indicating the presence of various ratios of soft and hard magnetic minerals. Some samples slowly acquired IRM in stepwise increasing ambient fields and, thus, do not show evidence for a significant content of a soft magnetic mineral. Magnetic mineralogy does not correspond to particular carbonate lithofacies.

Stable isotopic composition and carbonate mineralogy

Carbon isotopes have a narrow range and their values are typical for Lower Ordovician carbonate rocks. The range of $\delta^{18}O$ values of the limestones is −6 to −8.5‰, with the upper El Paso

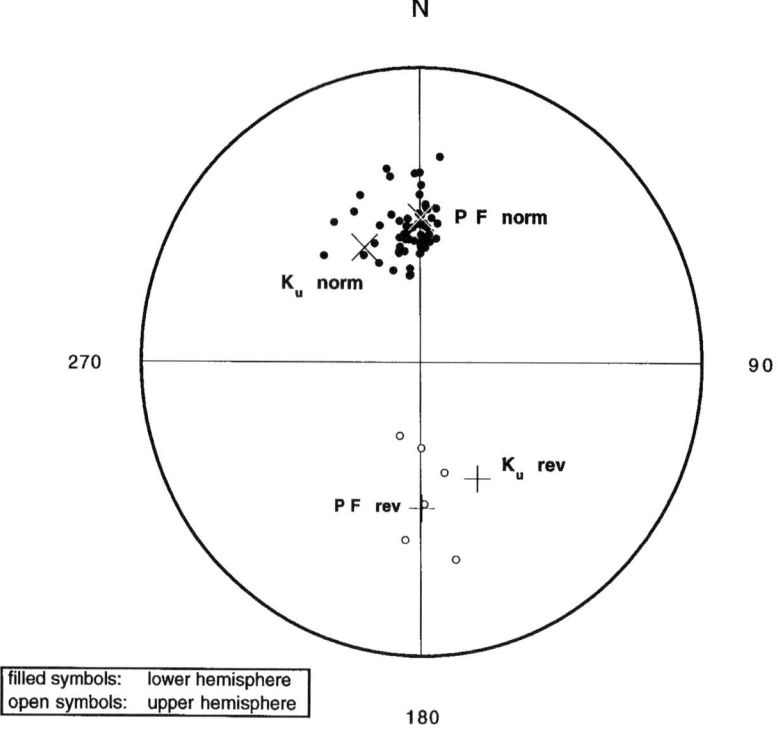

Fig. 16. Pre-tilt-corrected magnetic components, upper El Paso Group (Franklin Mountains). Most specimens contain a strong present field (PF) magnetization overprint but several components show a Late Cretaceous magnetization component before tilt correction.

Group dolomites, on average, being 3‰ heavier than their associated limestones. Magnetization age and stable isotopic signature do not correlate. Within the McKelligon Formation profile in McKelligon Canyon, only the lithoclastic beds facies is partly replaced by early-stage dolomite (Fig. 19). However, Cindy and Ranger Peak samples contain massive early-stage dolomite, with the dolomite content varying between 0 and 99.5%. There is no correlation between dolomite content and magnetization intensity or direction.

Discussion

Magnetic intensity and direction of samples from the Honeycut Formation (Llano Uplift) are contingent upon two sedimentological factors: grain size and mineralogy. Beds composed of fine-grained fabrics and limestones carry a weak Early Ordovician magnetization, whereas beds of coarse-grained fabrics and dolomite carry a strong Pennsylvanian magnetization, i.e. primary magnetizations are associated with comparatively low permeabilities whereas secondary magnetizations are associated with comparatively high permeabilities. Therefore, the magnetic properties are likely to reflect palaeohydrological conditions during regional or local fluid migration events. Samples carrying an Early Ordovician or a Pennsylvanian magnetization have moderate maximum NRM unblocking temperatures (450°C). During AF demagnetization, samples carrying a Pennsylvanian magnetization lose 90% of their NRM intensity after demagnetization to 90 mT. Either behaviour suggests that a soft to medium-hard magnetic mineral, such as fine-grained magnetite, is the predominant carrier of the ancient magnetization components observed in Honeycut samples. All samples with a present field magnetization lose this between 150 and 250°C, so, in most cases, the modern magnetization seems to reside in coarse-grained magnetite and, occasionally, goethite. Some samples that carry a streak magnetization

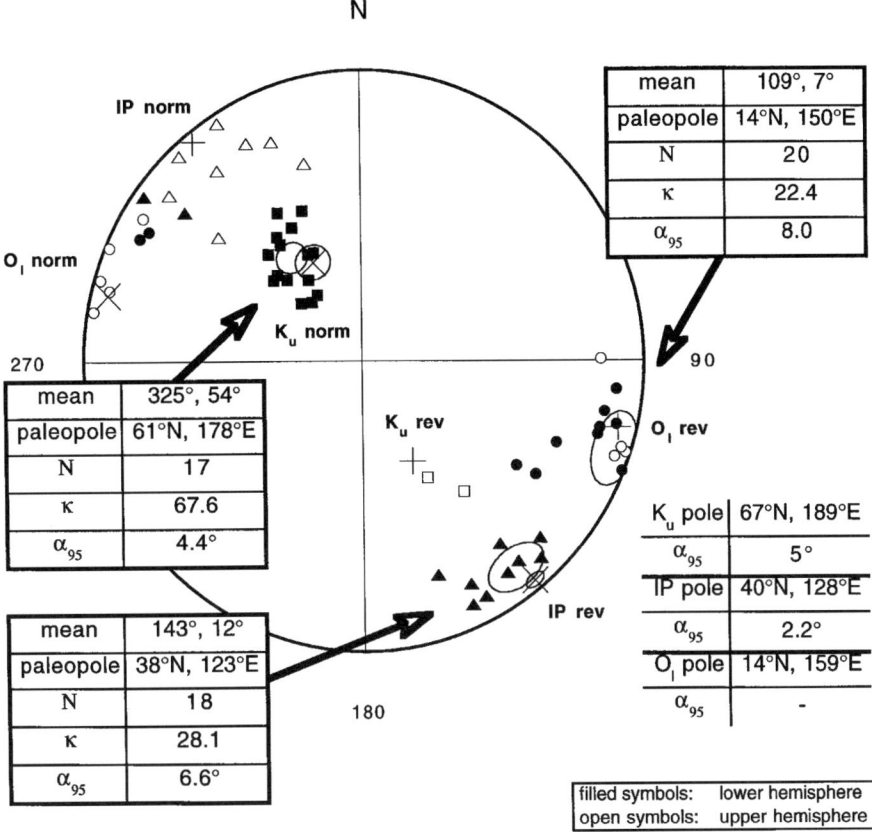

Fig. 17. Mean directions, after tilt correction, of components of remanence, upper El Paso Group, and Fisher (1953) statistics. ○, α₉₅. Normal polarity vectors were reversed and included in the calculations. The results indicate that these characteristic vectors reflect magnetization during early Ordovician (●), Pennsylvanian (▲), and late Cretaceous time (■).

(steep southeasterly, i.e. between reverse Pennsylvanian and present field) lose more than 90% of their initial NRM intensity after AF demagnetization to only 5 mT, indicating a very soft magnetic mineral such as coarse-grained (MD) magnetite, which will carry modern magnetizations and streak magnetizations of viscous origin, whereas other modern magnetization components reside in goethite, which is likely to be a product of recent weathering. These observations are supported by the behaviour during IRM acquisition and thermal demagnetization of three-component IRM. However, because most of the NRM is removed before 450°C or 90 mT, respectively, the high-coercivity magnetic minerals cannot be major contributors to the characteristic magnetizations.

Magnetite authigenesis is the most likely mode of Pennsylvanian remanence acquisition. A mere recrystallization of pre-existing detrital magnetite could not account for the up to 100-fold intensity variation between Early Ordovician and Pennsylvanian magnetizations. A thermoviscous overprint can also be ruled out because of shallow burial in the area and low diagenetic temperatures. This implies that fluids that penetrated the Honeycut rocks during the Pennsylvanian flowed preferentially in the dolo-grainstones, in which they triggered magnetite authigenesis. Therefore, magnetization intensity covaries with fluid/rock ratios, and the magnetization trends observed in this section actually document ancient fluid pathways. Aquitards were unaffected or less affected by Pennsylvanian fluid flow and hence retain most of their weak Early Ordovician magnetization, which is probably a primary magnetization residing in detrital magnetite.

These results are in accord with interpretations of the diagenetic history of Ellenburger carbonates based on their geochemistry. During

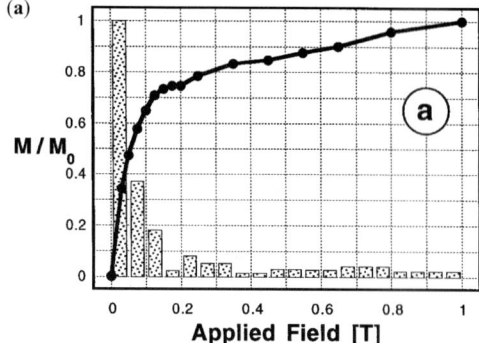

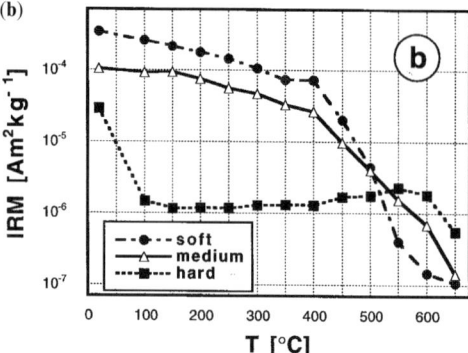

Fig. 18. Representative IRM acquisition and IRM thermal demagnetization curves, McKelligon Formation with Late Cretaceous magnetization. (a) IRM acquisition. The curve shows an initial steep remanence gain followed by a slow gradual increase. The magnetic phase is dominated by a low-coercivity mineral, but a high-coercivity mineral carries a considerable portion of the IRM. (b) Corresponding thermal unblocking spectrum of orthogonal component IRM (0.1, 0.4, 1 T). The soft magnetic carrier, coarse-grained magnetite, starts unblocking at 400°C, and is dominant, whereas fine-grained magnetite is the second most important carrier. It displays a slightly higher stability at temperatures above 500°C. The very low unblocking temperatures of a hard magnetic carrier imply goethite, and the gain in remanence during rise in temperature may be attributed to it partially converting to hematite.

regional tectonism, brines were expelled from basinal clastic sediments and penetrated the rocks in the subsurface, causing dissolution of metastable dolomite and the reprecipitation of a more stable phase (Kupecz 1989; Kupecz & Land 1991). The Pennsylvanian magnetization of the late-stage dolomites confirms that the Ouachita Orogeny caused these events. The range of $\delta^{13}C$ values observed in this study corresponds to their trend in the Cambro-Ordovician carbonate rocks of Oklahoma (Gao 1990). Moreover, Cambro-Ordovician carbonate rocks from several areas show comparable carbon isotopic compositions (Kupecz & Land 1991; Johnson & Goldstein 1993; Ripperdan et al. 1993; Montanez 1994). Therefore, these values can be considered typical for carbonate rocks of this age. However, the observed oxygen isotopic composition of both limestones and dolomites is considerably lighter than the presumed pristine Early Ordovician values. In general, depletion of heavy oxygen isotopes in carbonate rocks reflects diagenetic alteration involving water with a light oxygen isotopic composition, elevated temperatures, or both (Land 1983). Carbon isotopes, unlike oxygen isotopes, are not a stoichiometric constituent of both the rock and the water and, therefore, require extremely high water/rock ratios to be reset during diagenesis (Banner & Hanson 1990). For the same reason, the $\delta^{13}C$ range of carbonate rocks that undergo diagenesis is generally narrower than their $\delta^{18}O$ range (Allan & Matthews 1977). Thus, even entirely recrystallized rocks in this section may have inherited their carbon isotopic composition from precursor phases. However, the oxygen isotopic composition of the Lower Ordovician carbonate rocks from the Llano Uplift area reflects diagenetic alteration as it is heavy, in comparison with that for carbonate rocks of the same age from other areas, because the Llano Uplift has never been deeply buried. The temperature of the orogenic brines that affected the Honeycut rocks was up to 110°C. It was these elevated temperatures, rather than ^{18}O-depleted fluids, that caused the present oxygen isotopic signature in the late-stage dolomites in the Llano Uplift area (Kupecz & Land 1991). Limestones investigated in this study have $\delta^{18}O$ values from -4 to $-5‰$ PDB. It could be expected that those limestones which preserved an Early Ordovician magnetization also preserved their chemical properties acquired early in diagenesis, and, conversely, those limestones that were remagnetized during diagenesis in the Pennsylvanian should display stronger diagenetic alteration of their oxygen isotopic composition; however, this is not the case. This apparent contradiction possibly results from comparatively strong Pennsylvanian magnetization components residing in insignificant volumes of the rocks. Although these zones overwhelm the whole-rock magnetic data, they do not noticeably influence the whole-rock geochemical data. The oxygen isotopic composition of Honeycut Formation dolomites observed in this study is heavier than that of late-stage dolomites from other outcrops in the vicinity which show a $\delta^{18}O$ range between -7 and $-9‰$ (Kupecz & Land 1991). Possibly,

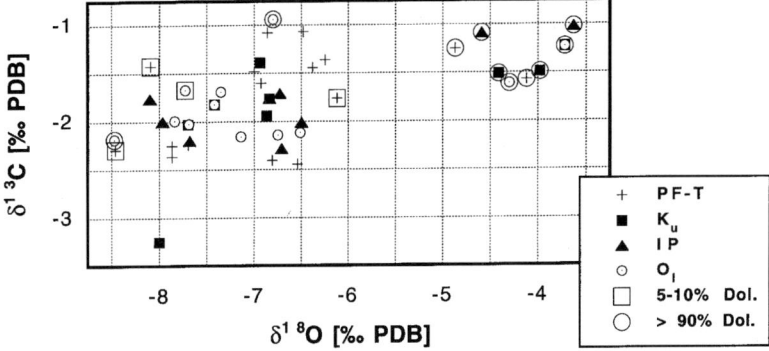

Fig. 19. Stable isotope cross-plot for the upper El Paso Group (Franklin Mountains). The range of $\delta^{18}O$ values is wider than that of the $\delta^{13}C$ values. The magnetization age and stable isotopic signature do not correlate.

local hydrological conditions can be invoked for causing relatively low temperatures of brines that interacted with these rock bodies. Small dolomite crystals along stylolites inside the cryptalgal laminites suggest that stylolites provided small-scale fluid conduits for dolomitizing fluids within what were otherwise almost impermeable strata. Stylolitization can generally enhance porosity and permeability of carbonate rocks, because sutured stylolites provide fluid pathways. Post-stylolitization dissolution can then result in the generation of fabric-selective secondary porosity (Von Bergen & Carozzi 1990). Dolomitization processes that coincide with the path of stylolites have been documented in other Palaeozoic carbonate units and are interpreted to have occurred during periods of decreasing pressure in the rock bodies (Major *et al.* 1990; Major & Holtz 1993). The Pennsylvanian magnetization components observed within the thick succession of cryptalgal laminites in the middle of the Honeycut section may record dolomite precipitation along stylolites during Pennsylvanian time. The volume of precipitated dolomite is small; the laminites typically contain less than 3% dolomite. However, in the lowest part of the section, two thin successions of cryptalgal laminites are entirely replaced by late-stage dolomite. Possibly, the magnesium source for these replacement processes was dissolution of early dolomite within the overlying dolo-grainstone strata. This hypothesis is supported by the negative covariance between dolomite content and NRM intensity of the dolomitic samples, which suggests that the dolomite content of these rocks was reduced during water–rock interaction. It is very unlikely that cryptalgal laminites were exposed to higher water/rock ratios than were dolo-grainstones. Therefore, the water–rock interaction path (Fig. 11) was in the direction of higher magnetization intensity and lower dolomite concentration. If some of the magnesium was mobilized during dolomite recrystallization within the dolo-grainstones, dolomitization along stylolites could occur, and some cryptalgal laminites in the middle part of the section acquired a Pennsylvanian magnetization; the two cryptalgal laminite successions that are sandwiched between grainstone aquifers were completely replaced by dolomite and were pervasively remagnetized during Pennsylvanian time.

The magnetic signature of upper El Paso Group samples is entirely different in that it does not record any pervasive remagnetization events. Importantly, lithologies that typically are relatively porous, such as grainstones in the McKelligon Formation or massive early dolomite in the Cindy and Ranger Peak Formations, do not preferentially carry Pennsylvanian magnetizations. Fluid–rock interaction during Pennsylvanian time did not affect these rocks to the degree observed in Honeycut Formation rocks from central Texas. Obviously, the Franklin Mountains area was less affected by fluid pulses associated with the Ouachita Orogeny because of the larger distance to the Ouachita front. However, the Franklin Mountains are in close proximity to the Laramide thrust front and the Rio Grande Rift. The fact that Late Cretaceous magnetization components could be isolated both before and after tilt correction was applied suggests that this magnetization was acquired during block uplift and tilting and, thus, documents the absolute timing of this tectonic event. During uplift, no rotation of the Franklin Mountains horst block around a vertical axis took place, because magnetization directions older than those of Cretaceous time still cluster around Early Ordovician or Pennsylvanian reference directions, respectively (Fig. 17). Therefore, Late Cretaceous tectonic activity, probably associated with Rio Grande rifting, caused a non-pervasive remagnetization in Franklin Mountains rocks. Unlike the tectonism-related remagnetizations observed in rocks from the Honeycut Formation, this remagnetization is not associated with strong magnetization intensities, particular magnetic mineralogies, or zones of high permeability. Thus, it does not indicate extensive influence by reactive fluids. Possibly, in this case, under an extensional regime during Cretaceous time, no fluids were expelled from subsurface strata.

Although the El Paso Group rocks were affected by a major karst event during Silurian or the Devonian time (Lucia 1969), this event is not recorded in the magnetic properties of the samples. Possibly, as a result of rapid development of major zones of collapse breccias, fluid flow was focused and hence even rock bodies adjacent to or within brecciated zones were not affected by meteoric diagenesis to a degree that would have caused remagnetization. Maximum unblocking temperatures during demagnetization of NRM suggest that the Early Ordovician magnetizations of the upper El Paso Group carbonate rocks predominantly reside in magnetite. Pennsylvanian and Late Cretaceous magnetizations show different maximum unblocking temperatures, implying that some of these magnetizations reside in magnetite, but other samples contain Pennsylvanian and Late Cretaceous magnetizations carried by hematite. Modern weathering-related overprints typically reside in goethite, as also indicated by rock magnetic experiments. The magnetic mineralogy of samples carrying an Early Ordovician magnetization is dominated by magnetite, whereas samples that carry a Pennsylvanian or Late Cretaceous magnetization are characterized by various magnetic minerals. The remanence acquisition mechanisms of the Pennsylvanian and Late Cretaceous magnetizations could not be resolved. Unlike the results obtained from Honeycut rocks from the Llano Uplift area, these results do not elucidate under which conditions NRMs were overprinted or preserved. However, considering the estimated maximum burial temperatures of the upper El Paso Group (Gao et al. 1992), a thermoviscous remagnetization is implausible. The $\delta^{18}O$ values of the limestones are c. 2‰ lighter than those of the Honeycut Formation and reflect deeper burial of the El Paso Group. The observed difference between the oxygen isotopic composition of limestones and dolomites from the El Paso Group is within a typical $\delta^{18}O$ range. Although a unique $\delta^{18}O$ may not exist at all (Land 1980) and values as low as 1.5‰ (Major et al. 1992) or as high as 5‰ (Land 1991) are possible, this finding suggests that both limestones and dolomites stabilized in the same ^{18}O-depleted (or hot) early-diagenetic fluid.

Rocks that preserved an Early Ordovician magnetization should conceivably also have preserved a comparatively heavy oxygen isotope signature. This is not the case, as observed in the Honeycut rocks from the Llano Uplift area. Therefore, remagnetizing diagenetic processes may not be recorded in the $\delta^{18}O$ values, or these values were reset after remagnetization took place, so erasing any previous correlation between $\delta^{18}O$ and magnetization direction. Both chemical stabilization and remagnetization are lithologically selective processes. Possibly, diagenetic events that affected the upper El Paso Group rocks were more selective to chemical alteration than to magnetic alteration. Consequently, the magnetic and chemical properties of a rock do not necessarily reside in the same lithological constituents. Little is known about the spatial distribution of magnetic grains within sedimentary rocks, but as the magnetic and lithological properties can hardly be correlated, the mode of remanence acquisition cannot be ascertained. These results illustrate a major limitation of sedimentary palaeomagnetic research, which probably results from an insufficient volumetric resolution. Consequently, in the absence of a major diagenetic event that involved a reactive fluid, magnetic properties seem to be randomly distributed through very diverse lithofacies.

The investigated Cambro-Ordovician platform carbonate rocks from Texas display evidence for regional remagnetization events similar to Kiaman overprints documented in many North American sedimentary rocks. Interestingly, although the polarity of Pennsylvanian magnetizations observed in this study is predominantly reversed, numerous samples, particularly from the Franklin Mountains, carry a normal polarity Pennsylvanian magnetization. The duration of the orogenic fluid pulses responsible for remagnetization events is not well constrained. Moreover, secondary magnetizations may be the result of multiple episodes of fluid–rock interaction. Nevertheless, this finding suggests significant normal polarity intervals within the Kiaman Reversed Superchron or Pennsylvanian remanence acquisition before or after the Superchron.

Conclusions

Late-stage recrystallization of dolo-grainstones in the Honeycut Formation is recorded by a strong, pervasive Pennsylvanian remagnetization. Reducing orogenic fluids expelled from basinal sediments during the Ouachita Orogeny triggered magnetite authigenesis that resulted in distinctive intensity trends covarying with fluid/rock ratios. Associated fine-grained cryptalgal laminites, owing to their low permeability, partially escaped from diagenetic alteration and retain a weak magnetization probably of syndepositional origin. However, rocks in the Franklin Mountains were not affected by orogenic fluids to the same degree because they are further from the Ouachita front. Even comparatively permeable strata, such as coarse-grained limestones or massive early dolomites, partially retain an early Ordovician signature. This result reflects the spatial limits of long-distance fluid migration during the Ouachita Orogeny. Detailed vertical correlation of NRM intensities and sedimentological variables can, thus, aid in determining pathways of reactive fluids and reconstruct palaeohydrological conditions. Measuring magnetic intensity is a fast and simple procedure, as opposed to the laborious determination of magnetization directions. Furthermore, as intensity measurements do not require the samples to be oriented, this method could be applied to core material. Large-scale subsurface mapping of magnetization intensity could semi-quantitatively elucidate the spatial distribution of regional orogenic fluid migration events. Provided that also magnetization directions and, thus, ages are available, this approach yields information on the spatial and temporal evolution of a rock body's ability to store and transmit fluids. However, this approach appears to be limited in that it is applicable only if a major diagenetic event affected the investigated sedimentary rocks that involved intensive interaction of the rock bodies with a reactive fluid. Future contributions of palaeomagnetic work to sedimentary research could be significantly improved if magnetic investigations have a volumetric resolution comparable with that of geochemical or petrographical work. The details of primary and secondary remanence acquisition are still poorly understood. Only integrated rock magnetic, petrographical, and geochemical work on a very fine scale could reliably clarify this problem. The sections investigated in this study allowed isolation of some magnetization vectors that probably are of primary origin. However, reliable criteria for identification of pristine magnetizations still have to be established. Because the Early Ordovician palaeopole for the North American craton is poorly constrained, these data may provide some improvement of the palaeomagnetic database for this craton.

This paper presents part of my PhD work at the University of Texas at Austin, which was co-supervised by L. S. Land and W. A. Gose. I gratefully acknowledge their advice and support. I wish to thank R. P. Major for invaluable conversations (regarding both scientific and other topics) and R. D. Elmore (The University of Oklahoma at Norman) for improving the interpretation of the data. Constructive comments regarding the manuscript by J. L. Banner, M. B. Lagoe, and P. Tecac were much appreciated. Fellowship support was provided by the Albert W. and Alice M. Fund in Geology. Fieldwork was supported by scholarships from the ARCO Oil and Gas Company Fund and the Ronald K. De Ford Field Scholarship Fund of the Geology Foundation at The University of Texas.

Reference

ALGEO, T. J. 1992. Continent-scale wrenching of southwestern Laurussia during the Ouachita–Marathon Orogeny and tectonic escape of the Llano block. *In*: CUNNINGHAM, B. K. & CROMWELL, D. W. (eds) *The lower Paleozoic of West Texas and Southern New Mexico – Modern Exploration Concepts.* Permian Basin Society of Economic Paleontologists and Mineralogists Publications, **89-31**, 115–129.

ALLAN, J. R. & MATTHEWS, R. K. 1977. Carbon and oxygen isotopes as diagenetic and stratigraphic tools: surface and subsurface data, Barbados, West Indies. *Geology*, **5**, 16–20.

AMTHOR, J. E. & FRIEDMAN, G. M. 1992. Early- to late-diagenetic dolomitization of platform carbonates: Lower Ordovician Ellenburger Group, Permian Basin, west Texas. *Journal of Sedimentary Petrology*, **62**(1), 131–144.

BANNER, J. L. & HANSON, G. N. 1990. Calculation of simultaneous isotopic and trace element variations during water–rock interaction with applications to carbonate diagenesis. *Geochimica et Cosmochimica Acta*, **54**, 3123–3137.

BARNES, V. 1982. *Geology of the Marble Falls Quadrangle, Burnet and Llano Counties, Texas.* Bureau of Economic Geology, The University of Texas, Austin, TX.

CANDELARIA, M. P. & REED, C. L. (eds) 1992. *Palaeokarst, Karst Related Diagenesis and Reservoir Development: Examples from Ordovician–Devonian Age Strata of West Texas and the Mid-Continent.* Permian Basin Society of Economic Paleontologists and Mineralogists Publications, **92-33**.

CLEMONS, R. E. 1989. The Ellenburger-El Paso connection: Lower Ordovician shelf carbonates. *In*: CUNNINGHAM, B. K. & CROMWELL, D. W. (eds) *The lower Paleozoic of West Texas and Southern New Mexico – Modern Exploration Concepts.* Permian Basin Society of Economic Paleontologists and Mineralogists Publications, **89-31**, 85–104.

CLOUD, P. E. & BARNES, V. E. 1946. *The Ellenburger Group of Central Texas.* Bureau of Economic Geology, The University of Texas, Publications, **4621**.

CROMER, J. B. 1991. *Stratigraphic Analysis of the Upper Devonian Woodford Formation, Permian Basin, West Texas and Southeastern New Mexico.* Bureau of Economic Geology, The University of Texas, Reports of Investigations, **201**.28.

DUNLOP, D. J. 1973. Theory of magnetic viscosity in lunar and terrestrial rocks. *Reviews of Geophysics and Space Physics*, **11**, 855–901.

EPSTEIN, A. G., EPSTEIN, J. B. & HARRIS, L. D. 1977. *Conodont Color Alteration – an Index to Organic Metamorphism.* US Geological Survey Professional Paper, **995**.

FARR, M. R., SPROWL, D. R. & JOHNSON, J. 1993. Identification and initial correlation of magnetic reversals in the Lower to Middle Ordovician of northern Arkansas. *In*: AÏSSAOUI, D., MCNEILL, D. F. & HURLY, N. F. (eds) *Applications of Paleomagnetism to Sedimentary Geology.* Society of Economic Paleontologists and Mineralogists Special Publications, **49**, 83–93.

FISHER, R. A. 1953. Dispersion on a sphere. *Proceedings of the Royal Society of London, Series A*, **217**, 295–305.

FLAWN, P. T., GOLDSTEIN, A., KING, P. B. & WEAVER, C. E. 1961. *The Ouachita System.* Bureau of Economic Geology, The University of Texas, Publications **6120**.

GAO, G. 1990. *Petrography, geochemistry, and diagenesis of the late Cambrian and early Ordovician Arbuckle Group, Slick Hills, SW Oklahoma.* PhD thesis, The University of Texas, Austin.

——, LAND, L. S. & FOLK, R. L. 1992. Meteoric modification of early dolomite and late dolomitization by basinal fluids, upper Arbuckle Group, Slick Hills, southwestern Oklahoma. *American Association of Petroleum Geologists Bulletin*, **76**(11), 1649–1664.

GEISSMAN, J. W., JACKSON, M., HARLAN, S. S. & VAN DER VOO, R. 1991. Palaeomagnetism of latest Cambrian–early Ordovician and Latest Cretaceous-early Tertiary Rocks of the Florida Mountains, southwest New Mexico. *Journal of Geophysical Research*, **96**(B4), 6053–6071.

GOLDHAMMER, R. K., LEHMANN, P. J. & DUNN, P. A. 1993. The origin of high-frequency platform carbonate cycles and third-order sequences (lower Ordovician El Paso Group, west Texas): constraints from outcrop data and stratigraphic modelling. *Journal of Sedimentary Petrology*, **63**(3), 318–359.

IRVING, E. & IRVING, G. A. 1982. Apparent polar wander paths Carboniferous through Cenozoic and the assembly of Gondwana. *Geophysical Surveys*, **5**, 141–188.

JOHNSON, K. 1991. Geologic Overview and Economic Importance of Late Cambrian and Ordovician Rocks in Oklahoma. *Oklahoma Geological Survey Circular*, **92**, 143–153.

—— & GOLDSTEIN, R. H. 1993. Cambrian sea water preserved as inclusions in marine low-magnesium calcite cement. *Nature*, **362**, 335–337.

KERANS, C. 1988. Karst-controlled reservoir heterogeneity in Ellenburger Group carbonates of west Texas. *American Association of Petroleum Geologists Bulletin*, **72**, 1160–1183.

—— & LUCIA, F. J. 1989. Recognition of second, third, and fourth/fifth order scales of cyclicity in the El Paso Group and their relation to genesis and architecture of Ellenburger reservoirs. *In*: CUNNINGHAM, B. K. & CROMWELL, D. W. (eds) *The lower Paleozoic of West Texas and Southern New Mexico – Modern Exploration Concepts.* Permian Basin Society of Economic Paleontologists and Mineralogists Publications, **89-31**, 105–110.

KIRSCHVINK, J. L. 1980. The least squares line and plane and the analysis of palaeomagnetic data. *Geophysical Journal of the Royal Astronomical Society*, **62**, 699–718.

KLEMENT, K. W. & TOOMEY, D. F. 1967. Role of the blue–green alga *Girvanella* in skeletal grain destruction and lime-mud formation in the lower Ordovician of West Texas. *Journal of Sedimentary Petrology*, **37**, 1045–1051.

KUPECZ, J. A. 1989. *Petrographic and geochemical characterization of the Lower Ordovician Ellenburger Group, west Texas.* PhD thesis, The University of Texas, Austin.

—— & LAND, L. S. 1991. Late-stage dolomitization of the Lower Ordovician Ellenburger Group, West Texas. *Journal of Sedimentary Petrology*, **61**(4), 551–574.

LAND, L. S. 1980. The isotopic and trace element geochemistry of dolomite: the state of the art. *In*: ZENGER, D. H., DANHAM, J. B. & ETHINGTON, R. L. (eds) *Concepts and Models of Dolomitization.* Society of Economic Paleontologists and Mineralogists Special Publications, **28**: 87–110.

——1983. The application of stable isotopes to studies of the origin of dolomite and to problems of clastic diagenesis. Society of Economic Paleontologists and Mineralogists Publications Short Course, **10**, 4.1–4.22.

——1991. Dolomitization of the Hope Gate Formation (north Jamaica) by seawater: Reassessment of mixing-zone dolomite. *Geochemical Society Special Publication*, **3**, 121–133.

LEE, Y. I. & FRIEDMAN, G. M. 1987. Deep-burial dolomitization in the Ordovician Ellenburger Group Carbonates, west Texas and south-eastern New Mexico. *Journal of Sedimentary Petrology*, **57**, 544–557.

LEMONE, D. V. 1968. The Canadian (Lower Ordovician) El Paso Group of the Southern Franklin Mountains, El Paso County, Texas. West Texas Geological Society Guidebook, **68-55**, 76–81.

LINDSAY, R. F. & KOSKELIN, M. 1991. Arbuckle Group depositional parasequences, southern Oklahoma. *Oklahoma Geological Survey Circular*, **92**, 71–84.

LOWRIE, W. 1990. Identification of ferromagnetic minerals in a rock by coercivity and unblocking temperature properties. *Geophysical Research Letters*, **17**(2), 159–162.

LUCIA, F. J. 1968. Sedimentation and Palaeogeography of the El Paso Group. West Texas Geological Society Guidebook, **68-55**, 61–75.

——1969. *Lower Palaeozoic History of the Western Diablo Platform of West Texas and South–Central New Mexico.* West Texas Geological Society Guidebook, **39-56**.

—— 1995. Lower Palaeozoic cavern development, collapse, and dolomitization, Franklin Mountains, El Paso, Texas. *In*: BUDD, D. A., SALLER, A. H. & HARRIS, P. M. (eds) *Unconformity and Porosity in Carbonate Strata*. American Association of Petroleum Geologists Memoir, 63, 137–152.

MAJOR, R. P. & HOLTZ, M. H. 1993. Depositionally and diagenetically controlled reservoir heterogeneity at Jordan field. (Reprint from: *Journal of Petroleum Technology*, 42(10), 1304–1309.) *Oklahoma Geological Survey Circular*, 95, 82–90.

——, LLOYD, R. M. & LUCIA, F. J. 1992. Oxygen isotope composition of Holocene dolomite formed in a humid hypersaline setting. *Geology*, 20, 586–588.

——, VAN DER STOEP, G. W. & HOLTZ, M. H. 1990. *Delineation of Unrecovered Mobile Oil in a Mature Dolomite Reservoir: East Penwell San Andres Unit Reservoir, University Lands, West Texas*. Bureau of Economic Geology, The University of Texas, Reports of Investigations, 194.

MCCABE, C. & ELMORE, R. D. 1989. The occurrence and origin of late Palaeozoic remagnetization in the sedimentary rocks of North America. *Reviews of Geophysics*, 27(4), 471–494.

MCDOWELL, K. 1983. *Lower Ordovician El Paso Group depositional systems and diagenesis, Southern Franklin Mountains, El Paso County, Texas*. PhD thesis, The University of Texas, Austin.

MONTANEZ, I. P. 1994. Late diagenetic dolomitization of Lower Ordovician upper Knox Carbonates: a record of the hydrodynamic evolution of the southern Appalachian basin. *American Association of Petroleum Geologists Bulletin*, 78(8), 1210–1239.

PULLAIAH, G., IRVING, E., BUCHAN, K. L. & DUNLOP, D. J. 1975. Magnetization changes caused by burial and uplift. *Earth and Planetary Science Letters*, 28, 133–143.

RANKEY, E. C., WALKER, K. R. & SRINIVASAN, K. 1994. Gradual establishment of Iapetan 'passive' margin sedimentation: stratigraphic consequences of Cambrian episodic tectonism and eustacy, southern Appalachians. *Journal of Sedimentary Research*, B64(3), 298–310.

RIPPERDAN, R. L., MAGARITZ, M. & KIRSCHVINK, J. L. 1993. Carbon isotope and magnetic polarity evidence for non-depositional events within the Cambrian–Ordovician boundary section. *Geological Magazine*, 130(4), 443–452.

SHEPARD, T. M. & WALPER, J. L. 1982. Tectonic evolution of Trans-Pecos Texas. *Transactions Annual Meeting, Gulf Coast Section, Geological Society of America*, 32, 74–83.

SPALL, H. 1971. Palaeomagnetism of Precambrian rocks from El Paso, Texas. *Physics of the Earth and Planetary Interiors*, 4, 329–346.

STACEY, J. K., JULIAN, F. E. & LEMONE, D. V. 1992. Structural and tectonic development of McKelligon Canyon, southern Franklin Mountains, El Paso County, Texas. *In*: CANDELARIA, M. P. & REED, C. L. (eds) *Paleokarst, Karst-Related Diagenesis and Reservoir Development: Examples from Ordovicean–Devonian Age Strata of West Texas and the Mid-Continent*. Permian Basin Society of Economic Paleontologists and Mineralogists Publications, 92-33, 195–201.

STEPANEK, B. E. 1987. *Dolomitization of Palaeokarst Collapse Breccias – the Ordovician El Paso Group, Franklin Mountains, West Texas*. SEPM Mid-Year Meeting Guidebook.

TOOMEY, D. F. 1970. An unhurried look at a Lower Ordovician mound horizon, southern Franklin Mountains, west Texas. *Journal of Sedimentary Petrology*, 40(4), 1318–1334.

VAN DER VOO, R. 1990. Phanerozoic palaeomagnetic poles from Europe and North America and comparisons with continental reconstructions. *Reviews of Geophysics*, 28(2), 167–206.

VON BERGEN, D. & CAROZZI, A. V. 1990. Experimentally-simulated stylolitic porosity in carbonate rocks. *Journal of Petroleum Geology*, 13(2), 179–192.

WIGGINS, W. D. 1982. *Depositional history and microspar development in reducing pore water, Marble Falls limestone (Pennsylvanian) and Barnett shale (Mississippian), central Texas*. PhD thesis, The University of Texas, Austin.

ZIJDERVELD, J. 1967. A.C. Demagnetization of rocks: analysis of results. *In*: COLLINSON, D. W., CREER, K. M. & RUNCORN, F. R. S. (eds) *Methods in Palaeomagnetism*. Developments in Solid Earth Geophysics, 3. Elsevier, Amsterdam, 254–286.

Glossary

anadiagenesis: A late diagenesis phase in which sediments become increasingly lithified during increasing burial (up to 10 km) causing interstitial fluids to migrate (usually upwards). It may be followed by tectonic metamorphism. Approximately equal to the phyllomorphic stage.

anchi-metamorphism: Extremely low, barely detectable, grade of metamorphism in which sufficient phyllosilicate minerals become oriented by horizontal (tectonic) forces to be detectable from the alignments caused by burial (phyllomorphism).

ankerite ($Ca(MgFe)(CO_3)_2$): hexagonal; a ferroan form of dolomite

apparent polar wander path: Sequential plot of positions of virtual geomagnetic poles of varying age from a particular continent reflecting the motion of a plate relative to the axis of the Earth's dipole field.

aragonite ($CaCO_3$): hexagonal; mostly as a skeletal mineral of many molluscs, corals, etc., and forming most modern calcareous muds. Chemically unstable, inverts to calcite.

ARM: Anhysteretic remanent magnetization. A laboratory induced remanence caused by a weak DC field imposed on a sample being exposed to alternating field (AF) demagnetization.

aulacogen: A continental rift system that failed to continue to open beyond an initial stage.

back-field coercivity (coercive force): The reverse field strength required to reduce the induced field of a sample to zero.

biogenic magnetite: Magnetite grown within living organisms, e.g. magnetotactic bacteria.

bioturbation: The disturbance of a sediment as a result of organic activity, e.g. burrowing.

bloch wall: see domain wall.

blocking temperature: Temperature at which a magnetic grain undergoes the transition from magnetic unstable to magnetic stable.

blocking volume: Grain volume at which the transition occurs from super-paramagnetic to stable single domain.

boundstone: Carbonate rock comprising organogenic components, either bound together in the position of growth to form a rigid framework, or deriving from the lateral accumulation of their fragments (e.g. biogenic reefs and associated bioclastic deposits).

Brunhes: The current normal polarity chron, commencing 0.785 Ma ago.

calcite ($CaCO_3$): orthorhombic structure; a major constituent of most limestones, largely of direct organic origin or derived from aragonite; rarely as a chemical deposit in hypersaline conditions.

Cartesian plot: see orthogonal plot:

cathodoluminescence: Emission of light resulting from the bombardment of a phosphor by electrons. In carbonates, cathodoluminescence correlates positively with the ratio Mn^{2+}/Fe^{2+}.

characteristic remanent magnetization (ChRM): Usually defined for an individual specimen by principal component analysis, or by Fisherian averaged ChRM specimens to determine a site mean ChRM.

chemical remanent magnetization (CRM): Magnetization acquired during crystal growth

chron: The basic unit of time, $\geq 10^5$ years, during which the Earth's magnetic field was predominantly of the same polarity.

coercivity: The magnetic field required to reduce the external magnetic field of an object to zero.

consistency index: A measure of the consistency in the direction of a vector during demagnetization that is independent of the change in intensity of remanence (see mda). It is given by the maximum value of (range of treatment^2)/ (circular standard deviation) for three or more successive demagnetization vectors. (Originally the Stability Index of Tarling & Symons (1967).)

crystalline remanent magnetization (CRM): Magnetization acquired during crystal growth.

Curie temperature: The temperature at which a ferromagnetic mineral ceases to have ferromagnetic properties and has only paramagnetic properties.

$\delta^{18}O$: The ratio of $^{18}O/^{16}O$ in a sample compared with that of the PDB standard.

depositional remanent magnetization (DRM): The remanence acquired by sediments as they are deposited from an aqueous or aeolian medium.

detrital magnetite: Allogenic (e.g. aqueous or aeolian) particles. The term is sometimes applied incorrectly to authigenic (e.g. bacterial) input at the time of sediment accumulation, as such particles can behave in the same way as truly deposited particles.

diamagnetic: Electrons in the ground state are paired and cause a zero net magnetic moment. Diamagnetic substances are repelled by magnets because an ambient magnetic field gives rise to a small net magnetic moment with a direction opposite that of the applied field.

diagenesis: All changes in physical and chemical properties in a sediment occurring between deposition and tectonic metamorphism. Diagenetic changes can form a continuum of modifications from the early stages when the sediment is still in or close to its ambient depositional environment (syndiagenesis), to those occurring during deep burial (anadiagenesis). The term also includes chemical and textural changes occurring when deeply buried sediments are raised to shallow burial con-ditions (epidiagenesis), including surface exposures where such changes merge into weathering processes.

dolomite ($CaMg(CO_3)_2$): hexagonal; normally as secondary replacement of other carbonates, usually caused by the entry of Mg-rich fluids; when it replaces calcite there is a reduction in volume of up to 13%.

dolostones: A carbonate rock mostly composed of dolomite and generally derived from profound diagenetic modifications (dolomitization).

domain: A region within a magnetic mineral in which the individual atomic magnetic moments are parallel to each other.

domain wall: A thin zone ($c.\,0.1\,\mu$m thick) between domains, usually magnetized in opposite directions, within which the electron spin magnetic vectors cant over between the vector directions in the two adjacent domains.

drift deposit: A sediment deposited in association with glacial ice, including glacial lake sediments.

epidiagenesis: Changes in mineralogy and crystalline forms in a sedimentary rock when it reached different environmental conditions, usually as a result of uplift, and may lead to increasing weathering effects.

epitidal: The upper part of the tidal zone; subject only to occasional brief submersion.

event (geomagnetic): A brief interval, $<10^5$ years, when the Earth's magnetic field was opposite to that of the chron in which it occurred.

Fisher statistical analysis: A statistic model which simulates a two-dimensional Gaussian distribution as a three-dimensional distribution of unit vectors on a sphere (Fisher 1953). This allows the estimation of the precision of the mean direction by k if $N > 7$ (which varies between unity and ∞) and by α_{95}, the radius of a cone centred on the mean direction and defining a cone within which there is a 95% probability that the true mean direction lies. The scatter can similarly be defined by a circle, centred on the mean direction, with a radius α_{63} (the circular standard deviation) which includes 63% of the vectors (if they have a Fisherian distribution).

flysch: Heterogeneous, mostly fine-grained sedimentary rocks in the Alpine region (Swiss dialect for slaty rocks subject to sliding).

goethite (α-FeOOH): Imbalanced anti-ferromagnetic mineral (lattice distortions or protohematite inclusions create spin imbalances that cause a net magnetic moment) with saturation magnetization <1 A/m, i.e. among the weakest magnetic minerals. However, it has very high coercivity, but is chemically unstable at temperatures above 100°C, usually transforming to hematite in oxidizing conditions and to magnetite in reducing conditions.

grainstone: Grain-supported carbonate sediment or rock without matrix.

halmrolysis: In clastic sediments, the early diagenetic processes involving changes in the clay minerals and including the formation of glauconite.

hematite (α-Fe_2O_3): Imperfect anti-ferromagnetic mineral (magnetic moments of atoms not exactly antiparallel, leading to a weak net magnetic moment) with saturation magnetization of 2×10^3 A/m, i.e. considerably weaker than magnetite but of much higher stability.

ilmenite ($FeTiO_3$): This mineral is normally paramagnetic at room temperature.

isothermal remanent magnetization (IRM): Remanent magnetization resulting from short-term exposure to strong magnetizing fields at constant temperature, usually room temperature.

Kiaman Reversed Superchron: A prolonged period of reversed polarity between about 320 and 250 Ma.

limnology: The study of all characteristics of lake and ponds (generally, but not exclusively, fresh water).

locomorphic phase: The phase of diagenesis when early cementation occurs.

loess: Wind-blown silts, fine sands and clay deposits.

mda: Mean diagonal angle (Kirschvink 1980) defining the degree of linearity of a vector; usually between 0° (perfectly linear) and 5° (reasonably linear).

MAD: See mda:

magnesite ($MgCO_3$): A hexagonal carbonate; usually occurs as a minor component in skeletal aragonite and calcite.

magnetite (Fe_3O_4): A cubic (spinel) ferrimagnetic mineral with a Curie point of 575°C. The most strongly magnetic of the common iron oxides, but of lower coercivity than hematite and goethite.

Matuyama: The last reversed polarity chron ending 0.785 Ma years ago.

micrite: Carbonate muds composed of clay-sized particles ($\geq 4\,\mu m$).

mudstone: A general term for a fine-grained sedimentary rock, including clay, shale, silt and argillites. A (lime) mudstone is a mud-supported carbonate sediment or rock containing <10% of grains >0.02 mm (see also micrite).

multi-domain (MD): Containing several magnetic domains, each with a domain wall that can usually move when a weak magnetic field is applied. Usually grains $>3\,\mu m$.

N: The number of observations, usually specimen, sample or mean site determination. In Fisherian statistical analysis, $N \geq 7$ is desirable for reliable statistics, and $N \geq 5$ is desirable in the analysis of individual vectors for linearity and consistency.

neomorphism: A phase of diagenesis involving changes in crystal properties including recrystallization with change in mineralogy and recrystallization into other minerals (polymorphism), e.g. aragonite to calcite, dolomite to calcite, etc.

NRM: The natural remanent magnetization of a sample or specimen. This usually comprises both magnetically soft and hard components, from which the soft components are removed by incremental partial demagnetization by heating or AF fields applied in zero external magnetic field.

orthogonal projection: During progressive demagnetization, vector endpoints are simultaneously projected on two orthogonal planes, horizontal and vertical. The length of the vector expresses the intensity of magnetization remaining after each demagnetization increment. (See also Zijderveld diagram.)

packstone: Grain-supported carbonate sediment or rock, with some intergranular matrix.

palaeo-direction: Expected magnetization at a particular place on the surface of the Earth calculated from a given pole position assuming an axial geocentric dipole.

PDB standard: A Cretaceous belemnite of the PeeDee Formation of South Carolina, USA, used as the standard for oxygen isotope ratios.

peritidal: In or around a tidal-flat environment.

phyllomorphic phase: Late-stage diagenesis during deep burial, usually associated with an increase in phyllosilicates; in this phase, the orientation of the new minerals is related to vertical (burial) pressures, whereas the onset of ancho-metamorphism is marked by their orientation relating to tectonic forces.

polymorphism: A change of crystal shape, e.g. aragonite to calcite, without necessarily a change of chemical composition.

post-depositional remanent magnetization (PDRM): The remanent magnetization acquired following deposition, usually considered to occur by rotation of magnetic grains, within interstitial water or air, by the Earth's magnetic field. The name could include other remanences, but is generally confined to those acquired immediately after deposition.

principal component analysis (pca): The analysis to derive the average direction of three or more vectors isolated successively by partial demagnetization. The linearity of the mean vector is defined by the mean diagonal angle (see mda) (Kirschvink 1980).

pseudo-single domain: A magnetic domain that behaves as a single-domain particle. There are many possible forms, such as a partially pinned microstructure within a multi-domain grain.

redoxomorphic phase: The early phase of diagenesis when deposited materials begin to equilibrate with their environment

relaxation: Decay of the remanent magnetization of a magnetic grain or an assemblage of magnetic grains.

relaxation time: The time taken for $1/e$ of a magnetization to relax into the new direction of a weak field, usually at ambient temperatures. It is a function of mineralogical composition, grain size and temperature.

remanent magnetization or remanence: Stable magnetization recording past action of magnetic fields that have acted on the rock.

rutile (TiO_2): Titanium oxide.

secular variation: Variation of the Earth's magnetic field over time scales greater than 1 year, but excluding the 22 year sunspot cycle. Attributable to processes interior to the Earth.

sediment: Solid material settled down from a state of suspension or solution and composed of grains (detrital particles or grains) and generally of matrix (fine detritus).

siderite ($FeCO_3$): Hexagonal, iron carbonate, often as concretions and ooliths.

single domain (SD): A grain containing only one magnetic domain, usually around $1\,\mu m$ dimensions.

stability index: See consistency index.

superchron: Long period, >20 Ma, of unchanged geomagnetic polarity.

superparamagnetic: A ferromagnetic grain that has strong paramagnetic properties, but loses any remanence over a few minutes. Usually associated with sub single-domain sized grains.

syndiagenesis: The diagenetic processes affecting sediments before deep burial (i.e. at depths <100 m), including alterations before the grains finally settle as well as the early stages of compaction and cementation. In clastic sediments, it includes changes in the clay minerals and glauconization. In carbonate sediments, it includes a range of mineralogical and textural changes caused by variations in the original depositional environment or by the passage of fluids through pore spaces.

syngenesis: The same as syndiagenesis.

Tesla (T): Unit for magnetic flux density; $T = Wb/m^2$ (Weber/metre^2).

TRM: Thermal remanent magnetization. The magnetisation acquired during cooling, in a weak magnetic field, usually from a temperature at or above the Curie temperature.

varves: Laminated sediments such as in glacial lakes where freezing-over results in fine clay winter laminae and coarser summer layers.

virtual geomagnetic pole (VGP): Pole of a geocentric dipole corresponding to an observed direction of magnetization.

viscous remanent magnetization (VRM): The magnetization acquired by a substance lying in a weak magnetic field for prolonged periods. The intensity is generally a function of log time.

wackstone: A mud-supported carbonate sediment or rock with >10% grains.

weak magnetic field: Generally a field comparable in strength with that of the Earth, i.e. less than or about $100\mu T$.

Zijderveld diagram: A Cartesian (orthogonal) plot of changes in a vector during partial incremental demagnetization. The conventional plot illustrates the horizontal component in terms of the north (x) component of remanence against the east (y) component, and the vertical (z) component against either x or y (Zijderveld 1967). The so-called 'modified' Zijderveld diagram shows the same horizontal component, x v. y, but plots z against $h = \sqrt{(x^2 + y^2)}$ so that the vertical part of the plot illustrates the inclination.

References

FISHER, R. A. 1953. Dispersion on a Sphere. *Proceedings of the Royal Society*, **A217**, 295–305.

KIRSCHVINK, J. L. 1980. The least-squares line and plane and the analysis of palaeomagnetic data. *Geophysical Journal Royal Astronomical Society*, **62**, 699–718.

TARLING, D. H. & SYMONS, D. T. A. 1967. A stability index of remanence in palaeomagnetism. *Geophysical Journal Royal Astronomical Society*, **12**, 443–448.

ZIJDERVELD, J. D. A. 1967. A.C. demagnetization of rocks: analysis of results. *In*: RUNCORN, S. K. CREER, K. M. &, COLLINSON, D. W. (eds) *Methods in Palaeomagnetism*. Elsevier, Amsterdam, 254–286.

Index

Pages numbers in *italics* refer to Figures or Tables.
Page numbers in **bold** refer to glossary definitions.

AARM *see* anisotropy of anhysteretic remanent magnetism
Adriatic Sea sediments
 magnetic properties and environmental proxies 78–80
 methods of analysis 72–3
 results 75–7
AF demagnetization *see* alternating field demagnetization
Albano Lake sediments 72
 magnetic properties and environmental proxies
 methods 72–3
 results 73–5
 results discussed 77–8
algal magnetite production 152
alternating field (AF) demagnetization 29
 Chalk 32
 Monte Raggeto carbonates 150–2
 Texas carbonates
 method of analysis 184
 results 196, 197
AMS *see* anisotropy of magnetic susceptibility
anadiagenesis 2, **205**
 Trubi Formation 65–7
Ancaster Limestone 31
 VRM tests 36–7
ancho-metamorphism 4, **205**
Andilská Hora Formation 132–3
anhysteretic remanent magnetism (ARM) 45, 139, **205**
 New Jersey slope sediments 89
 relation to environmental proxies
 methods of analysis 72
 results *73, 75, 76, 77, 81*
 results discussed 77, 78, 80–1
 Trubi Formation 58, 59, 64
anisotropy of magnetic susceptibility (AMS)
 Matinenda Formation 142–4
 New Jersey slope sediments 89
 use in fabric analysis 126
 Bohemian massif study 132–6
 Carpathians study 129–31
 waterlain sediment
 method 10
 results 11
anisotropy of anhysteretic remanent magnetism (AARM) 139
 Matinenda Formation 143–4
ankerite **205**
aragonite 3, **205**
archaeological studies 16
 York
 methods of analysis 10
 results 10–13
 results compared 13–18
ARM *see* anhysteretic remanent magnetism
aulacogen **205**

back-field coercivity **205**
bacteria *see* magnetotactic organisms
biogenic magnetite *see* magnetotactic organisms
bioturbation **205**
block wall **205**
blocking temperature **205**
blocking volume **205**
Bohemian massif 127
 magnetic fabric study 132–6
boundstone **205**
Brunhes chron (epoch) 30, 32, **205**

$\delta^{13}C$ and magnetic properties 77
Calabrian Ridge 90–1
calcite 3, **205**
Caldicot archaeological site 16
Callisto gas field 109
carbonates
 clay content effects 167
 relation to susceptibility 173, 176–7
 concentration effects 77
 crystalline interactions 3
 depositional cycles
 methods of analysis 150–2
 results 152–3
 results discussed 153–4
 diagenesis and magnetism
 Tata Kálvária Hill
 methods of analysis 158–61
 results 161–3
 results discussed 163–4
 Texas carbonates 181–93
 diagenetic history 196–201
 lithofacies analysis 184–6, 190–2
 magnetic properties 183, 186–9, 192–5
 mineralogy 189, 195–6
 sedimentary characteristics 183
 syndiagenesis 147
Carpathians 127
 magnetic fabric study 129–31
cathodoluminescence **205**
cementation reactions 2
Chalk of eastern England
 isothermal remanent magnetization 31
 natural remanent magnetization 32
 viscous remanent magnetization 35, *36, 37, 38, 40*
characteristic remanent magnetization (ChRM) **205**
 Leman Sandstone Formation 118, 121
 Trubi Formation 54–5, *56–7*, 59, *62, 63,* 64–5
chemical remanent magnetization (CRM) **205**
 Leman Sandstone Formation 120
 Texas carbonates 181
China loess 46
chlorite formation 4
ChRM *see* characteristic remanent magnetization
chromite *99, 100,* 101, 106
chron **205**
clays
 effect on magnetic record 167
 porosity of 2
 relation to susceptibility 173, 176–7
cluster analysis *see* fuzzy c-means
Cochiti subchron 54

INDEX

coercive force **203**
coercivity 10, 18, **205**
   relation to environmental proxies 76, 80
   Trubi Formation 58, 64
conglomerate test 35, *36*
consistency index **205**
Cretaceous rock studies *see*
   Chalk of eastern England
   Monte Raggeto, Italy
CRM *see* chemical remanent magnetization
crystalline remanent magnetization **205**
Curie temperature **205**
cyanophyta in magnetite production 152

dating methods 17
Dead Sea *see* Lisan Lake
debris flows *34*, 35, *37*
declination record
   Lisan Formation *48*, *49*, *51*
   Monte Raggeto carbonates *153*
   Trubi Formation *62*, *63*
deformation
   association with diagenesis 127
   magnetic analysis
      Bohemian massif study 132–6
      Carpathians study 129–31
depositional remanent magnetization (DRM) **205**
detrital magnetite **205**
   see also magnetite
diagenesis 2, 147, **205–6**
   carbonates
      Tata Kálvária Hill
         methods of analysis 158–61
         results 161–3
         results discussed 164
      Texas 196–201
   magnetic analysis
      Bohemian massif study 132–6
      Carpathians study 129–31
   role of deformation 127
   Trubi Formation 65–7
   Wanganui Basin sediments 106
diamagnetic **205**
diamagnetism 139
diatoms and magnetic properties 77–8, *79*
dolomite 3, **206**
dolomitization and magnetic properties
   Estonian carbonates 174–5
      relation to susceptibility 173, 176–7
   Honeycut Formation 189
   McKelligon Formation 195–6
   Monte Raggeto carbonates *154*
dolostone **206**
domain **206**
domain wall **206**
   remanence 30
drift sediments **206**
   PDRM measurement
      method *21*, 23
      results 24–5
      results discussed 25–6
DRM *see* depositional remanent magnetization

El Paso Group *see* McKelligon Formation
Ellenburger Group *see* Honeycut Formation
environmental proxies *see* Quaternary
epidiagenesis 4, **206**
   Trubi Formation 67
epitidal **206**
Estonian carbonates 167–9
   methods of analysis 169–70
   results 171–6
   results discussed 176–7
Europa gas field 109
event **206**

fabric in sediments
   magnetic characterization
      Bohemian massif study 132–6
      Carpathians study 129–31
Farnley archaeological site 16
ferrimagnetic *v.* paramagnetic effects 104
field excursions 50, *51*
Fisher statistical analysis **206**
flysch **206**
Flysch Belt 127
   magnetic fabric study 129–31
foliation, magnetic 129, 139
foraminifera and magnetic properties 79–80, *81*
francolite 168
Franklin Mts 182–3
   see also McKelligon Formation
fuzzy c-means clustering (FCM) 85, 86–8
   application 89–91

Ganymede gas field 109
Gauss chron 54
geomagnetic field excursions 50, *51*
Gilbert chron 54
glauconite 167, 169
goethite **206**
   Estonian carbonates 168
   river sediments 13, 18
   Tata Kálvária Hill 162
   Texas carbonates 194, 195
grain shape effect on AMS 139
grain size effects on unblocking 30
grainstone **206**
greigite 3
   bacterial production 43
   conditions of formation 72
   Trubi Formation 58

halmrolysis **206**
Hartlepool archaeological site 16
hematite **206**
   AF demagnetization 29
   blocking temperature *28*
   Estonian carbonates 169
   Leman Sandstone Formation 118, *119*
   Red Chalk 31
   river sediments 13, 18
   Tata Kálvária Hill 162
   Texas carbonates 189
   Trubi Formation 63
   Wanganui Basin sediments *99*, *101*, *105*, 106
hemo-ilmenite 106

Honeycut Formation 181–82
  carbonate mineralogy 189
  diagenetic history 196–201
  lithofacies analysis 184–6
  magnetic properties
    methods of analysis 183
    results 186–9
  sedimentary characters 183
Horní Benešov Formation 132
Hradec-Kyjovice Formation 132, 133, *134*, *136*
Hungary *see* Tata Kálvária Hill
Huronian Supergroup *see* Matinenda Formation
hydrogoethite 169
hydrohematite 169
hysteresis
  limestones of eastern England 32
  New Jersey slope sediments 89
  Wanganui Basin sediments 103

ilmenite **206**
  Wanganui Basin sediments *99*, 101, 102, 103, 105, 106
imbrication 129, 139
inclination error 1
inclination record
  Leman Sandstone Formation *114*, *115*
  Lisan Formation *48*, *49*, *51*
  Monte Raggeto carbonates *153*
  Trubi Formation *62*, *63*
Indian Ocean magnetotactic bacteria 45
isothermal remanent magnetization (IRM) **206**
  limestones 31
  New Jersey slope sediments 89
  relation to environmental proxies
    methods of analysis 73
    results 74, *80*
  Tata Kálvária Hill 162
  Texas carbonates
    methods of analysis 184
    results 188–9, 194–5
  Trubi Formation 57, 58, *59*, *60*, 61, 64
  Wanganui Basin sediments
    methods of analysis 97
    results 102
  waterlain sediment study
    methods 10
    results 12
Italy
  Cretaceous carbonates
    depositional history 148–50
    palaeomagnetism
      methods of analysis 150–2
      results 152–3
      results discussed 153–4
    syndiagenesis 148
  lake sediment environmental proxy analysis
    methods 72–3
    results 73–5
    results discussed 77–8

Jupiter Fields *see* Leman Sandstone Formation
Jurassic rock studies *see*
  Ancaster Limestone
  Lincoln Limestone
  Tata Kálvária Hill

Kaena reversal 54
Kiaman Superchron 120, **205**

lake sediments PDRM measurement
  method *21*, 23
  results 24–5
  results discussed 25–6
Laschamp excursion event 50, *51*
Leman Sandstone Formation 109
  magnetostratigraphy
    methods of analysis 110–13
    results 113–18
    results discussed 118–23
limestones of eastern England 30
  isothermal remanent magnetization 31
  natural remanent magnetization 32
limnology **205**
Lincoln Limestone 36–7, *38*, *40*
Lisan geomagnetic event 50, *51*
Lisan Lake Formation NRM study
  methods 47
  results 47–52
Llano Uplift *see also* Honeycut Formation 179–80
locking-in of remanence 50–2
locomorphic diagenesis 2
locomorphic phase **205**
loess **205**
  magnetic extracts 46
  PDRM measurement
    method *21*, 23
    results 24–5
    results discussed 25–6

McKelligon Formation
  carbonate mineralogy 195–6
  digenetic history 196–201
  geological setting 182–3
  lithofacies analysis 190–2
  magnetic properties
    method of analysis 183
    results 192–5
  sedimentary characters 183
mackinawite 72
maghemite 43
magnesite **205**
magnetic fabric *see* fabric
magnetite **205**
  acquisition coefficient 27
  bacterial production 43
  blocking temperature *28*
  effect on AMS 139
  low-temperature demagnetization 30
  regional studies
    Eastern England limestones 31, 32
    Estonian carbonates 169
    river sediments 13, 18
    Texas carbonates 181, 187, 194, 195
    Trubi Formation 55, 63
    Wanganui Basin sediments 104
  transdomain remanence 30
magnetobacteria 3
magnetotactic organisms 43–6, 152
MAGPORE 126
marcasite 168, 169

Mascarene Plateau sediments 45
masonry test 35, *37*
Matinenda Formation
 depositional setting 140–2
 magnetic properties 142–4
Matuyama chron 54, **205**
Mediterranean Sea 90–1
Messinian evaporites 67
metamorphism/diagenesis boundary 4
micrite **203**
micro-organisms and diagenesis 3
Milankovitch cycles 148, 153–5
mineralogy 18
Monte Raggeto carbonates
 depositional history 148–50
 palaeomagnetism
  methods of analysis 150–2
  results 152–3
  results discussed 153–4
 syndiagenesis 148
Moravice Formation 132, *134*, *135*, *136*
mudstone **206–7**
multi-domain (MD) **207**
multivariate analysis *see* fuzzy c-means *also* non-linear mapping
Myslejovice Formation 132, 133, *134*, *135*

natural remanent magnetization (NRM) **207**
 Calabrian Ridge sediments 91
 factors affecting 53
 Leman Sandstone Formation
  method of analysis 112
  results 113
 Lisan Formation
  methods of analysis 47
  results 47–52
 Monte Raggeto carbonates 153
 New Jersey slope sediments 89
 Tata Kálvária Hill 161–2
 Texas carbonates 183
  methods of analysis 184
  results 186–8, 192–4
 Trubi Formation 58, 59–61, *62*, 65
 Wanganui Basin sediments 96–7
 waterlain sediment study
  methods 10
  results 11, 12, 14
Nemi Lake sediments 72
 magnetic properties and environmental proxies
  methods 72–3
  results 73–5
  results discussed 77–8
neomorphism **207**
New Jersey slope 904 89–90
New Zealand *see* Wanganui Basin
non-linear mapping (NLM) 85, 88
 application 89–91
North Sea *see* Leman Sandstone Formation
NRM *see* natural remanent magnetization
Nunivak subchron 54

$\delta^{18}O$ **205**
 relation to dolomitization 189, 195–6, 198–9
ODP sediments 45, 89–90

Ontario *see* Matinenda Formation
Ordovician rock studies *see*
 Estonian carbonates
 Texas carbonates
orthogonal projection **207**
Ouachita orogeny 198
Ouse River flood deposits
 methods of analysis 10
 results
  modern sediment 12–13
  Roman sediment 10–12
 results compared 13–18
Owen Ridge sediments 45

packstone **207**
palaeo-direction **207**
palaeocurrent analysis
 Matinenda Formation
  depositional setting 140–2
  magnetic properties 142–4
Palaeogene rock studies *see*
 Bohemian massif
 Carpathians
paramagnetism 104, 139
PDB standard **207**
PDRM *see* post-depositional remanent magnetization
peritidal **207**
Permian rock studies *see* Leman Sandstone Formation
phyllomorphic phase 4, **207**
Pliocene rocks studies *see*
 Trubi Formation
 Wanganui Basin
polymorphism **207**
pore geometry
 analysis by AMS 126
 role in reservoir characterization 125
post-depositional remanent magnetization (PDRM) **207**
 relation to wetting and drying
  method of measurement 23–4
  results 24–5
  results discussed 25–6
principal components analysis (PCA) 32, *33*, **207**
Protivanov Formation 132, 133, *134*, *135*
pseudo-single domain (PSD) **207**
pyrite
 conditions of formation 72
 Estonian carbonates 168, 169
 Trubi Formation 63, 67
 Wanganui Basin sediments *99*, 100–1, 106
pyrrhotite 58, 65

Quaternary environmental proxies
 Adriatic Sea sediments
  methods of analysis 72–3
  results 75–7
  results discussed 78–80
 Italian lake sediments
  methods of analysis 72–3
  results 73–5
  results discussed 77–8

R-mode factor analysis 175
red beds 2, 29

Red Chalk
  isothermal remanent magnetization 31
  natural remanent magnetization 32
redoxomorphic diagenesis 2
redoxomorphic phase **207**
relaxation **207**
relaxation time **207**
  factors affecting 28
remanence **207**
remanence consistency index 10
remanent magnetization **207**
reservoir characterization
  role of pore geometry 125
  use of AMS 126
reworking effects of sediments 106
Rheno-Hercynian Zone 127
  magnetic fabric study 132–6
rutile **207**

St Petersburg varved clay *21*
salinity effect on carbonates 3
saturation isothermal remanent magnetization (SIRM)
  Estonian carbonates
    methods of analysis 170
    results 174
  relation to environmental proxies
    methods of analysis 72
    results 74, *75, 76, 77*
    results discussed 78, 80–1
  waterlain sediment 13
secular variation (SV) 47, 48, 50, 153, **207**
sediment **207**
shale diagenesis 2
Sicily *see* Trubi Formation
siderite **207**
Sidufjall subchron 54
Silurian rock studies *see* Estonian carbonates
single domain (SD) **207**
Sinope gas field 109
SIRM *see* saturation isothermal remanent magnetization
specularite 118
stability index **207**
stable isotope analysis 189, 195–6, 198–9
sulphate chemistry, Trubi Formation 67
superchron **207**
superparamagnetic **207**
susceptibility 17–18
  Estonian carbonates
    methods of analysis 170
    results 171–3
  Leman Sandstone Formation 113
  relation to environmental proxies
    methods of analysis 72
    results *74, 75*, 77, 78
  Trubi Formation 64
  Wanganui Basin sediments 103
  see also anisotropy of magnetic susceptibility (AMS)
SV *see* secular variation
syndiagenesis 2, **207–8**
  Monte Raggeto carbonates 147
  Trubi Formation 65
syngenesis **208**

Taman Peninsula loess *21*
Tashkent loess *21*
Tata Kálvária Hill 157–8
  palaeomagnetic study
    methods of analysis 158–61
    results 161–3
    results discussed 164
temperature and remanent magnetization 27, 28
Tesla **208**
Texas carbonates
  diagenetic history 196–201
  geological setting 181–3
  lithofacies analysis 184–6, 190–92
  magnetic properties
    methods of analysis 183
    results 186–9, 192–5
  mineralogy 189, 195–6
  sedimentary characteristics 183
Thebe gas field 109
thermal remanence 1
thermal remanent magnetization (TRM) **208**
  relation to VRM 28–9
thermoviscous remanent magnetization (TVRM) 27, 183
Thvera subchron 54
titanomagnetite 29, 31, 74
total organic carbon (TOC) and magnetic properties *75, 76*, 78
transdomain remanence 30
Triassic *see* Tata Kálvária Hill
TRM *see* thermal remanent magnetization
Trubi Formation
  depositional environment 54
  magnetostratigraphy 54–6
    CMD profile 59–64
    LCM profile 56–9
    PMD profile 59
  role of diagenesis 64–7
Tupholme Limestone *38, 39, 40*
TVRM *see* thermoviscous remanent magnetization

unblocking temperatures 29–30
USA *see* Texas carbonates study

Väo Formation 168
  dolomitization 175
  susceptibility 172, 173, 176
Variscan orogeny 127
varves and varved clay 1, **208**
  PDRM measurement
    method *21*, 23
    results 24–5
    results discussed 25–6
virtual geomagnetic pole (VGP) **208**
viscous remanent magnetization (VRM) 27–8, **208**
  anomalous stability 38–40
  Leman Sandstone Formation 118, 120
  relation to TRM 28–9
  tests
    conglomerate 35, *36*
    debris flow 35, *37*
    masonry 35, *37*
Voronezh *21*
VRM *see* viscous remanent magnetization

wackstone **208**
Wanganui Basin 96
  Pliocene sediment history
    methods of analysis 96–7
    results 97–104
    results discussed 104–6
waterlain sediment analysis
  methods 10
  results
    modern sediment 12–13
    Roman sediment 10–12
  results compared 13–18
weak magnetic field **208**
West Heslerton archaeological site 16
wetting/drying cycles
  effect of 22
  experiment to measure
    method 23–4
    results 24–5
    results discussed 25–6

White Chalk
  isothermal remanent magnetization 31
  natural remanent magnetization 32
Wood Hall Moated Manor archaeological
  site 16

York archaeological study
  methods of analysis 10
  results
    modern sediment 12–13
    Roman sediment 10–12
  results compared 13–18

Zanclean neostratotype 54
Zijderveld diagram **208**